ELIZA SCIDMORE

Praise for *Eliza Scidmore*

"To the canon of women explorers like Gertrude Bell and Nellie Bly, add journalist Eliza Scidmore to the list of exceptionals. Diana Parsell's meticulously researched biography brings to life the woman whose curiosity and passion for travel bought the wonder of distant lands in words and pictures to American readers."

—Cathy Newman, author of *Women Photographers at National Geographic*

"Parsell has brilliantly rescued Eliza Scidmore, a celebrity journalist and travel writer, from obscurity. Her family background and character are intriguing, and the biography is packed with cultural and historical detail that positions Scidmore as a professional with friends in the highest positions in many fields, both in the United States and the Far East. In addition to the long and complicated saga of her role in securing Japanese cherry trees for Washington D.C., she is especially esteemed for her books about Alaska and Japan, and her instrumental work for *National Geographic* magazine."

—Susan Schoenbauer Thurin, author of *Victorian Travelers and the Opening of China 1842–1907*

"Diana Parsell's meticulous biography of the important, intrepid, though still sadly under-researched and insufficiently known, Eliza Scidmore, will be an invaluable resource for travel writing scholars and students. The interweaving of the author's own biography with Scidmore's history makes for a wonderful connecting of two women writers' stories more than a century apart."

—Julia Kuehn, The University of Hong Kong

"Parsell writes in a clear and lively style and makes thorough use of primary sources, effectively blending narrative drive with evocative detail."

—Michelle McClellan, Bentley Historical Library, University of Michigan

"A riveting read, this comprehensive biography of Eliza Scidmore is full of surprises, demonstrating a legacy that extends far beyond her role in bringing the now-iconic cherry blossoms to Washington, D.C. Through prodigious research and vivid writing, Diana Parsell brings to life the dynamic period from America's Gilded Age into the 1920s, when Scidmore was an eyewitness to major world events. I highly recommend this book."

—Ann McClellan, author of *Cherry Blossoms* and *The Cherry Blossom Festival: Sakura Celebration*

"One part writer, one part adventurer, one part cultural ambassador, and 100% tenacious—and at a time when women were supposed to linger in the shadows—Eliza Scidmore literally changed the landscape of the nation's capital. In this terrific biography, Diana Parsell's obsessive quest to piece together Scidmore's extraordinary life moves this forgotten journalist from footnote to center stage."

—Lisa Napoli, author of *Susan, Linda, Nina & Cokie: The Extraordinary Story of the Founding Mothers of NPR*

"On the Way Home," by Kazumasa Ogawa in *The Hanami* (Flower-picnic), A. B. Takashima, 1899.

(Special Collections and Archives, The Claremont Colleges Library)

ELIZA SCIDMORE

The Trailblazing Journalist Behind Washington's Cherry Trees

Diana P. Parsell

Great Clarendon Street, Oxford, OX2 6DP,
United Kingdom

Oxford University Press is a department of the University of Oxford.
It furthers the University's objective of excellence in research, scholarship,
and education by publishing worldwide. Oxford is a registered trade mark of
Oxford University Press in the UK and in certain other countries

Impression: 1

Published in the United States of America by Oxford University Press
198 Madison Avenue, New York, NY 10016, United States of America

British Library Cataloguing in Publication Data
Data available

Library of Congress Control Number: 2022945703

ISBN 978–0–19–886942–9

DOI: 10.1093/oso/9780198869429.001.0001

Printed and bound by
CPI Group (UK) Ltd, Croydon, CR0 4YY

To Bruce,

for being there

Who knows how circumstance and character conspire to write the narrative we come to recognize as our life? Of course, one part of the answer is that we bear responsibility for our own narrative.

Cathy Newman,
Women Photographers at National Geographic
(National Geographic, 2000)

Contents

Author's Note

To respect the integrity of the nineteenth-century texts on which this work is drawn, I give many of the place names in their original published form. Thus, today's Beijing appears as *Peking*. In cases where there are minor variations in spelling, as in *Tokio* and *Kioto*, I have adopted the more familiar modern terms (*Tokyo* and *Kyoto*). As for the names of people, I have converted many of Eliza Scidmore's original spellings, such as for the emperor and empress dowager of China, to more recognizable spellings used today. Inconsistencies in stylistic guidelines reflect my intent to aid readability.

About the illustrations: Most of the images used in this book consist of historical photographs out of copyright and in the public domain. In the case of those from various collections for which I received reprint permission, I have made every effort to properly credit the original photographers and sources. Any errors or omissions will be redressed in future editions.

Prologue: A Grave in Yokohama

> "If one should inquire of you concerning
> the spirit of a true Japanese,
> point to the wild cherry blossoms
> shining in the sun."
>
> —Motoori Norinaga

They visit her grave site every year, always in the springtime. If nature cooperates, the cherry trees are at peak bloom. Clutching bright bouquets of flowers, several dozen Japanese citizens and their guests gather inside the tall iron gates of the Foreign General Cemetery, which spills across a high cliff overlooking the harbor of Yokohama. Tea shops and gingerbread-trim bungalows line the streets of this leafy neighborhood where many Western residents lived in the late-nineteenth century, when the area was known simply as "the Bluff."

The cemetery, lush with trees and shrubs, has the air of a private garden. Tombstones instead of statuary line the paths. On a Sunday afternoon in early April, an enormous crow-like bird alights on a stone monument, its oily black feathers burnished with a dull sheen. Overhead an airplane putters across a patch of blue sky visible through the treetops.

Eliza Ruhamah Scidmore lies in Section 11, on an upper terrace near the main road. In this region prone to earthquakes and violent storms, the forces of nature have upended many of the 4,200 graves. Hers is intact, marked by a thick rectangular slab of granite with a peaked top (Figure 0). Nearby stands a flowering cherry tree, its fifteen-foot canopy in full bloom, a graceful study of symmetry and strength. Planted three decades ago by those who come to pay tribute and descended from one of the 3,000 cherry trees that Japan sent to Washington a century ago, it provides a fitting monument to the American woman whose

Figure 0. Scidmore family grave, center, at the Foreign General Cemetery in Yokohama.
(Courtesy of Shinya Oto)

dream of seeing cherry blossoms every spring along the Potomac River indelibly shaped the public face of the U.S. capital.

Of a popular variety called Yoshino, the cherry tree has snowy-white blossoms tinged in pink. The flowers appear early in the season and last barely two weeks, disappearing before the tree gets its leaves. When the petals drop, they fall onto Eliza Scidmore's grave. A small plaque near her name bears the brush-like strokes of Kanji characters. The inscription explains in Japanese, with the power of a haiku poem: "A lady who loved Japanese cherry blossoms rests here in peace."[1]

***

Eliza Scidmore was born in the American Midwest in 1856, three years after Commodore Matthew Perry of the U.S. Navy sailed his fleet of coal-powered "black ships" into Edo (Tokyo) Bay with a letter from President Millard Fillmore demanding that Japan open its ports to American vessels.

A year later, during Perry's follow-up visit, a young marine died aboard ship while the squadron sat anchored in the bay.

When Perry requested a burial spot overlooking the sea, Japanese officials offered the grounds of the Buddhist Zotokuin temple, above the tiny fishing village of Yokohama.[2] Over the years the site became a cemetery for foreigners who died in the area: Russian sailors and Dutch sea captains; merchants and military officers; Western diplomats, missionaries, scientists, and engineers. Some were struck down in violent early clashes between the Japanese and the foreigners.[3] Others succumbed to diseases like cholera and typhoid. Many went to Japan as technical experts and teachers—invited by the government to help build the country's modern infrastructure and institutions—and ended their days there. The gravestones show that a tragically high number buried there lost their lives in the Great Kanto Earthquake of 1923 that leveled Yokohama.[4]

A few years after that cataclysmic event, Eliza Scidmore's ashes were deposited in the cemetery, next to the remains of her mother and her brother George. Three Americans self-exiled far from their native land, intimately tied to one another in death as they were throughout their lives. Eliza's name is inscribed on the narrow side panel of the shared gravestone.

She went to Japan for the first time in the summer of 1885, making the long voyage across the Pacific with her mother to visit George, an American consular official who would spend most of his career in Japan. Though still in her twenties, Eliza had just published her first book, based on her sightseeing adventures in the remote U.S. territory of Alaska.

Japan thrilled her from the start. The soft strokes of summer were washing across the land when she arrived, and Eliza found the country "beautiful from the first green island off the coast to the last picturesque hill-top."[5] The landscape unfolded in a succession of terraced hills, cleft by narrow valleys and ravines. Tiny villages of thatched cottages dotted the countryside, framed by dark groves of pine, palm, and bamboo.

With outside knowledge of Japan still quite limited, Eliza arrived, as other visitors did, with images of the country informed by common stereotypes and prejudices of the day. Under patronizing attitudes in the West that regarded Eastern cultures and peoples as inferior and less developed, Japan was seen as picturesque but backward.[6] A "flowery fairyland inhabited by little people in kimonos, carrying fans and parasols," as one U.S. minister put it.[7] Though traveling with an attitude of openness toward cultural discovery, Eliza expressed similarly simplistic and clichéd views in her initial reporting, while revealing what would become her lasting admiration for the artistic sensibility of the Japanese. "The houses seem toys, their inhabitants dolls, whose manner of life is clean, pretty, artistic, and distinctive," she wrote. Everything looked "too theatrical" to be real. "Half consciously the spectator waits for the bell to ring and the curtain to drop."[8]

Despite the studied effect, change was rippling through Japan.[9] Before Perry's arrival, a long line of military rulers, or *shoguns*, had tightly restricted Japan's interaction with the outside world for more than two centuries to insulate the country and its people from the menace of Christian missionaries and the colonizing grasp of European nations. Trade occurred chiefly with the Chinese and a few Dutch merchants confined to an island off Nagasaki. Most Japanese people lived as their ancestors had for many generations, working the land under powerful feudal lords known as *daimyo*. A well-educated class of *samurai* served the governing elite as both warriors and skilled bureaucrats.

Perry's voyage forced open the gates. Other countries were bound to follow. With the world suddenly at their doorstep, a faction of Japan's leaders concluded that for their country to protect its sovereignty and ancient culture, it would have to abandon its isolationist policies and adopt practices and technologies like those that had made nations in the West so strong. A turning point came in 1868, after a decade of civil war, when the shogunate lost its dominance and governing power was restored to the emperor. The period that followed—known as the Meiji era, for "enlightened rule"—brought a sweeping program of reforms across society and in the military that would transform Japan, in a half-century, from a feudal state into a modern world power.

Among Westerners, pent-up interest in the little-known island empire gave way to a craze for all things Japanese. Globetrotters streamed across the sea for a firsthand look. Back home they stuffed their Victorian parlors with Japanese objects—fans, porcelain, woodblock prints, and painted screens. In a wave of cultural imitation, "Japonisme," as the trend was known, influenced Western art and design in works ranging from Tiffany luxury goods to Impressionist masterpieces by artists such as James McNeill Whistler, Edgar Degas, and Mary Cassatt.[10]

Eliza Scidmore returned to Japan many times, chronicling the ways of old and new Japan in prolific writings that included her best-known book, *Jinrikisha Days in Japan*. Living there on and off for four decades, she came to feel a deep affinity with the Japanese and their cultural values. She shared their passion for the ethereal beauty of cherry blossoms—*sakura*—and loved the ancient ritual of *hanami*, when huge crowds of Japanese turned out every spring to stroll among the trees in bloom. Washington, she decided, needed something like it. Though a series of men in charge of Washington's parks rebuffed her suggestion, she kept her idea alive for more than two decades, until she finally saw her vision become a reality in 1912 with the support of First Lady Helen Taft and a gift of 3,000 cherry trees from Japan, bestowed as a gesture of friendship at a time when its relations with the United States had grown tense.

***

Before I ever heard Eliza Scidmore's name, I crossed her path many times halfway around the world. In the late 1990s I lived for two years with my husband in Jakarta, and over the next decade I returned often to Southeast Asia as a science writer and editorial consultant. On one of those trips, I bought a paperback reprint of an 1897 travelogue, *Java, the Garden of the East*. Back in America, I laid it aside, only to come across it again a few years later. Dipping into the book for first time, I was quickly engrossed. One passage that caught my eye was a description of the author's visit to Buitenzorg (now Bogor), forty miles south of Jakarta. A popular "hill station" in the nineteenth century where colonial officials sought relief from the tropical heat, Buitenzorg had rich volcanic soil, frequent showers, and abundant sunshine that offered ideal conditions for a world-class botanical garden.

I had worked in Bogor and knew it well. My early morning commutes were the best. The route to the office took me along an avenue lined with banyan and kenari trees, towering palms, giant ferns, and thickets of bamboo; past a whitewashed presidential palace, where a large herd of slender deer grazed on the lawn. Orange and magenta bougainvillea spilled across the walls of the neighboring houses and shops. With noxious car exhaust not yet fogging the air, I breathed the loamy, rainforest scent of the tropical garden at the end of the street.

Back home in a suburb of Washington, D.C., the book brought it all back to me. The writing was vivid, and I found the voice beguiling. Who was this author, "E. R. Scidmore," I wondered, and what had taken him out to Java so long ago?

What I discovered in a quick online search astonished me. Not only was the author an American woman, but she had led a remarkable life a century earlier as a prolific author and world traveler. Among her impressive achievements, Scidmore became the first female board member of the National Geographic Society (in 1892) and an early writer and photographer for its now-iconic magazine. Most surprising to me was her key role in bringing Japanese cherry trees to Washington. I had lived in the D.C. area more than three decades. I went every spring to see the city's cherry trees in bloom. Yet I had never heard of Eliza Scidmore.

As I pondered the limited details about her, I realized Scidmore must have been in her late twenties when she first pitched her idea of the cherry trees to the men in charge of the city's parks. Thinking about myself at that age, I begged to know: what would have given a young woman of that era such audacity and confidence? Who was she and where did she come from? And what was it about Japanese cherry trees that made her so taken with the idea in the first place? I headed downtown to the Library of Congress in search of answers. There I found copies of Scidmore's seven books and some of her articles, as well as entries in two dozen biographical indexes. But there was little about her personal life—and no biography. So began my quest to learn all I could about Scidmore and ultimately to tell her story.

It came as a great disappointment to learn early on that a relative destroyed most of Scidmore's personal papers soon after her death, reportedly at her wish. A final, decisive act, it would seem, of self-editing her life and controlling her own narrative. What fascinating secrets those records must have held. Evidence of romantic relationships, perhaps, or revelations of regrets and disappointments she experienced along with her many triumphs. The paucity of records helped explain why she remained little known over the years despite her historical significance.

My growing obsession with uncovering her life story took me on a zigzagging course: from Washington to New York and Boston; across the Midwest; as far as Japan and Alaska. To my delight, research at two dozen libraries and archives turned up a surprisingly rich trove of information about Scidmore, much of it hidden in plain sight for decades in historical newspapers and magazines, letters collections, photographs, public records, and other primary and secondary sources. Digitization and other advances in research tools helped me uncover about 800 of Scidmore's published articles, an important but little-examined body of work that shows she deserves greater recognition as one of America's pioneering female journalists (a contemporary of the early newspaperwomen whom historian Alice Fahs describes in *Out on Assignment*).[11] Story datelines and several hundred letters to friends and editors, many never revealed before, enabled me to piece together the course of Scidmore's life for the first time. The findings offer new insight into her environmental advocacy and her role in Washington's cherry trees and magnify her legacy by revealing her importance as one of the Western travel writers who "opened" China to mass tourism in the late-nineteenth century.

A driving question I had from the time I began pursuing Scidmore's story was how she managed to achieve all she did at a time when few women had careers and society frowned on female ambition. Happily, the historical record provided some answers. Gradually, a picture of Scidmore came into focus like a photograph that takes form in a chemical bath. Despite gaps in her story, the evidence points to a woman who was clearly exceptional: a trailblazer who melded the traits of journalist, explorer, geographer, and ethnographer; environmentalist; advocate of world peace; and collector of Oriental art.[12]

The late writer and editor Charles McCarry captured Scidmore deftly, I think, in an anthology of excerpts from *National Geographic*. Calling her "the best pure *National Geographic* writer the magazine ever had," he saw her "fluently confident" style as an expression of her personality: that of an independent, intelligent, and highly principled woman of the late-Victorian period "whose visible passions were those of the mind." Scidmore reminded him, McCarry said, of his maternal aunts—schoolteachers, travelers, suffragists. "Reading one of her stories was like reading a letter from my Aunt Carolyn, a teacher of geography who, every summer, set off for some interesting foreign destination aboard a tramp steamer."[13]

Though a maverick in many ways, Scidmore was not exceptional in her day as a woman who cherished her independence. A large number of women in the decades after the Civil War showed little interest in marrying. With gender constraints loosening, they discovered they liked the freedom of doing as they pleased, of having agency over their own lives.

Because Scidmore did not have family wealth or a spouse to depend on for support, she might be thought of today as a "self-made woman" who owed her success to the force of her own talent and drive. A freelancer for all of her forty years in journalism, she spent much of her adult life living out of apartments and hotel rooms and ship's cabins, packing and unpacking her steamer trunk for life on the road. Thanks to a wide circle of friends and professional contacts, she was often the guest of prominent people, yet she also endured, as a matter of course, the rigors of travel that could be quite challenging in many of the places she visited.

Though her personal circumstances were relatively modest, in the context of her day Scidmore traveled from a position of privilege as a white, well-educated American woman who enjoyed the advantages of her race, nationality, and family ties. She benefited from favorable foreign exchange rates in her travel costs and acquisition of Asian artifacts. Her brother's consular status ensured a measure of security and eased her access to places off the beaten track. Deference from local people opened the door to her acquaintance with "natives" up to the highest levels of power.

A real-life "Forrest Gump," Scidmore was an eyewitness to many historical events and rubbed elbows with famous people around the globe, from John Muir and Alexander Graham Bell to U.S. presidents and Japanese leaders. One of the most accomplished women of her day, she became so widely known that newspapers routinely reported the comings and goings of "Miss Scidmore." In the scope of her adventures, she was an American equivalent of intrepid British travelers like the famous Isabella Bird.

As with any life, Scidmore was full of contradictions. Though generally enlightened in her attitudes, she made derogatory and racist observations about some nationalities that echoed cultural biases and stereotypes typical of her era. As Nicholas Clifford writes in his study of British and American travel writing in China, Scidmore conveyed "a decidedly unflattering picture of hundreds of millions of Chinese, from one end of the empire to the other, as alike as peas in a pod."[14] Culturally insensitive language she used throughout her reporting, as in her references to boatmen in India as "black man-apes" and to an elderly Mexican woman in Santa Fe as "a shriveled old crone," makes us wince today.[15] Such passages are especially jarring from a woman who came to see her writings as a bridge to increase understanding between people of different cultures.

Even when she expressed abhorrent ideas, however, Scidmore grounded her writing in diligent reporting and research that won her the respect of scientists and other experts. As Clifford says of her reporting on China, Scidmore spoke, like Isabella Bird, "with the kind of authority thought to be reserved to men."[16] For all her faults, Scidmore's peripatetic life and keen reporter's eye make her a fascinating guide to the dynamic period of U.S. history from the Gilded Age to the Progressive Era, as seen from an all-too-rare female perspective.

This book is not a work of literary analysis or critical theory. Rather, my aim has been to write the first deeply researched biography of Eliza Scidmore, chronicling her life in a narrative style that makes her story accessible to a wide range of audiences. My research began not long before Scidmore was "rediscovered" during the centennial of Washington's Japanese cherry trees in 2012. Blogging about her on my website, I fielded inquiries from scores of people keen to know more about her: journalists, scholars, and authors; cherry blossom enthusiasts and Japanophiles; U.S. park rangers; women's studies bloggers; students and educators. Over a decade, interest has expanded such that even the Girl Scouts now include information about Scidmore in their Cherry Blossom Patch program. By providing a well-documented account of her life based on the best available evidence to date, I hope to satisfy the curiosity of general readers like these while giving scholars and other specialists a foundation for further investigation of Scidmore's life and deeper analysis of her work.

Any biography raises the implicit question of why the subject matters. Scidmore certainly matters as a "hidden figure" of historical importance whose life was long overlooked. Despite my diligent efforts in piecing together her story, much about her remains unknown. One thing that became clear in my research is that she zealously guarded her privacy and avoided personal references in her published work. In most cases it's not even clear who she was traveling with at various times.

Biographical writing is an iterative process of sleuthing, to connect the many dots of evidence that add up to a life. Over time, with additional research and scholarship, our knowledge of Scidmore will grow. In the meantime, I believe her achievements speak for themselves and merit attention. It is through her impressive body of published work that Eliza Scidmore tells us who she was.

PART 1
FOUNDATIONS

Child of the Frontier

It must have shocked people in Madison, Wisconsin, to hear that Mrs. Scidmore planned to up and move to Washington, D.C., in the middle of the Civil War. That she had two small children made it seem all the more fool hardy. But what was she to do? Three times married, and still she lacked the security that women of her era were led to expect in the traditional role of wife and mother. Remaining in Madison meant accepting family charity and being subject to pity and gossip. Unfortunate circumstances handed her a chance to take control of her own destiny, and in a leap of faith, she acted.

For her five-year-old daughter, Eliza—"Lillie," the family called her because mother and daughter shared the same first name—leaving Madison would be the latest in a series of moves that made the idea of home a fluid concept. So it would be for much of her life. Born October 14, 1856, in the Mississippi River town of Clinton, Iowa, during an itinerant period of her parents' marriage, she came into her early consciousness of the world in Madison, romping with her brother George and a brood of Sweeney cousins in her mother's large clan. Later in life, the writer Eliza Scidmore would feel such an association with Madison as her place of origin that she listed it at times as her birthplace on U.S. passports.[1]

Her roots on the American frontier became a part of her self-identity. Her forebears, who hailed from England and Ireland, included patriots who fought in the Revolutionary War and the War of 1812.[2] Later generations on both sides of her family left their homes in New York State to join the Great Westward Migration that expanded the contours of the United States in the middle of the nineteenth century. The American "West" of the time encompassed the vast region of the continent stretching from the Allegheny Mountains to the Pacific Ocean—an area where Native Americans were being pushed off their ancestral lands to make way for white settlers. To hundreds of thousands of Americans in the East and newly arrived immigrants, making a home in the western territory offered the prospect of freedom and self-sufficiency, and room for families to spread out. In an unpublished and heavily autobiographical novel Eliza Scidmore wrote late in life, one central character refers to residents of the West, only half-mockingly, as the "real Americans."[3]

To those seeking land, the rich soil and bountiful forests of Wisconsin loomed as a paradise. "The natural resources of Wisconsin are almost unlimited and nothing is wanted but the hand of cultivation to make it the garden of the world," one man wrote to his wife back East.[4] The Sweeneys arrived in Wisconsin during a period of heavy migration to the area in 1848, the year the territory joined the Union as the thirtieth state.

Madison, where Lillie Scidmore spent a formative period of her childhood, was an uncommonly pretty town on the frontier. Named for the fourth president shortly after his death, Madison grew from a thousand acres of primeval forest that the town's founders saw as an alluring spot for a state capital. Under their city plan, the dome-topped capitol building of white stone stood atop a rise of land between lakes Mendota and Monona, overlooking pristine waters whose shores of pebbles were fringed by rows of cedars. Avenues of white and burr oaks radiated outward from the capitol building, giving Madison the sylvan charm of an English park. Editor Horace Greeley of the *New-York Tribune*, America's most widely read newspaper, declared on a lecture tour that Madison had "the most magnificent site of any inland town I ever saw."[5] An "Interlaken of the West," some people called it.[6]

Lillie Scidmore's parents, George Bolles Scidmore and the former Eliza Catherine Sweeney, married in Madison a few weeks before Christmas in 1851. They likely met through Eliza C.'s brother-in-law, David Atwood, a talented political reporter who was soon to become the founding editor of the *Wisconsin State Journal*. The two men were both strong backers of the Free Soil Party, a short-lived but influential populist movement—later absorbed into the new Republican Party—that sought to prevent the extension of slavery into the western territories.[7] Eliza C.'s own views as an "ardent abolitionist" undoubtedly factored into the couple's mutual attraction, at a time when Northerners were still far from united on the issue of slavery.[8] At a deeper emotional level, George and Eliza C. shared an understanding of how hard and lonely life on the frontier could be without someone to share the burdens. Each was marrying for a third time.

George Scidmore, of Herkimer County, New York, had moved west in 1847 with his first wife to farm forty acres in Indiana's LaGrange County, near a homestead staked out by his parents.[9] By the time he married Eliza Catherine Sweeney four years later, George had buried two wives at the United Methodist Cemetery in Plato. At thirty-two years old, and with no record of surviving offspring, he may have ached for a family.

Eliza C. came into the relationship with two young children from her first marriage, Eddie and Fanny Brooks. To the pretty widow of twenty-eight with lively blue-gray eyes—the eyes her daughter Lillie would inherit—the marriage offered more than just the prospect of a making a new home. It gave her the chance of a fresh start, after a scandalous incident that had shaken her world a few years earlier, soon after she first arrived in Wisconsin with her family.

***

When the Sweeneys packed up their wagons in Ohio and moved north to Wisconsin in 1848, they all migrated together: five adult siblings and their elderly parents. The family patriarch, Connor

Murray Sweeney, was one of five sons of an Irishman who emigrated to America just after the Revolutionary War. He learned the hatter's trade as a young man and ran a shop in his native Tonawanda, New York, along the Niagara River, until he was widowed. After remarrying, he settled with his second wife at Canton, in northern Ohio's Stark County. There, the couple ran a store where Connor peddled felt and wool hats while his wife, Susan, did "a brisk business" selling fresh gingerbread and "pop beer."[10] The couple raised their five children—a son and four daughters—in rooms next door. Eliza Catherine Sweeney, the future mother of the writer Eliza Scidmore, was born in 1823 as their middle child.

Eliza C. developed her antislavery ideas while growing up in Ohio, and later passed those convictions on to her own children. The Sweeneys' neighbors in northern Ohio included Quakers, German-Americans, Yankees, and free blacks whose opposition to slavery made the area a hotbed of abolitionism.[11] As early as 1827, during the presidency of John Quincy Adams, Connor Sweeney joined town folk who called publicly for the freeing of slaves.[12]

After leaving home to marry at the age of nineteen, Eliza C. found herself widowed a few years later. When the Sweeneys looked to make a new home in Wisconsin, Eliza C. and her two young children joined them. The family settled initially at Potosi, in the southwest corner of the territory, where land opened up in a former lead-mining area. Gradually, they drifted sixty miles east to Madison.[13]

The "dearth of womankind" that plagued many settlements across the West made the Sweeney household a popular place after the family's arrival in southern Wisconsin.[14] A stream of suitors came calling, and in 1849, three of the four Sweeney sisters married within several months of one another. Eliza Catherine went first to the altar, remarrying in early May when southern Wisconsin vibrated with the greening shoots of spring. Her new husband, William W. Wyman, was a prominent businessman and a widower twice her age with several grown children. Originally a printer from New York, he started one of Madison's first newspapers, the *Express*, and owned a hotel.[15] A month later, Eliza's favorite sister, Jane, wed a local man named George Oakley. In late summer, the oldest of the girls, Mary Ann, married the newspaperman David Atwood, who bought the *Express* from Wyman and remade it as the *Wisconsin State Journal*.

The glow of the wedding season had not yet dimmed when a family crisis broke out: something turned sour in Eliza C.'s marriage to William Wyman. That fall, she left him and fled to her family, agonizing over whether to seek a divorce. The measure was drastic; divorce carried such a stigma that many women put up with a lot to avoid it. No court records exist that describe her complaints, but most divorces by women in the nineteenth century were granted on the grounds of a husband's cruelty, habitual drunkenness, desertion, or failure to provide for the family.[16] Her family must have found her concerns reasonable, as the Atwoods and a friend helped Eliza seek a divorce. Wyman fought the move, but eventually consented, on the condition that no "attempt was made to furnish proof [against him] of an infamous character."[17]

Eliza Catherine obtained the divorce in a neighboring county, possibly to minimize public exposure. To erase other traces of the marriage, she legally restored the use of her earlier married name; henceforth she would be known again as "Eliza C. Brooks."[18] With no home of her own, she had to depend on her family to take her in. She and her children lived alternately with her parents in Potosi and the Atwoods in Madison.[19]

The situation shattered her expectations of a secure future. And it was troubling to think what a divorce could mean for her social standing. She knew herself to be a good and moral woman, but reputation was everything in middle-class respectability.

***

As Eliza C. embarked on her new marriage to George Scidmore in 1851, the couple decided to move west of the Mississippi, where the region was booming under a fast-growing lumber industry, busy steamboat traffic, and the coming of the railroads. George sold his farmland in Illinois and took up insurance work.[20] In Dubuque, Iowa, Eliza C. gave birth to a son, George Hawthorne, on October 14, 1854. The couple's daughter Eliza arrived two years later, when the family was living sixty miles south in Clinton. The name they gave her, Eliza Ruhamah Scidmore, was a weighty one for a tiny baby, though she would grow fully into it in time. The distinctive middle name, derived from her paternal grandmother, was a biblical name meaning "having obtained mercy."[21]

To Eliza C., starting a second family in her early thirties, a baby girl must have seemed a special gift. Two years earlier, she had suffered the death of her older daughter, Fanny Brooks, at the age of nine.[22] Now Lillie would become the vessel of her mother's hopes and dreams.

Through their father's lineage, Lillie and her brother George were born ninth-generation Americans. The family's earliest ancestor in America, Thomas Skidmore (or Scudamore), emigrated from England to Boston during the period of the Massachusetts Bay Colony. The Scudamore family—whose ancestral home at Holme Lacy, in Herefordshire, survives today—dated its ancestry in England to the time of the Norman conquest (Figure 1.1).[23] As descendants of Thomas Skidmore branched out across America, they adopted many variants of the name. Some people who met the writer Eliza Scidmore as an adult recalled her tendency to correct anyone who addressed her as "Miss Skidmore" instead of pronouncing her name correctly as SID-more, with a silent "c."[24]

Around Lillie's first birthday, in the fall of 1857, the Scidmores' situation grew precarious when a banking crisis sparked a financial panic that spread across the United States. Thousands of businesses and private fortunes were wiped out. George found work managing an inn in Decatur, Nebraska, a new town on the upper Missouri River. Presuming the family accompanied him, Lillie, a blond and round-faced child, would have learned to walk

Figure 1.1. Holme Lacy, the Scudamore manor house in Herefordshire, England, as shown c. 1890.
(Copyright Francis Firth Collection)

in the long corridors of Brown's Hotel, a square frame building with banks of windows on every side. The views overlooked a still-raw settlement where many of the dwellings were cabins built along a creek, surrounded by woods that ran thick with turkeys and other wildlife. When the local railroad boom failed to materialize, the hotel went bust. By then, George Scidmore had likely exhausted local goodwill in a long-running dispute centered on "wrongful detention of property." In the 1859 trial, the court ruled that the plaintiff could seek to recover from Scidmore "one yoke of oxen valued at $85" and the costs of the suit.[25]

By the eve of the Civil War, the Scidmores were back in Madison, living on a farm. George and Eliza apparently took in boarders to raise cash, as the 1860 census for Dane County shows George heading a household of eighteen people. One of the other couples living with them had a little girl of three, the same age as Lillie.[26]

Eliza Catherine oversaw the cooking, cleaning, laundering, and other chores; given her later ease in handling money, she may have managed the household budget as well. All the work she did made her, in effect, an equal partner of George in providing for the family. Yet any fruits of her labor could never be her own. Legally a husband had proprietary claim to his wife's property, and even any wages she might earn. Victorian America valued women's work in the "domestic sphere," as wives, mothers, and caretakers. But money and politics were strictly the purview of men.[27]

A decade earlier, in 1848, activists at a convention in Seneca Falls, New York, had launched a movement calling for women to be given the same civil and social rights that men had, including the right to vote. Eliza Catherine Scidmore left no record of her views on the issue, though in later years she and Lillie would both act in ways that aligned with today's notions of feminism. At the least, Mrs. Scidmore would have understood that when it came to looking after their own interests, women had little power over their own lives and remained dependent on men. The unfairness of it must have rankled, for her own sake and her daughter Lillie's future.

***

Upon their return to Madison, Eliza C. and her family enjoyed the advantages of close ties to the Atwoods, now one of the most prominent families in town. Besides running the *Wisconsin State Journal*, David Atwood had influence stemming from a range of business activities and his activist role in the founding of the Republican Party in 1854. The grand home he built for his family in Madison's "silk-stocking district" of Mansion Hill was a meeting place for powerful men from all walks of life. Atwood had the kind of connections that opened doors.[28]

As part of his kinship obligations, Atwood hired his sister-in-law Mrs. Scidmore's older son, Edward P. Brooks, as an apprentice at the *Wisconsin State Journal*. The family's avid readership of newspapers gave the child Lillie her first glimpse of what would become her life's work. None of them could ever have imagined the journalistic success she herself would one day achieve, though her precociousness was evident at an early age.

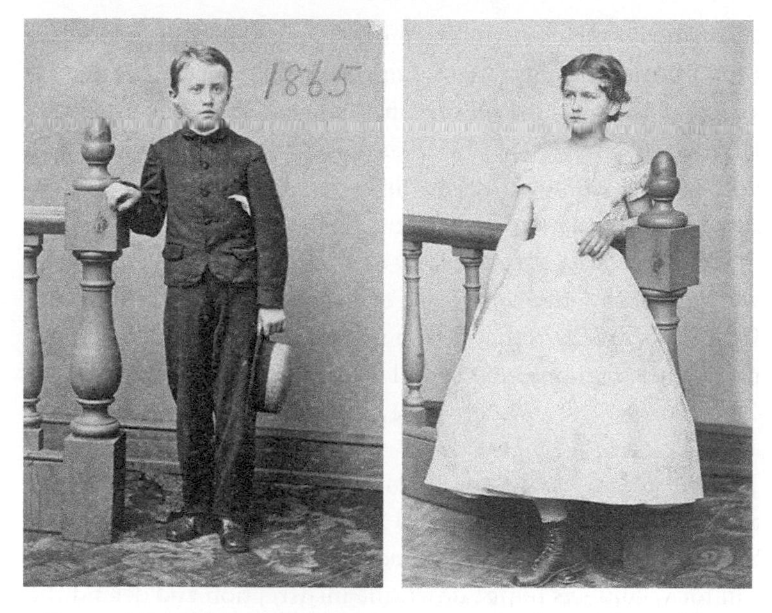

Figures 1.2 and 1.3. "Lillie" Scidmore at around age eight, and her brother George, two years older.
(Wisconsin Historical Society, WHI-91614 and WHI-90043)

Twin portraits of young Lillie and her brother George, two years older, taken around the end of the Civil War, suggest their temperamental differences, and offer clues to the adults they would become (Figures 1.2 and 1.3). The photo of George shows a sweet-looking child with soulful eyes and a pensive expression; the kind of boy who would grow into a man beloved for his courtesy, kindness, and "dignified, yet unassuming nature."[29] Lillie, in contrast, looks high-spirited, and possibly stubborn. Most striking in the slender little girl with an assertive pose and a firm-set mouth is what appears to be a great deal of self-possession for one so young. A strong sense of herself would be a lasting trait. Throughout her life, Lillie Scidmore would show little inclination to regard herself as her brother George's inferior, by age or by sex.

When it came to a true big brother, Eddie Brooks fit the bill. Blond, boyishly handsome, and thirteen years older than Lillie, he was a joy to be around. "Everyone liked Ed," a cousin recalled.[30] His comrades in the Civil War would laugh recalling how "Little Ed Brooks"—so called for his slight stature—amused everyone in camp with his pranks.[31]

Lillie was several months shy of her fifth birthday when she watched her brother march off to fight in the summer of 1861. Little did she realize that before the second year of the war had passed, her own childhood would be turned upside down.

***

The outbreak of the war in the spring of 1861 turned Madison into a military town. After the April attack by Confederates on Fort Sumter in South Carolina, President Abraham Lincoln issued a call for volunteers to put down the insurrection and defend the Union. Men and boys from across Wisconsin joined the rush to enlist. By summer, 2,000 recruits were in training at Camp Randall, on the former county fairgrounds just outside Madison. As they drilled for hours every day, the faint beating of drums floated across the meadow at the edge of town. Families and town folk often rode out to the camp late in the day to watch the men preparing for war: sons, husbands, brothers, and sweethearts.[32] Lillie's brother Edward Brooks joined up in mid-July, soon after turning eighteen.[33] He was assigned to the Sixth Wisconsin Infantry, a regiment that would gain fame at Gettysburg and other major battles as part of a hard-fighting unit of men from several western states who came to be known as the Iron Brigade.

On the evening of July 13, Eliza Catherine Scidmore was among a group of Madison residents who met in Assembly Hall to plan a banquet for the departing soldiers. "On a motion of Mrs. Scidmore," a local newspaper reported, several prominent women—including her sister Mary Ann Atwood—were elected to head the organizing committees.[34] Mrs. Scidmore herself offered to help raise funds. The plans called for farm families to provide chickens, boiled ham, and vegetables; women from town would donate sweets and decorate the dining hall. A week later, banners waved, and bands played as wagons rumbled through Madison on their way to the campgrounds. By late afternoon, 6,000 people from across the county had turned up for the supper, one of the largest patriotic gatherings held in the North.[35]

As the men prepared to set off, news came of the disastrous defeat of 35,000 Union troops in the First Battle of Bull Run, in Northern Virginia. Eddie and his comrades expected to be sent to the area but were assigned instead to help guard a line of fortifications ringing the U.S. capital. Eddie wrote home in late October, not long after Lillie and George both celebrated their birthdays, just two days apart. Lillie had turned five and George seven. Eddie and his comrades were camped in the wooded heights of Virginia overlooking Washington and the Potomac River. A mile away stood Arlington House, the stately former home of General Robert E. Lee that now provided a base of defense for the Army of the Potomac. The fine discipline among the Wisconsin men had done the state proud, Eddie reported, when Major General Irvin McDowell arrived with Secretary of State William Seward and other dignitaries to review the troops. "On the whole we are satisfied with our situation save that we wish to be nearer the rebels," Eddie wrote. The closest Confederate outpost was seven miles away.[36]

That winter, as Lillie felt her brother's absence, how much did she detect of the tensions at home that were about to shatter her world? Children who grow up to become writers tend to be observant; they notice things, even though the meaning of what they perceive may not be clear at the time.

The record is blank on exactly what happened. But around that time, George and Eliza Catherine Scidmore went their separate ways. The estrangement apparently became permanent, leaving Lillie and her brother George to grow up essentially fatherless.

George Bolles Scidmore remains the biggest enigma in his daughter Eliza's story. Though he is noticeably absent from the Sweeney family's photo albums, military pension records describe him as five-feet-ten-inches tall, with gray eyes and brown hair. His written statements in the files attest to his intelligence, and character references vouch for his "judicious" nature; there is nothing that suggests he was the kind of man who would abandon his family.[37] Still, at some point, George left home and struck out on his own for parts farther west.

A Scidmore family historian would later lay much of the blame for the failed marriage on Eliza C., describing her as "a dedicated social climber" for whom George Scidmore "never amounted to the kind of successful husband she envisaged." Yet George, the author conceded, was "hard on wives," and Eliza C. was no doubt unhappy with "all the moving and removing" in her life.[38] Social ambition may in fact have been a source of strength and aspiration as she strove to make a better life for herself and her children. Those hopes would not have accorded with the prospect of starting over in the harsh environment of the far West, a region regarded as "a heaven for men and dogs but a hell for women and oxen."[39]

For George, the siren call of the West may have derived in part from his political ideals as a "free-soiler." In May 1862, Congress fulfilled one of the movement's chief aims by passing a homestead law that allowed settlers to claim 160 acres of federal land. That summer, records show, George tried acquiring land in Colorado, though the deal fell through.[40] Later that year, while living in Pueblo, he enlisted at age forty-three in Colorado's Third Infantry Regiment. In subsequent transfers to cavalry units in Colorado and Missouri, he became a first lieutenant and worked as a recruiting officer. But military service left him a broken man. During a grueling Tennessee campaign in the fall off 1864, he contracted a severe case of dysentery that lasted several months, leaving him in a debilitated state that would affect his health the rest of his life. After several hospitalizations, he was discharged as "unfit for military duty."[41]

George married again after the war, but he and his fourth wife, a homeopathic doctor, remained together only two years.[42]

Another attempt to stake a homestead in Colorado failed when he relinquished his claim after a few months, likely because of poor health.[43] Records show that he moved often in the last three decades of his life, holding a string of jobs in a half-dozen states.[44] He died in March 1898, at age seventy-nine, of "old age and la grippe" (influenza) at a home for disabled soldiers in San Diego, California. His military pension records mention no family except for a sister he lived with in Indiana for a while after the war. It is not clear whether his children Eliza and George stayed in touch, though his obituary identified them as his survivors and described their careers.[45]

Even in his absence, George Bolles Scidmore had a major influence on his daughter Eliza's life. Children, and the adults they become, bear the marks of genetics and family dynamics. Eliza and her brother George owed their intelligence in part to their father. For Eliza, growing up outside conventional family life likely contributed to the maverick sensibility that led her to follow an independent life of adventure highly unusual for a woman of her day. The workaholic habits she exhibited in her many years as a travel writer may have stemmed in part from a deep longing for security.

In total, available records on George Bolles Scidmore suggest a man who, despite his obvious intelligence and talents, found it hard to ground himself and get a firm foothold in life. The chief legacy he left his daughter may have been a restless spirit that made her pine for the novelty of little-known places.

***

The ceaseless travel that the writer Eliza Scidmore pursued as an adult was a classic quest of leaving home to explore the world. Her mother's motivation in striking out from Madison seems more an act of desperation—or maybe liberation. Scholars have noted that the Civil War had a transforming effect on the lives of many women by making them more self-reliant and confident of their own abilities.[46]

Perhaps, in the end, Madison was too small for someone of Eliza Catherine Scidmore's energy and ambition. Growing up, Lillie Scidmore would have in her mother a model for how a white middle-class woman could prevail against the gender conventions of Victorian America through vision, pluck, resourcefulness, and charm.

It was a cold, wet spring in Madison in 1862 when Mrs. Scidmore sold off the family's household goods—a table here, a chair there.[47] Sometime in the weeks that followed, life changed radically for young Lillie. The turn in her destiny probably occurred one day at the height of summer, when grass in the meadow was high and horses' hooves kicked up dust in the road. One can imagine the scene as the little girl, wearing her Sunday best and maybe a straw hat, arrived at the train depot with her mother and climbed aboard a coach. Lugging parcels, they would have made their way down the aisle and settled into the plush seats, then leaned into the window to wave goodbye to family members who came to see them off.[48]

With the blasts of a shrill whistle and great puffs of white steam, the trip began that would carry them 900 miles across America's heartland. At the other end lay Washington, D.C., and a future far different from the life the little girl would have led had she never left Madison.

At the age of five, Eliza Scidmore set off on the first big journey of her life.

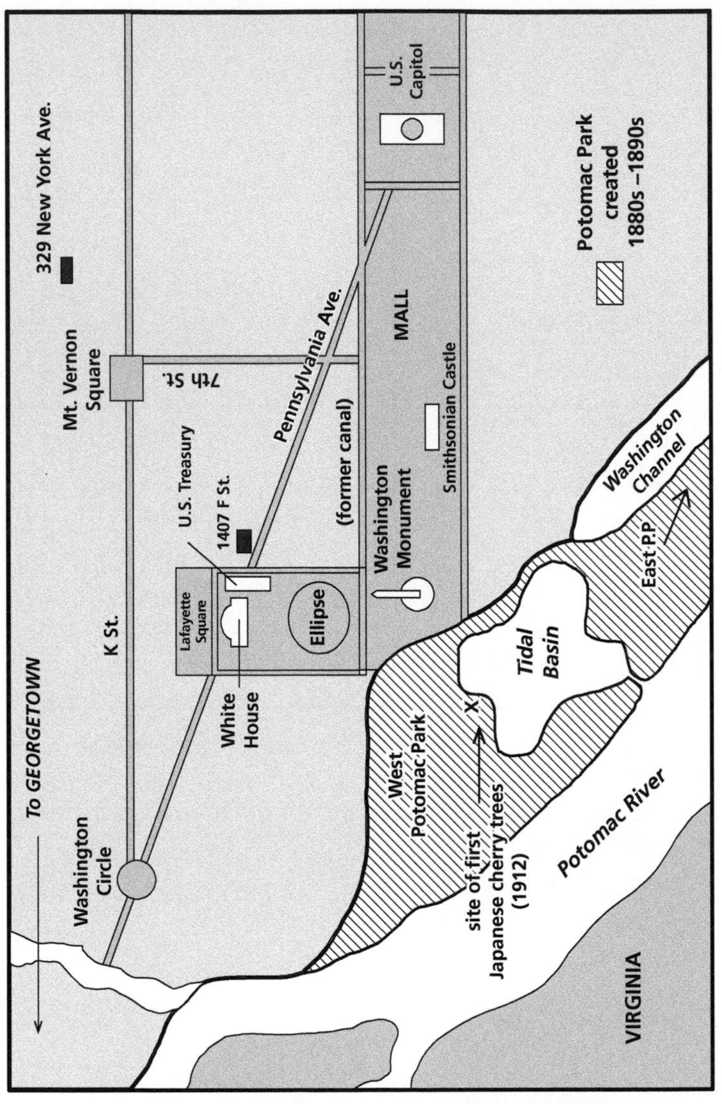

Washington, D.C., and National Mall

2

A Fresh Start

Eliza Catherine Scidmore needed a way to provide for herself and her children, and no place in America offered more opportunities than wartime Washington. In the summer of 1862, she joined the huge stream of people flowing into the capital, an influx so heavy the number of residents would triple to almost 200,000 during the war.[1] With thousands of troops assigned to guard the city, young Lillie Scidmore saw soldiers in blue jackets everywhere she looked. The crowds filling the streets included men and women from all around the country—young and old, rich and poor, and from every walk of life. They came to drive the wagons, run the stables, cook for the soldiers, and build the fortifications. Reporters came to cover the war; volunteers to nurse the casualties; family members to look for loved ones. Blacks fleeing slavery in the South came seeking sanctuary under the protection of Union troops, adding to the city's sizable community of free blacks.[2] As one federal official observed, "No other city ever presented such a conglomeration of humanity as may be found in Washington just now."[3] Under the crush, a local newspaper reported, the nation's capital was bursting at the seams, "crammed almost to suffocation."[4]

For Mrs. Scidmore, the crowded and chaotic conditions held the seeds of a way forward. To her daughter Lillie, the sights and sounds were an endless source of fascination in a strange new city that was about to become her adopted home.

The initial views must have been a letdown, as the city possessed little of the grandeur one would expect of a place that had been the nation's capital for more than sixty years. President George Washington himself had chosen the scenic ten-mile-square site along the Potomac River on which to build a permanent seat of government for the young nation. He and his fellow planners envisioned a capital that would one day be as beautiful as the great cities of Europe. The man they called on to render the concept into a city plan, Pierre Charles L'Enfant, a Paris-born military engineer who served with George Washington in the Revolutionary War, produced an elegant design featuring wide, diagonally intersecting avenues, a tight geometric grid of streets and neighborhoods, and lots of parks, both large and small. In one of the plan's most prominent features, he anchored the city's federal district on a spacious rectangle of green space—the National Mall—which stretched for a mile west of the U.S. Capitol, ending along the banks of the Potomac.[5]

After decades of sluggish and haphazard development, however, the grand plan of a handsome capital remained mostly a lofty idea. Like the experiment in democracy that America represented, Washington in the summer of 1862 was still a work in progress.

When newcomers like the Scidmores arrived by train at the depot a block north of the U.S. Capitol, they stepped into a city that seemed more a provincial Southern town than a vibrant urban center. Clusters of houses and shops stood surrounded by empty lots and large areas of blight. Most of the roads were rutted and unpaved. Washington looked rough and unfinished, with war operations magnifying the sense of disarray. At the far western end of the Mall, the tall stone monument intended as a memorial to George Washington himself had been abandoned midconstruction, prompting the author Mark Twain to mock the ugly stump as "a factory chimney with the top broken off."[6] Pens of cattle grazed in its shadows, awaiting slaughter to feed the hundreds of thousands of men now fighting on battlefields around the region. Cries of pain rang out from horse-drawn ambulances that rumbled through the streets carrying the sick and wounded to makeshift hospitals around town.

Even the U.S. Capitol, the most iconic building in America, looked pitifully sad and denuded. During a recent period of

Figure 2.1. West front of U.S. Capitol under renovation, c. 1860, with silted-up Washington Canal running along the National Mall.
(War Department Records, National Archives; via Wikimedia Commons)

national growth, wings had been added at each end to expand the chambers. Now, the top was sheared off to accommodate the installation of a larger dome made of cast iron, and a giant hole gaped at the center. President Lincoln had ordered construction to continue during the war, as an expression of faith in the nation's endurance (Figure 2.1).

Apart from its bedraggled appearance, Washington was a city that smelled at the height of summer. As mother and daughter navigated their way through the city, Mrs. Scidmore had to lift her skirts above the filth in the streets—piles of livestock manure and streams of slop water, urine, and rotting garbage. The most fetid smell came from the area south of the White House, where the former Washington Canal ran parallel to the northern edge of the Mall.[7] Built decades earlier as an innercity waterway to promote commerce, the canal had silted up over the years. Now it functioned essentially as an open sewer, a "slimy trench" that spilled its runoff onto the mud flats of the Potomac, just beyond the base of the Washington Monument.[8] When the foul waste festered under the hot sun, a sickening stench drifted north to the White House.

Even before the current onslaught of people, the city's antiquated sanitary system and water supply had been woefully inadequate. Tragically, President Lincoln and his wife, Mary, suffered the consequences. In February 1862, their eleven-year-old son Willie died of typhoid fever, leaving his mother inconsolable and the president to conduct the war under a dark cloud of personal grief.[9] Mr. Lincoln moved his family for the summer to a thirty-four-room Gothic Revival "cottage" at the Soldier's Home, three miles away, where the air was healthier. The president rode his horse back and forth every day to conduct the affairs of state and follow military dispatches in a room at the War Department, next door to the White House.

Confronting the city's miserable conditions must have given Mrs. Scidmore pause. What had she been thinking, in deciding to move her family to such a squalid place? Madison, in contrast, seemed idyllic. But there was no going back. At the age of thirty-nine she had made a clean break. In dreary and tumultuous Washington, Eliza Catherine Scidmore was determined to do whatever it took to make a new life.

***

With the city's public-education system still in its infancy, most families in Washington who sent their children to school enrolled them in private academies.[10] That fall, Mrs. Scidmore decided only one place would do for her gifted little girl: the Academy of the Visitation, in Georgetown, which many people considered the best girls' school in the country.[11]

Georgetown, a quaint and prosperous port town of 9,000 residents, lay three miles west of the White House.[12] Fortunately, getting there was now easier, thanks to a new horse-drawn streetcar service that started from the Navy Yard on Capitol Hill, ran on rails through the center of Washington, and reached the end

Figure 2.2. Visitation Academy in Georgetown.
(Photography Collection, New York Public Library)

of the line in Georgetown.[13] After getting to Georgetown, Mrs. Scidmore and Lillie had to climb a hill to get to the school, which was just down the street from Georgetown College. The Visitation complex consisted of the girls' academy, a monastery for eighty nuns, a chapel, an orchard, and spacious grounds cobbled together from several former estates. A high brick wall secluded the residents from outside distractions (Figure 2.2).[14]

The girls' school had been founded in 1799 by the Jesuit founder of Georgetown College to give Catholic families a place where they could send their daughters for a first-rate education grounded in Christian values. Before the war, its pupils came from all over the United States and even a few other countries. The Sisters of the Visitation who ran the school belonged to an order with roots in seventeenth-century France, where intellectual women led popular salons. In a like-minded spirit, Visitation aimed to turn out well-rounded young ladies who would use their talents in service to family and society. "Be who you are and be that well," the nuns counseled in the words of St. Francis de Sales, one of the order's founders.[15]

Lillie Scidmore was joining a community of girls who learned to regard themselves as exceptional. The academy taught practical skills and "feminine arts" expected of well-bred girls in the nineteenth century. But the school's fame came chiefly from a rigorous curriculum that included algebra, history and mythology, poetic composition, elocution, moral philosophy, languages, and several sciences. The notion that girls could master such subjects made Visitation a curiosity. Many foreign visitors toured the school, and U.S. presidents often attended the end-of-year ceremonies to watch the young ladies demonstrate their proficiency.

Lillie, at not quite six years old, was registered among 166 girls who hailed from sixteen states and the District of Columbia.[16] In deference to divided loyalties between North and South, the nuns banned all talk of the war except for expressions of concern for family members engaged in the conflict. Lillie's days at the school were highly regimented. A number assigned to her dictated where she sat, which washstand to use, what peg on which to hang her cloak and bonnet. She and her classmates sat through three hours of lessons every morning, followed by specialized instruction and study halls in the afternoon. Though Mrs. Scidmore was a practicing Lutheran, Lillie was expected to join the rest of the school for Catholic services in the cream-colored chapel. Except for hours set aside for socializing, silence prevailed—undoubtedly a challenge for a lively little girl who would grow into a garrulous young woman.

In the classroom, Lillie clearly blossomed. By the end of her first year, she showed aptitude in areas that presaged her future career. At the awards ceremony that year, she stepped forward to receive "premiums" in two subjects: reading and orthography (the rules of spelling), and geography.[17]

To a child with a curious mind, classroom geography offered a framework for making sense of the world—its places, people, physical features, and natural wonders. In an educational trend of the midnineteenth century, geography lessons included the copying of maps. Textbooks presented countries as being at different levels of "civilization," on a scale from "savage" to "enlightened," with white Christian nations on top.[18] So prevalent was such thinking in America and Europe that even decades later, as an experienced journalist, Eliza would convey some of that racist and culturally insensitive perspective in her own reporting on foreign countries.[19]

In an interview at the height of her success as a celebrated writer who roamed the world, Eliza attributed her passion for travel to her early fascination with geography. While other children enjoyed toys and games, she liked to amuse herself with maps and a globe—"always planning journeys," she said. "As a child I always cared for geography to the exclusion of almost everything else," she recalled. "My great pastime was to study maps and locate places. My daydreams were always of other countries."[20]

***

By 1864, Mrs. Scidmore had installed the family in a house at 329 New York Avenue NW, just east of Mount Vernon Square. In a half-dozen moves over the years, the family would remain in the heart of the city, living in the fifteen-block area that extended from Capitol Hill to the White House.[21] Young George and Lillie roamed the area around Seventh Street, a lively neighborhood of stores and offices, churches, restaurants, and hotels. Washingtonians came every day to shop at the Center Market on Pennsylvania Avenue, a jumble of food vendors on the present-day site of the National Archives. Mrs. Scidmore joined St. Paul's Lutheran Church, at Eleventh and H Streets.[22]

The household expanded when Mrs. Scidmore's youngest sister, Harriet, came to stay during the war. In her thirties, and still unmarried, Harriet Sweeney had used her connections in Wisconsin to acquire a job at the U.S. Treasury.[23] With male employees off fighting, a few federal managers were hiring women for the first time, as clerks and copyists, press and bindery workers, and weapons makers at the Arsenal and Navy Yard.[24] Women generally earned half of what men were paid for the same work.

But the money—in Harriet's case, fifty dollars a month—was still more than teachers and factory girls made, leading women around the country to seek work in Washington. Though meant to be temporary, the federal hiring of women would persist. By the time Lillie came of age, female workers like her Aunt Harriet were a common sight in the streets of the capital, going to and from their jobs and exploring the sights on their days off.

To earn a living, Mrs. Scidmore took in boarders, a line of work that would remain a financial lifeline for the family over the next decade.[25] Washington's perennial housing shortage forced many residents, even members of Congress, to live in boardinghouses or hotels. Widow-run boardinghouses were especially popular for their homey environment. Eliza Catherine Scidmore obviously concluded that widowhood offered certain advantages; at some point, she started passing herself off as a widow. Besides the practical benefits, the move may have helped her deflect awkward questions about her marriage at a time when her estranged third husband was still alive and living in the West.[26]

In the summer of 1864, the Scidmores shared the tension that gripped all of Washington as the city faced the threat of direct attack. A large force of troops under the Confederate General Jubal Early advanced on Washington, bent on an all-out assault. Local militias managed to repel them at Fort Stevens, where President Lincoln narrowly missed being shot when he rode out to watch the fighting.[27] Meanwhile, the family worried about the fate of George and Lillie's half-brother, Eddie Brooks. Now a lieutenant and adjutant of the Sixth Wisconsin Infantry, he had fought valiantly with his comrades at Gettysburg and other major battles before being captured by Confederate sympathizers in Virginia and imprisoned in Richmond. Mrs. Scidmore appealed personally to President Lincoln to obtain her son's release.[28] On March 18, 1864, the president sent a telegram to Major General Benjamin Butler requesting "if practicable" that Brooks be exchanged for "a rebel prisoner of same rank." Butler reported back five days later that Brooks had been freed.[29] Eddie had since rejoined his regiment and was off fighting again in Virginia.[30]

To support the war effort, Mrs. Scidmore visited sick and wounded soldiers. In August 1864, she aided Clement Warner, a young soldier from Windsor, Wisconsin, who lost an arm in a battle near Richmond. During his convalescence at the Armory Square Hospital, a thousand-bed complex on the site now occupied by the National Air and Space Museum, he dictated a letter home asking his younger sister to come care for him. He arranged for her to board with Mrs. Scidmore. As Sabra Warner recalled many years later: "Mrs. Scidmore received me with open arms saying: 'I know all about you. I have been to see your brother and am ready to receive you into my home and heart and be your mother while you are here.'"[31]

Because of its location near rail lines and a steamship landing, Armory Square received many of the worst casualties. The poet Walt Whitman spent hundreds of hours at the hospital comforting its patients.[32] Another frequent visitor was President Lincoln. One of the nurses recalled the "sad, far-away look" in the president's eyes as he passed through the wards, grasping the hand of each man and uttering "God bless you" to those in the greatest pain.[33]

Several blocks north of the hospital, the tall, gangly figure of Abraham Lincoln became familiar to Lillie and George when they played in the meadow near the White House and made a game of watching for the president as he came and went. On October 24, 1864, they rushed off with autograph books they had received as birthday presents from their mother. Lillie had just turned eight and George ten. When they approached the president to request his autograph, he pulled them onto his lap, according to one family account, and signed his name in their albums.[34] In his spidery script, he wrote in Lillie's small red-leather book:

For Miss Lillie Scidmore
A Lincoln.

The following spring, during Easter weekend, it shocked the family to get the news on April 15, 1865, that President Lincoln was dead. He had been shot the previous evening while attending a play at Ford's Theatre. Only two weeks earlier the city had erupted in celebration at the news that General Robert E. Lee and his Confederate army had surrendered at Appomattox Court House in Virginia, effectively ending the Civil War. Now, a pall of numbness fell across Washington.

A week after the president's death, a funeral train left town carrying his body home to Springfield, Illinois. On April 27 the train reached Buffalo, New York, where a hundred thousand people turned out to file past the coffin.[35] That same day, at some unidentified event back in Washington, Lillie managed to get the new president's autograph. Andrew Johnson entered his name in her album just below that of Abraham Lincoln.

***

As Lillie was outgrowing her adolescence in the early 1870s, Washington itself underwent a transformation that permanently altered its character. For months, the clanging and pounding of construction disrupted city life for the Scidmores and their boarders in the house they occupied on F Street NW, a few blocks east of the White House.[36]

The war had left the city in tatters. Visitors and residents alike were appalled by its shabby state. When a group of politicians and their supporters seized the opportunity to try moving the nation's capital farther west—preferably to an up-and-coming city like St. Louis—President Ulysses S. Grant and others demanded physical improvements.[37] In a blitz of public-works projects directed by city alderman Alexander "Boss" Shepherd, crews graded and paved the streets, laid 128 miles of sidewalks, installed 3,000 gaslights, modernized the water and sewer systems, and filled in the noxious Washington Canal to form a new roadway (today's Constitution Avenue). They planted 60,000 trees along the avenues and in public parks—maples, elms, lindens, and poplars, twenty varieties in all.[38] The runaway costs bankrupted the city and led to a loss of local control over the

District of Columbia. But everyone agreed the results were astonishing. President Grant boasted in his annual address to Congress, "Washington is rapidly becoming a city worthy of the Nation's Capital."[39]

The city's physical progress paralleled the Scidmores' improving status. A decade after arriving in Washington, they were moving up the ladder of social mobility, like millions of other Americans in the postwar era. A job Eliza Catherine Scidmore held for many years at the Treasury Department helped stabilize the family's financial situation. Hired initially in 1870 as a "matron," she reported for work every afternoon at four o'clock to supervise a crew of seventy-five women who came to scrub and polish the marbled interior.[40]

Lillie's half-brother Edward Brooks also contributed to the household while living on and off with the family. After the war, he and a partner had tried publishing a newspaper in Raleigh, North Carolina, the *Journal of Freedom*, catering to the region's huge population of black freedmen. Although the venture started off promising, the paper folded after white-owned businesses refused to buy ads.[41] Now Eddie was back in Washington, working as a correspondent for several dailies.[42] His work gave Lillie an insider's view of the city's newspaper industry, a scene she herself would soon be part of. The Scidmores lived just around the corner from Newspaper Row, on Fourteenth Street NW, where many of the country's leading papers had bureaus and a telegraph office stood on the corner. Two nearby hotels, the Willard and the Ebbitt, served as popular watering holes where cigar-chomping lobbyists and politicians fraternized with male reporters.[43]

By her late teens, Lillie had grown into a striking, dark-haired young woman whose final height of five foot seven inches made her fairly tall for a woman of her day. Though not conventionally pretty, she exuded a confidence and vitality that left strangers impressed by her social poise. "Sparkling in conversation," a Wisconsin reporter said after meeting her at a reception she attended with her mother.[44]

Ambitious for herself and her children, Mrs. Scidmore was pleased to discover that Washington society had a welcoming attitude toward newcomers. As the *Evening Star* boasted, there would always be a place in the capital for anyone "who may be so intelligent, entertaining, and well behaved as to prove an agreeable acquaintance."[45] Mrs. Scidmore threw herself into charity work and social-reform causes. She joined the temperance movement and supported the black activist Sojourner Truth in her efforts to secure government land in the West for formerly enslaved families in the South.[46] With her son George, Mrs. Scidmore became a charter member of a new congregation, Luther Place Memorial Church, that dedicated itself to improving conditions in the city's black communities.[47]

Lillie and George understood the hard work and sacrifices their mother made for the family's welfare, a devotion they would repay later in life. From her strong-willed and enterprising mother, Lillie learned important lessons that would serve her well: how to cultivate connections and straddle different worlds with ease; how to seize opportunities and push gender conventions without going outside the bounds of social respectability. In perhaps the strongest lesson, she learned from the example of her mother that marriage offered no guarantee of security and emotional fulfillment.

While George and Lillie were both intelligent, it must have seemed obvious to their mother that George's more reserved nature held him back in comparison with his gregarious younger sister. In the summer of 1873, a couple of years after George completed his education at a boys' school in Washington, Mrs. Scidmore obtained passports for herself and George. In July, they sailed out of New York on their way to Europe, Mrs. Scidmore carrying a signed letter from President Grant that served as an introduction to U.S. ministers and consuls abroad.[48] They landed in Liverpool, where the U.S. vice consul, Charles Atwood, was a close relative, as the son of Mrs. Scidmore's brother-in-law,

David Atwood.[49] In what may have been his mother's intention, the trip gave George a preview of what would become his life's work.

Upon returning home, George enrolled in law school at National University, a new institution that held evening classes to accommodate government clerks and other young men with full-time jobs.[50] In 1876, soon after being admitted to the D.C. Bar, he passed an exam that led to his appointment as a clerk in the U.S. Consular Service, under a new program that offered job security outside the whims of political patronage. That summer, George followed in the footsteps of his cousin Charles by crossing the Atlantic to take up his first consular assignment, in Liverpool.[51]

***

Mrs. Scidmore also sent her daughter to college, something women of Lillie's generation were pursuing in ever greater numbers.[52] In the fall of 1873, a few weeks before she turned seventeen, Lillie boarded a train with her trunk and rode 400 miles to Oberlin College, in northern Ohio. The choice of school may have been in part sentimental; Mrs. Scidmore had grown up nearby, in Canton, Ohio. Oberlin also represented values the family held dear. Oberlin College had been the first institution of higher learning in the country to admit black and female students as well as white men. The town of Oberlin played an active role in the Underground Railroad movement that aided the flight of enslaved people heading to freedom in Canada.

Oberlin had been founded by two Presbyterian ministers as a religious community to train Christian missionaries for the American frontier. The college's educational mission fused the zeal of evangelicalism with a dedication to humanitarian causes and social reform. Oberlin taught its students that success in life meant doing good works and being "deeply engaged in the great moral concerns" of the day.[53]

The college, with its campus of ivy-covered brick buildings and a student population of 1,400, dominated the leafy village of Oberlin.[54] Lillie had plenty of opportunities to make new friends, as nearly half the students were women. While the majority of female students attended the music conservatory or completed a shorter course of studies to become teachers, she intended to pursue the standard four-year academic program in the College of Arts and Sciences. To meet the admission requirements, which included basic proficiency in areas of English and Latin, classical texts, algebra and geometry, history, and government, she spent her first year in the Preparatory Department.[55] As a routine part of their studies, students were also required to study the Scriptures and attend daily prayer services.[56]

Though most female students at Oberlin lived in the three-story Ladies' Hall, Lillie boarded at a house on Forrest Street, several blocks away. Living off campus did not allow her greater freedom. Young women at Oberlin had to follow strict rules of conduct, designed to curb unsanctioned mingling of the sexes. They had to be in their rooms by 8 p.m., could use the library only during certain hours, and were not permitted to walk with a man on campus except with special permission. Oberlin admitted women and welcomed their presence as a socializing influence. At the end of the day, however, their education played second-fiddle to the mission of training men—"the leading sex"—as future ministers and Christian gentlemen.[57]

Lillie entered the collegiate program in the fall of 1874, one of six women in a freshmen class of 157 students.[58] Then, in a move that demonstrates her emerging independence and self-identity, she decided in 1875 to leave Oberlin. A hand-written note in her academic files explained that "Lillie Scidmore is going to Smith College." That fall, the *Oberlin Review* reported, "Lillie Scidmore is studying Spanish and French privately at her home in Washington, D.C. She intends completing her studies at Smith College."[59]

She had grown enamored with Smith, a new college in Northampton, Massachusetts, that was getting ready to open its doors. Sophia Smith, a New England woman of wealth, had decided to use her family inheritance to create an institution that would give young women access to an education equal to that at leading men's colleges. Smith College departed from the model of other women's colleges, like Mount Holyoke and Vassar, where students lived and studied in close quarters under the guidance of their teachers. Convinced that cloistered education made young women "unfeminine" and poorly socialized, Smith's founders created a residential campus involving few regulations and limited supervision. The college planned to offer a rigorous classical curriculum rather than a literary or "ladies" course.[60]

To Lillie, coming into a sense of her own potential, it must have sounded heavenly: an environment that offered the camaraderie of a girls' school while treating students like grown-ups. Smith would give her a chance to test her intellectual mettle in the company of some of the brightest young women of her day.

But it was not to be. For unknown reasons—related perhaps to the college's rigorous admission standards, as well as her family's financial situation after the Panic of 1873—Lillie Scidmore was not among the select group of women who entered Smith College after it welcomed its inaugural class in the fall 1875. The situation was a great blow to her pride and expectations. More essentially, it left her to face a burning question: what was she to do with her life?

World's Fair

It rained in Philadelphia on May 9, 1876, the day before the Centennial Exhibition opened its doors to the world.[1] Eliza knew that on this most consequential of days, she had no choice but to forge ahead amid the showers for the walk to Fairmount Park, on the western bank of the Schuylkill River, where the exhibition was being held. As she drew nearer, the scene ahead brought her to a halt: the pedestrian path was covered in slippery mud. Fortunately, she spotted a line of railroad tracks leading toward the park. Clutching her skirts and wielding her umbrella, she hopped onto the tracks and continued on her way, pausing now and then until the screech of an approaching train forced her to jump aside.[2]

She had come up from Washington to preview, on this Tuesday, the most-talked-about event in America. A century earlier, on July 4, 1776, the Founding Fathers had signed the Declaration of Independence at the statehouse in downtown Philadelphia, two and a half miles away. To celebrate a hundred years of nationhood, the country was now holding a grand exposition in "friendly competition" with three dozen other nations.[3] Though officially the "International Exhibition of Arts, Manufactures and Products of the Soil and Mine," everyone called it simply "the Centennial."[4] It was the first major world's fair held in the United States.

Twenty-five years earlier, Great Britain had dazzled the world with its innovative Crystal Palace exhibition in London's Hyde Park. There, Victorians gawked in wonder at steam engines, wrought-iron fireplaces, stained-glass windows, and other technological marvels, as well as displays of "exotic" objects and people from across the British empire. A few other European cities copied the concept with mixed success. Now it was America's turn to tout its progress, and the Centennial's planners had pulled out all the stops, determined to hold the biggest and splashiest event of its kind the world had ever seen. At Fairmount Park they built a mini city that combined the air of a country fair and an English village. More than 200 buildings and pavilions spread across 285 acres held a mind-boggling array of some 40,000 displays, featuring objects from the quirky to the sublime.[5]

Eliza would not have missed it for anything. Because her Uncle David Atwood served on the Centennial Commission, appointed by President Grant to represent the state of Wisconsin, her family had the benefit of VIP privileges.[6] But Eliza had come on a mission of her own. She was in Philadelphia to write about the exposition for the *National Republican* newspaper in Washington. Her presence at the historical event marked her debut as a reporter, at the age of nineteen.

***

Inside the park, Eliza found that the soggy weather had not dampened the public's enthusiasm. Crowds of people wandered around, too excited to wait until opening day. The Main Exhibition Hall, a massive but graceful structure of iron and glass, towered above the site like a modern cathedral. Said to be the largest building in the world, it covered twenty-one acres and stretched the length of six football fields; by one account it could hold half the population of Philadelphia. Next door stood Machinery Hall, the heartbeat of the exposition with its state-of-the-art industrial technology.[7] Seven miles of walkways crisscrossed the fairgrounds, and a narrow-gauge rail line circled the park to ferry visitors from one attraction to the next.

Touring the site to get oriented, Eliza discovered that many of the buildings and displays were not yet finished. She paused beneath her umbrella to watch carpenters hammering away in the rain. Outside the Art Gallery, workmen were hauling giant crated paintings and other works of art, many sent from as far away as Europe. In its reporting on the preparations, the *National Republican* noted that fifty cars of the Pennsylvania Railroad sat backed up on the tracks and several ships idled at the docks in Delaware, all waiting to unload hundreds of exhibits that were still arriving.[8]

Eliza owed her big break in journalism to her half-brother, Edward Brooks, now a senior editor at the *National Republican*. It may have been Brooks himself who provided the paper's traditional news coverage in Philadelphia while acting as his sister's chaperone and tutor. Eliza spent her first day reporting on the Women's Pavilion, an assignment the editors no doubt thought well suited to a young female correspondent, especially one as yet untested.

The pinkish-brown Women's Pavilion was a handsome building designed in the shape of a Maltese cross. To thousands of American women, it was a great source of pride. Never before had they had such a prominent forum in which to showcase the fruits of their labor. They themselves commissioned the building, after the fair's organizers refused to allocate space for women's work in the main hall, and they raised the $30,000 cost through Centennial tea parties and other fundraisers. But not everyone approved. Some women resented the segregation of their work from the main exhibits and refused to participate. Other critics dismissed the endeavor as a glorified church bazaar. Under Susan B. Anthony and Elizabeth Cady Stanton, a group of suffragists put out at being denied official space to advance their cause were secretly planning a demonstration at the exposition on July 4 to demand that women be given the vote.[9]

Inside the airy, light-filled building, Eliza found much to admire. Yes, she conceded, the exhibits included a great deal of ephemera: painted seashells, flowers fashioned from fish scales, even a whistle made from a pig's tail. If that was not ridiculous enough, the wife of an Arkansas dairy farmer would draw huge crowds all summer to her bas-relief bust of the literary heroine Iolanthe, carved in a giant block of butter. Eliza failed to see the appeal. "It seems a pity," she wrote a few weeks later upon viewing the good lady's artistry, "that a person with her evident talent should not work in some better or more enduring material than butter."[10]

The sillier items notwithstanding, Eliza told readers, women distinguished themselves well in the displays. Young ladies from the Cincinnati School of Design were exhibiting gorgeous furniture crafted from walnut, oak, cedar, and ebony. There was a lot of needlework, and some of it looked amateurish—"the usual worsted monstrosities of coarse stitches and horrible design," Eliza noted—but most of the textiles were exquisite.[11] Queen Victoria had sent many pieces from the Royal School of Needlework; nuns in Canada, gold- and silver-threaded ecclesiastical garments; Japanese women, embroidered silks; and Belgian women, fine lace work, including an ensemble of Brussels point lace that made Eliza swoon. "No description could give an adequate idea of its grace, beauty and delicacy," she reported. In an amusing contrast, glass display cases also featured dress-reform garments offered as an alternative to tight corsets and heavy Victorian clothing.

Down another gallery, Eliza studied a display of seventy-five patented inventions by women, ranging from household labor-saving devices to a night-signal system used by the federal government. She stood entranced for some time in front of one working model, designed to show the operation of a life-preserving mattress for steamboats. Inside a small aquarium, the miniature device floated around "rescuing three distressed-looking dolls from a watery grave."

On her swing through the building, Eliza came upon the latest flap involving Elizabeth Gillespie, head of the Women's Committee and the great-granddaughter of Benjamin Franklin. Mrs. Gillespie, Eliza informed readers, managed to offend many of the people she dealt with. Among her edicts, she had banned nudes in the women's building, prompting some of the country's leading female artists to boycott the Centennial. "Today she again distinguished herself," Eliza wrote, "by utterly refusing to allow Vinnie Ream to exhibit any of her work in the Women's Pavilion; her reason, she said, being that she personally disliked Miss Ream."

Eliza knew the incident would tickle readers back in Washington, where Vinnie Ream was a local darling. A young women of modest birth, Vinnie had taken sculpting lessons in her teens from an eminent artist on Capitol Hill, then stunned everyone a few years later by winning a $10,000 commission from Congress to sculpt a life-size statue of President Lincoln not long after his death.[12] The contretemps between Miss Ream and Mrs. Gillespie was happily resolved, Eliza reported, when Vinnie—"all smiles and dimples"—was promised a spot for her latest creation in the main Art Gallery.

***

Eliza proved to be a quick study as a reporter. Later that day, she wrote up a 1,500-word piece on what she saw, then sent it off the next morning to the *National Republican*'s offices in Washington. The article appeared May 12 on the front page, the first of a handful of "letters" on the Centennial that Eliza filed that week from Philadelphia.[13] They introduced readers to a new "special correspondent" writing under the pen name "Ruhamah." Within a few years the distinctive byline would be familiar to readers around the country, as editors excerpted passages from her work.

Those early dispatches from Philadelphia exhibited the characteristics that would distinguish Eliza's substantial record of journalistic writings. They showed her to be a fast and fluent writer with a keen eye for detail and a flair for description. In a quality that no doubt reflected both her youth and personality, she also conveyed an insouciant attitude in her witty asides and strong opinions—a style readers must have loved. During the week in Philadelphia, Eliza managed to complete her work so

Figure 3.1. *Harper's Weekly* illustration of opening-day ceremonies at the 1876 Centennial Exhibition in Philadelphia. (Library of Congress)

efficiently that she had time to enjoy the exhibition with friends who came up from Washington. One evening, they all climbed onto a "wagonette" and rode around town to see the patriotic displays.[14]

Eliza's reporting on the Centennial included an account of the opening-day ceremonies on Wednesday, May 10 (Figure 3.1). At dawn that morning, a great pealing of bells rang out in central Philadelphia, where Independence Hall and other public buildings were swaddled in red, white, and blue bunting. Residents had hung out the Stars and Stripes to welcome visitors, and flags of the guest nations lined the main avenue.

The rain had continued through the night but was tapering off to a light drizzle when Eliza joined the throngs of people streaming to the park. Arriving since dawn, they came by train, trolley, carriage, and on foot, some by steamboats that docked along the riverbank. By the time the gates opened at nine o'clock, more than a hundred thousand people were massed outside the seventeen entrances, ready to rush forward, plunk down fifty cents, and push their way through the turnstiles.[15]

Inside the park, Eliza dashed to claim a spot near the grandstand. By half past ten o'clock, when the program got underway, the sun was breaking through the clouds. As President Grant and his wife appeared and took their seats, an orchestra struck up a "Grand March," which the Women's Committee had commissioned from the famous German composer Richard Wagner. Eliza found it difficult to hear the music above the loud cheers that erupted from the crowd every time a dignitary crossed the stage. Benedictions and speeches followed, along with a hymn by John Greenleaf Whittier and a cantata sung by a chorus of 800 voices.

As the program droned on, Eliza studied the beaming figure at President Grant's side. He was Pedro II, the emperor of Brazil. An affable Renaissance man with a passion for science and technology, he had come to Philadelphia as the president's special guest. Back in Washington, "Dom Pedro" cut a dashing figure in the diplomatic community when he entered drawing rooms in his brilliant white uniform adorned with gold braid, oversized epaulettes, and a chest full of ribbons and medals.[16] On this day, Eliza reported, the forty-one-year-old emperor had donned "a plain suit of black, without tinsel or gold trimmings, and looked very much like an American citizen."[17] As the program concluded at noon—in a rousing hundred-gun salute and a choral rendition of Handel's "Hallelujah Chorus," accompanied by chimes, church bells, and factory whistles—Eliza watched the president and his entourage hurry off to Machinery Hall. Mrs. Grant, a short, bosomy woman in a pale gray dress, "never looked so well in her life," Eliza observed, and "seemed radiant from the honor of having a live Emperor as an escort."

The main attraction in Machinery Hall, the giant Corliss engine, offered a muscular symbol of the Centennial and all it was meant to celebrate. Named for its Rhode Island inventor and custom built to power the exhibits, the behemoth stood four-stories high, weighed seventy tons, and had a flywheel thirty feet in diameter. The husky, bearded President Grant mounted the steps of a high platform, Dom Pedro at his heels, and a sea of faces gazed upward to watch as the two men turned the switches to kick the mighty engine to life. Steam poured into the powerful cylinders, and within minutes the entire room vibrated with the pumping and squeaking and groaning of long rows of machinery—lathes and saws, drills and looms, presses and pumps, all cranking away.[18] To Eliza and the ten million other fairgoers who would witness the demonstration over the fair's six-month run, the moment was electrifying. Nothing better epitomized the talent for science and ingenuity that would fuel America's industrial age and turn the country into the world's technological powerhouse.[19] As a *Times of London* reporter put it: "The American invents as the Greek sculpted and the Italian painted: it is genius."[20]

A promise of great things to come permeated the air in Philadelphia. Two decades earlier the Civil War had torn the nation apart. Now the country was working to regain its footing. Despite the upheavals of Reconstruction, political scandals in the Grant administration, and the lingering effects of an economic depression, Americans were starting to feel optimistic again.

Eliza too felt buoyed by possibilities. At a major crossroads in her life, she had found direction. The path that took her to Philadelphia as a reporter was not the course she had pursued in aiming for admittance to Smith College. Had that worked out, the elite education would probably have led her to a more intellectual life; teaching other young women, perhaps, like the astronomer Maria Mitchell, who gained world fame for discovering a comet and went on to become a much-loved professor at Vassar College. It seems unlikely, however, given her strong streak of independence, that Eliza would have found fulfillment in a classroom. She hungered for learning and discovery, but books alone would never be enough to satisfy her curiosity.

***

"In answer to several inquiries, let me state that 'Ruhamah' still lives," Eliza wrote that fall in the *National Republican*.[21] She had returned to Philadelphia and was staying at the Globe Hotel, a thousand-room temporary hotel across the street from

Fairmount Park. Every day she roamed the fairgrounds making a systematic study of the exhibits and other attractions. The range and amount of stuff was stupefying, from Italian nudes and Norwegian codfish to Russian sealskins and South African diamonds. Little went unnoticed. As Eliza told readers: "I feel like a walking encyclopedia from the amount of knowledge that has been let in upon me through guide books, catalogues, circulars, cards, signs and disinterested parties."

Just as the Centennial marked a new phase in Eliza's life, it signaled America's transition from an agrarian past to a modern and more urbanized society. The goods and gadgets on display would create a mass craving for consumer products that promised to make everyday life easier and more pleasant. Eliza was quite taken with a sewing machine that stitched names in a wink; orders were flying off the shelf. She and the millions of other fairgoers sampled novelty foods such as bananas, which came wrapped in tinfoil and sold for ten cents apiece. Among the newfangled machines, she watched demonstrations of Otis elevators, which would pave the way to skyscrapers, and heard the *clackety-clack* of Remington typewriters, which would fill offices in those buildings.

As much as the exposition celebrated American progress, there were also harbingers of hard challenges facing a young nation that was still struggling to define itself and its values. In one fresh attraction that fall, Eliza viewed a peculiar piece of sculpture rising out of the ground beside an artificial lake. Installed in September, months behind schedule, the colossal copper arm holding aloft a torch offered a preview of the monumental work that French sculptor Frederic Bartholdi was building in his Paris studio.[22] Mounted atop its pedestal in New York Harbor a decade later, the Statue of Liberty would loom as a beacon of hope for millions of immigrants arriving on U.S. shores.

Another soon-to-be-iconic object had been unveiled that summer, on a day that represented America at its best and worst. On the steamy Sunday morning of June 25, a young Scottish-born teacher of the deaf named Alexander Graham Bell—a man who would one day become a friend and colleague of Eliza back in Washington—met with a committee of judges to demonstrate a newly patented device he called the telephone. Meanwhile, 2,000 miles away, Gen. George Armstrong Custer and his U.S. cavalry troops were fighting Northern Plains Indians in a fatal encounter at the Little Bighorn River in Montana. At the exposition, the Smithsonian Institution and Department of the Interior sponsored displays intended to educate visitors about Native American cultures. Perversely, the bows and arrows, beaded moccasins, tomahawks, and teepees only reinforced the public's views of America's native people as "primitive" and savage.[23]

For Eliza and other Americans, most knowledge of foreign people and cultures came from maps, schoolbooks, and newspapers. Now, the scene in Philadelphia offered a microcosm of the world. Some fairgoers complained that many of the foreigners, having abandoned their native costumes, did not look foreign enough. As the *New York Times* reported, a number had switched to Western-style coats, vests, and trousers after being harassed by crowds of men and boys "who hooted and shouted at them as if they had been animals of a strange species instead of visitors who were entitled only to the most courteous attention."[24]

In spite of such incidents, it gladdened Eliza's heart to see the overall mingling and spirit of tolerance that prevailed. "A more motley audience probably never attended a concert," she noted admiringly after turning up one afternoon for a performance of the popular Gilmore's Band in Memorial Hall.[25] She relied on prevailing stereotypes in her snapshot description of the many nationalities represented—French, Turkish, Egyptian, Spanish, Chinese, and Japanese. But her larger point stressed the marvel of the scene as the foreigners and all classes of Americans sat peaceably intermixed.

Of all the guest nations at the Centennial, none intrigued Americans more than Japan.[26] All winter, crowds had gathered in Fairmount Park to watch in amazement as Japanese carpenters constructed a house without the use of nails. Fairgoers who arrived with views of the long-isolated country as backward came away surprised by the design and craftsmanship of Japanese objects on display. The "Japonisme" craze that had caught on in Europe after Japan's participation at the 1867 Paris Exposition Universelle was about to sweep America as well.[27]

Eliza's own encounters with the Japanese in Philadelphia left her impressed by their manners and quiet dignity. In the Women's Pavilion, she had watched several Japanese commissioners remove their jackets to help their countrywomen arrange their embroideries and silk paintings. All the while, they answered her questions politely in soft, low voices—a sharp contrast to the brashness that seemed typical of American men and of those of some other nationalities.[28]

Clearly, the world had much to learn from the Japanese, Eliza concluded. In one of the most defining aspects of her long career as a reporter and travel writer, she herself would become, by the end of the century, one of the best-known interpreters of Japan and its people for American readers.

4

"Lady Writer"

Back in Washington that fall, Eliza found the capital in a strange state of suspense. The presidential election that pitted Republican Rutherford B. Hayes of Ohio against Democrat Samuel J. Tilden of New York had ended in a tight vote margin, clouded by allegations of fraud, and with no clear winner. "Every one concedes that the election could not have been worse managed," Eliza wrote in mid-December, in the first Washington column she filed for the *St. Louis Globe–Democrat*. Congress had recently reconvened, and the country "will be happy when they have set their mighty brains to work to untangle the election affairs."[1]

She had started writing occasionally for the *Globe–Democrat*, a prosperous and rapidly growing newspaper in the West, during the final weeks of the Centennial.[2] Her brother Edward Brooks may once again have been her steppingstone to the assignment. Brooks had been a colleague of the *Globe–Democrat*'s talented editor, Joseph McCullagh, in the small, tight-knit circle of men who reported from the capital after the war.[3] Soon after returning home from Philadelphia, Eliza supplanted another young woman as the *Globe–Democrat*'s society columnist in Washington.[4]

The holidays passed on a somber note, Eliza told readers in St. Louis. Under the lingering effects of the country's financial crisis, "a great deal of suffering and distress exists in the city which charity cannot alleviate."[5] The day before Christmas, she climbed Capitol Hill to take the pulse inside the marbled seat of government. Under the gray sky, in the biting winter chill, the building looked "cold and forlorn." The grounds had been plowed up and many old trees felled for an ambitious landscaping design by

Frederick Law Olmsted, the creator of Central Park in New York. The interior of the Capitol also felt dreary, its once-blue carpets now worn and faded, the halls reeking of stale tobacco smoke. In the public gallery of the House chamber, Eliza came upon vagabonds and unemployed men hanging out to keep warm.

The year 1877 opened with a snowstorm. Powdery flakes were falling fast at midmorning when Eliza left home at 1407 F Street NW and hurried two blocks west to the White House. She was about to cover one of the most important social events of the year, the presidential "levee" on New Year's Day. Thousands of people attended. The cold and snow did little to dampen the enthusiasm, Eliza reported, as it would be the last such reception hosted by President Grant.[6]

Eliza slipped into the mansion entrance on Pennsylvania Avenue, where the doormen had unfurled a cloth to protect the ladies' gowns and slippers. Inside, a Marine band played in the vestibule. By ten o'clock, members of the diplomatic community stood clustered in the Red Room, waiting to be announced and passed on to the president in the Blue Room. Eliza scribbled notes as the first family took their places in the receiving line. Mrs. Grant was wearing diamonds and black velvet with point lace, while her daughter had chosen a princess-style dress of maroon velvet, and her daughter-in-law a pale-blue silk gown with an overskirt of Brussels lace. As Eliza and the city's other society reporters knew, female readers followed such details closely to stay current with the latest fashions.

Stylish in his own way was a Chinese official who arrived in a lemon-yellow brocade robe lined in white fur. As he and the other ministers filed through the room, Eliza admitted defeat at trying to figure out their ranks from "the strips, straps and stripes of gold lace that make them what they are." U.S. army and navy officers strutted into the room like "gilt-edged roosters." Judges and members of Congress rounded out the dignitaries—"and a very dingy, ordinary set of blackbirds they seemed after the splendor that had preceded them."

Following a brief rest for the first family, policemen took up their posts, and the doormen opened the mansion to the public. Throngs of citizens poured inside for a perfunctory handshake with the president. The egalitarian nature of the event warmed Eliza's heart. "They could never be called the unwashed," she wrote, "for never did a crowd give plainer evidence of soap and water diligently applied." One young man with a pair of ice skates slung across his shoulder impressed her with his "cool self-possession." He had been skating on the river and decided to drop by the White House on his way home.

The reception was one of many visits, both personal and professional, that Eliza would make to the White House in nearly a dozen administrations stretching into the next century. The following January, she had a brush with her future legacy when she returned to report on the first New Year's Day levee hosted by President Hayes and his wife, Lucy. The contested presidential election had finally been resolved in Hayes's favor by a special commission, in a much-maligned compromise that effectively ended postslavery Reconstruction in the South.

The Hayeses had old friends staying with them over the Christmas holidays. John and Harriet Herron had come from Cincinnati with three of their eight children. Sixteen-year-old Helen—known as "Nellie"—was such a delight, the president noted in his diary, that she made the White House "alive with laughter, fun, and music."[7] Because she had not yet "come out" in society, Nellie spent much of the week cloistered in the first family's private quarters on the second floor. An exception was allowed on New Year's Day when she joined the reception downstairs. Holiday decorations made the mansion warm and cheery, and Nellie found the event, animated by thousands of well-wishers, a magical experience. She would fantasize for years of returning to the White House one day—and not as a guest. Back in Cincinnati later, she confided to girlfriends that she would only consider marrying a man with the talent to become president.[8]

As she worked the room that day, Eliza may have noticed Nellie, four years her junior, among the president's family and friends. Unbeknownst to the two young women, they were kindred souls: strong-willed, proudly nonconformist, and imbued with great curiosity about the world, which fueled a yearning to travel. Their paths would cross again many times over the years, sometimes in far corners of the globe. On a spring afternoon in 1909, Eliza would arrive at the White House to meet with the former Nellie Herron—by then, Mrs. William Howard Taft—to discuss a scheme that had captured their imagination: the planting of Japanese cherry trees in the new park that had taken shape south of the White House, near the grounds of the Washington Monument.

***

As she took up society reporting, Eliza joined a couple dozen women who wrote from the capital for newspapers around the country, in New York, Boston, Philadelphia, and Chicago; Syracuse, Sacramento, New Orleans, and Cleveland.[9] A few had worked in Washington for years. Sarah Jane Lippincott ("Grace Greenwood") and Mary Abigail Dodge ("Gail Hamilton") were both veterans with well-known bylines. Mary Clemmer Ames, who penned a long-running Washington column in the *New York Independent*, had such a devoted following that she became the highest-paid female journalist of the day. The most colorful of the lot was Emily Briggs, who wrote as "Olivia" for the *Philadelphia Press*. A tiny bird-like woman who seemed to turn up everywhere around town, she set a standard for society reporting from the capital with her long, colorful descriptions and a peppery style that veered at times into venomous attacks on people and things she despised, like the railroad lobby.[10]

Some of the women who were now her colleagues Eliza recognized from trips to Capitol Hill over the years. More than two decades earlier, Jane Swisshelm, a political correspondent for

Horace Greeley's *New-York Tribune*, had become the first woman admitted to the congressional press galleries. Now a dozen women had accreditation.[11] Sitting with friends in the public galleries to watch proceedings on the chamber floor, Eliza would have seen the female reporters in the balcony across the room, hunched over inkpots as they scribbled away next to the men.[12]

The small but growing corps of society reporters that now included Eliza owed their jobs to the rise of ostentatious living in Washington during the Grant administration. Under an onslaught of wealthy Republican politicians and influence-seekers, "the smell of new money" wafted through the halls of Congress.[13] The competition to impress grew so intense that society life was like "living over again in the last luxurious days of Louis XV," Emily Briggs reported.[14] *Nouveaux riche* families flush from overnight fortunes in railroads, mining, and manufacturing began spending the winter in Washington, where the social elite were more tolerant of newcomers than well-entrenched "bluebloods" in major cities like Boston, Philadelphia, and New York. Mark Twain and Charles Dudley satirized the city's "parvenus" in their 1873 novel *The Gilded Age*, which gave the era its name. "It doesn't need a crowbar to break your way into society there as it does in Philadelphia," one character explains to another. "It's democratic, Washington is. Money or beauty will open any door."[15]

"Lady writers," the press called the female reporters.[16] Editors hired them chiefly to cover society events and "women's news." But in a town where the social and political overlapped, the lines blurred. Eliza had the diligence that female reporters in Washington needed to make it as freelance correspondents. She also had an advantage in having grown up there. She knew the city like the back of her hand—its streets, its politics, its local culture and notable people. She knew its very pulse. Though her editors in St. Louis labeled her columns "Washington Gossip," she covered whatever caught her fancy. Readers west of the

Mississippi devoured it all, endlessly fascinated by the goings-on in the nation's capital.

The long, free-form columns Eliza wrote three times a week for the *Globe–Democrat* show tenacity in reporting more than artfulness of prose. A playful undertone suggests her fondness for Washington and its characters. A keen eye for detail that became a hallmark of her published work is illustrated in a March 1877 column describing President Hayes's arrival at the Senate to take the oath of office. The soles of his boots, Eliza wrote, "betrayed the creamy pinkness of new leather."[17]

***

Eliza's apprenticeship with the *Globe–Democrat* came as a lucky break. Its editor, Joseph McCullagh, who had started his own career at the paper, called the *Globe–Democrat* "the best school of journalism in the country."[18] Many of his colleagues regarded the former Civil War reporter and Washington correspondent as the best newsman in the business. His sharp news instincts and innovative practices helped make the *Globe–Democrat* a trendsetter in journalism.[19] The paper spent huge amounts on telegraph transmission to aid newsgathering. To cover news wherever it broke out—and scout out original stories—McCullagh dispatched a network of correspondents far and wide.[20] Eliza and other women benefited from his willingness to hire talent regardless of sex. A strong hands-on editor, McCullagh demanded clear writing and utmost accuracy, and he insisted that his correspondents refrain from interjecting themselves into their stories. Above all, he valued enterprise in reporting. Like most dailies, the *Globe–Democrat* generally paid six to ten dollars a column. But McCullagh pledged that the paper would cover expenses and award bonuses to any correspondents who went out of the way to get valuable or exclusive news.[21]

Eliza rose to the challenge. She craved a chance to travel; now the *Globe–Democrat* offered a way. She joined the paper when its editors were pushing to expand news coverage and gain new subscribers west of the Mississippi. To a young woman whose own character and identity had been forged in part by the ethos of the frontier, the prospect was irresistible. Was there any place in the country more exciting to explore than the American West?

On a late July day in 1877, Eliza and a female companion—possibly her mother—rolled westward by train. They left Washington only two days before railway workers shut down the tracks near Baltimore, sparking a wave of strikes and labor riots that would spread to other cities. During the journey to Chicago, the women made conversation—"in bad German and worse French"—with a blond Danish girl traveling out to California to meet her sweetheart.[22] The train went on to St. Paul, arriving the same day that Barnum's circus came to town. That afternoon, Eliza organized a side trip to see Minnehaha Falls, a tourist site made famous by Henry Wadsworth Longfellow's poem "The Song of Hiawatha." Eliza described for readers the glee she felt, during the return, when she brought the train to an unscheduled stop by pinning a handkerchief to the end of her umbrella and waving it out the window to flag the crew's attention.

An even bigger thrill occurred the next day, after they left St. Paul. Their "military escort," whom Eliza dubbed "Brass Buttons" in her newspaper account of the trip, arranged for them to meet the engineer in his cab. Eliza pressed for a ride at the front of the train, and the men indulged her. The officer offered his cloak, the engineer placed a cushion on the metal bench above the cowcatcher, and the two women hopped onto the seat. "For fifty miles," Eliza wrote, "we rushed over the country ahead of the train, getting fine views that were totally lost when viewed from the last platform." The experience was transcendent, filled with the joy of feeling at one with the sky and the vast landscape and the natural beauty of the surroundings. In the cool and cloudy afternoon, the smell of pines perfumed the air as forests and lakes sped by.

The women changed trains at the Northern Pacific junction in Brainerd, Minnesota, where frontiersmen, Black Hills miners, and a chatty party of Canadians and Englishmen climbed aboard.

At a stop further on, several Native American men in blankets and otter-skin hats—tomahawks at their sides—stood waiting to greet an arriving passenger. The women enjoyed the comfort of a Pullman sleeper as the train crossed the Mississippi River and entered the Red River Valley, the fertile prairie land of the Dakotas. Thousands of acres of wheat lay ready to be harvested. At one point Eliza climbed down to gather wildflowers.

Bismarck, their destination, was a four-year-old town along the Missouri River. The local mosquitoes were so "savage," Eliza told readers, that people walked around wearing head nets. Most of the town's 1,200 residents were men, who reminded Eliza of characters in a Bret Harte story.[23] Saloons and a few other businesses lined the dirt street, but the busiest activity came from mining (Figure 4.1). Long wagon trains rumbled along a 300-mile trail carrying supplies to Deadwood, the raucous gold-boom town in the Black Hills.

The women spent all of autumn in the Dakota Territory. They stayed in Bismarck and made trips to Standing Rock, Fort

Figure 4.1. Main street of Bismarck in 1877.
(Photo by Jay Hayes, Institute for Regional Studies, North Dakota State University, Fargo. No. 2029.9.4)

Abraham Lincoln, and Fort Rice. On an excursion to Fort Rice, twenty-eight miles away, they rode for several hours in a spring wagon, accompanied by ten armed men. It was Sioux country, Eliza reported, playing up the risk of danger that readers expected. "A little to my disappointment we rode the whole way without encountering a single hostile Indian, or having our blood run cold at the sound of a screeching war whoop."[24]

Fort Lincoln, just across the Missouri River from Bismarck, hardly looked like a fort at all; not so much as a rail fence surrounded its dozens of wooden buildings. Fort Lincoln was home to the U.S. Army's Seventh Cavalry, which had moved into the area a few years earlier to secure the expansion of the railroad. General George Custer had been the post commander until he departed on the campaign against the Sioux that led to his death at the Little Bighorn River.

During her trip, Eliza witnessed the effects of another aggressive displacement of Native Americans. That summer, newspaper headlines had riveted readers around the world with reports of battles between U.S. Army troops and a band of Nez Perce they were pursuing in present-day Idaho and Montana.

For years, the Nez Perce had lived peacefully, alongside white settlers, on their ancestral homelands in the Pacific Northwest. Under a treaty with the federal government, they occupied a reservation carved from the land. Then, after gold was discovered nearby, the government broke the treaty by ordering the Nez Perce to leave the region and move to a smaller reservation in Idaho. But a large number resisted. When open warfare broke out, a group of several hundred tribal members—including women, children, and old people—trekked north toward Canada, hoping to find sanctuary among a group of Lakota who had fled there with Sitting Bull after the Battle of the Little Bighorn. For three months, across a distance of more than a thousand miles, the Nez Perce fought the U.S. Army and eluded capture in a conflict that won them the admiration of much of the reading public. Finally, after an exhausting five-day battle in October 1877, Chief Joseph and his fellow Nez Perce surrendered in the Bear Paw Mountains of Montana, only forty miles south of the Canadian border.

As Eliza reported in one of several dispatches from her trip that fall, the 400 or so survivors who were taken prisoner passed through Bismarck and Fort Lincoln. They were on their way to Fort Leavenworth in Kansas, where they would spend the winter before being sent onward to the "Indian Territory" in Oklahoma. "In that abode of unhappy spirits," she wrote sympathetically in the *Globe–Democrat*, "their individuality and history as a race will cease. Mixed with other tribes and lost sight of in that unknown country, they will soon be forgotten, in spite of ills they suffered at civilized hands and the brave fight they made when peaceful endurance had ceased to be right."[25]

She described the scene as the Nez Perce arrived in Bismarck under military guard. Everyone in town turned out in the streets for a look at the great leader himself, Chief Joseph. Nobly handsome in a war bonnet and robes, and "bearing a melancholy weight," Eliza wrote, he passed by "like a captive king in a Roman triumph."

Late on a Saturday in mid-November, she watched as a group of about seventy-five tribal members set up camp at Fort Lincoln. With sleet and snow moving in, the women hurried to assemble the lodge poles and cover them with canvas and skins, while the children ran and played nearby. Eliza secured an introduction to an elderly medicine man. She visited him inside the smoky lodge, sitting on a pile of robes and blankets as a young member of the tribe served as interpreter. She described for readers the sadness and sense of defeat in the old man and his people. Unable to return to their valley, they felt the weight of impending extinction. They were also "maddened by the dishonesty and injustice of whites," Eliza wrote. Public sentiment was largely behind them, as the *New York Times* made clear in a scathing 1877 editorial declaring that the "war was, on the part of our Government, an unpardonable and frightful blunder."[26]

One day during her stay in the area, Eliza had a formal photograph of herself taken in Fort Lincoln, possibly as a record

Figure 4.2. Eliza Scidmore, around age twenty-one, photographed in Orlando Goff's studio at Fort Lincoln in the Dakota Territory, c. 1877.
(Wisconsin Historical Society, WHI-90045)

of a milestone birthday: she turned twenty-one that October (Figure 4.2). She posed in the studio of Orlando S. Goff, who took the first portrait of Chief Joseph around the same time.[27] The image of Eliza shows a young woman who appears fully at ease, her direct gaze and guileless pose a contrast to the coyness seen in many female portraits of the era. The unfussy, slapdash styling of her dark hair seems a nod to practicality in the rough travel conditions of the West. For the photo, Eliza pinned a silver horse-shaped brooch to her collar. A more powerful symbol is the book she cradles in her arm. The portrait seems to capture her emerging identity as a writer.

***

The resourcefulness Eliza exhibited as a roaming "special correspondent" set the pattern of her working life for the next decade. During the winter and spring, she reported from Washington. In the off months she hit the road for long periods, pausing every few days to mail travel dispatches to the *Globe–Democrat*.

In 1878 she went to Europe for the first time. She attended the Exposition Universelle in Paris, where her brother George was working as a consular clerk for the U.S. government, and went on to Berlin, Dresden, Venice, and Milan before returning home. A year later, on a trip to the West Coast, she described watching a hateful incident in which the labor agitator Denis Kearney harangued crowds in San Francisco with his message that "the Chinese must go."[28] In the spring of 1881 she traveled widely around Colorado and New Mexico with a group of friends, returning the last week in June in time to report from Newport and Saratoga.[29] For three years in a row she made sweeping trips along the Eastern seaboard, where a string of coastal resorts attracted many well-heeled Washingtonians and other members of America's elite.

When in Washington, Eliza worked out of the news bureau of Charles T. Murray, the *Globe–Democrat*'s chief correspondent in the capital. A highly accomplished journalist, Murray had an office on Newspaper Row and wrote for several newspapers with the help of a handful of reporters, including his wife. Eliza furnished most of the bureau's society material.[30]

Eliza scrambled to keep up in the winter of 1882, when Washington enjoyed one of the busiest and most brilliant social seasons the city had ever seen.[31] Oscar Wilde came to town, causing a sensation in his outlandish dress. The American author Henry James arrived as well, on his first visit to the capital. After the huge success of his latest novel, *The Portrait of a Lady*, Eliza noted, everyone in town wondered whether his prolonged stay meant that "he meditates some etchings of Washington life."[32] Accustomed to the cosmopolitan scene of Europe, where he spent most of his time, James sniffed at the provincialism of Washington's art and culture. But he found himself unexpectedly charmed by daily life in the capital, especially in contrast to the money-centered hustle of New York. A "City of Conversation," he called Washington, after passing time there among witty friends such as Henry and Clover Adams.[33]

June brought the start of the summer exodus. Like many other residents, Eliza postponed her departure until after the biggest event of the year: the June 30 hanging of Charles Guiteau for the murder of President James Garfield. Guiteau, a deranged office-seeker, had shot the popular new president a year earlier at the city's train depot, only a few months after the inauguration. The president's slow, agonizing death from his bullet wounds gripped the nation for months. Mrs. Scidmore secured passes to attend Guiteau's murder trial.[34] Eliza's boss, Charles Murray, covered the hanging, and she herself described the mood in Washington that day. At half past noon, a "strange quiet of suspense" hung over the city, she wrote.[35] The silence was broken a short time later when newsboys shouted that the execution had been carried out.

A few days later, Eliza set off on a long working vacation. She was traveling in a party of seven young women, including two from St. Louis, chaperoned by a railroad executive who was undoubtedly a relative of one of the girls. In Atlantic City, New Jersey, where they spent the July Fourth holiday, Eliza described the surf "black with bobbing heads" and young girls strolling the boulevard "buttoned and braided" in fashions inspired by the heroines of Henry James's novels. At Coney Island, New York, festive with electric lights and the strains of a brass band, Eliza tried soft-shell crabs and champagne for the first time.[36]

In Rhode Island, they visited Narragansett Pier, a favorite destination of many Washington people and certain families from St. Louis. They also spent time in aristocratic Newport, where wealthy Americans like the newspaper magnate James Gordon Bennett and Mrs. William Astor occupied behemoth "cottages."[37] After posh Newport, the former whaling colony of Nantucket seemed especially dreary with its sparse accommodations and grim New Englanders. "Hawthorne's characters," Eliza called them. The one large hotel was a boardinghouse that let rooms for twenty-one dollars a week. Eliza went shark hunting and guiltily described the pleasure: "It may be a cruel and murderous sport, but there is a strange excitement in waging war against such wild beasts."[38]

The itinerary included a stay in Saratoga Springs, New York, known for its horse racing and European-style mineral baths. "All the world seems bound to getting to Saratoga in August," Eliza said of the famed resort.[39] She found the place stuffy and pretentious. Its three grand hotels were full of overdressed women who came to "first drink the water to lose weight," she wrote, "then put on extra layers like tree rings."[40]

Her whirlwind travels ended that fall with a camping trip in the Adirondacks. Eliza and her companions had caught the craze for camping that many young women were now pursuing under a "physical culture" movement that promoted robust health and rigorous exercise, in defiance of society's assertions about the delicacy of the female constitution.[41] For two weeks, the young women stayed at a surveyor's camp, enjoying the freedom from laced-up Victorian life. Clad in flannel blouses and awning-cloth skirts, they forged through the woods on ten-mile hikes, climbed the mountains, and rowed on the lake with the vigor of athletes, Eliza reported in the *Globe–Democrat*.[42] They ate pancakes with maple syrup for supper and stayed up late, laughing and telling stories by the campfire as loons wailed in the blackness of the star-studded nights.

***

By the mid-1880s, Eliza had won the admiration and affection of many people in Washington—and almost certainly the envy of colleagues. "A charming and piquant young lady, beloved in society for her personal traits," one newspaper described her.[43] In an article on the local press corps, a reporter for the *St. Paul Daily Globe* sang her praises: "I mention Miss Scidmore first because I regard her as one of the very best lady correspondents at the capital in this line of business." Through a mix of society reporting and travel writing, the writer noted, "she earns more money with her pen every year than half the regular newspaper men."[44] The flattering portrait continued:

> Besides being a correspondent Miss Scidmore is a lady, and is received and treated as such wherever she goes. She is on intimate social relations at the White House, and with the family of every cabinet officer, and has the entrée to the most select social circles of the capital. Where other women are compelled to seek invitations for the purpose of doing journalistic work Miss Scidmore is invited as a guest. The consequences of this is her letters and dispatches on social matters may be taken as authority.

Her early years as a journalist were a period in which Eliza navigated her emerging identity. Like many women after the Civil War, she tested the boundaries of social conventions in a quest to find fulfillment outside traditional female roles. Before the war, women had looked for guidance from authors such as Catharine Beecher, an education reformer who supported the Victorian notion that women achieved their greatest purpose through domestic obligations as wives, mothers, and teachers. Now, Eliza and her contemporaries were getting a different message. In one such voice, the spinster author Louisa May Alcott, whose 1868 novel *Little Women* became a bestseller, urged young women to travel for intellectual enrichment.[45]

There is nothing in Eliza Scidmore's few surviving papers that makes it clear whether she remained single all her life by choice or circumstances. Yet clues to her way of thinking come from an unpublished novel she wrote late in life. A character named Eleanor, who shares many of Eliza's traits, bristles at the suggestion that widowed or unmarried women traveled "to find a husband." She defends the action of a young niece by declaring, "American women don't have to marry if they don't want to."[46]

In the growing sense of her individuality and independence, Eliza was coming to embody the "American girl" that Henry James portrayed in his novels and short stories. In *The Portrait of a Lady*, James presented the newspaperwoman Henrietta Stackpole, a close friend of the main character Isabel Archer, in a way that conveyed the public's general distaste toward women who made their living as journalists. But social attitudes were changing. While Henrietta seems at times repulsive for her pushiness and smug self-satisfaction, she also comes across as a sympathetic figure, admirable for her pluck, loyalty, self-sufficiency, and good heart.[47]

James modeled Henrietta in part on Kate Field, a celebrity journalist whom the *New York World* called "the brainiest woman in the United States."[48] A prolific and versatile writer, Field staged dramatic readings of her work and befriended powerful people like Mark Twain and the *New-York Tribune* editor Whitelaw Reid. She even started her own newspaper in Washington. Eliza met Field socially at gatherings of female reporters in the capital. Though Miss Field's flamboyance may have been a step too far for Eliza, she greatly admired Field, whose published work included a considerable amount of travel writing.[49] Kate Field offered a model for the heights of success women could achieve in journalism.

Travel writing would come to define Eliza's own career. Her future was about to be sealed in that regard when she set off on her latest travels in the summer of 1883. A trip quite out of the ordinary, it would help raise Americans' awareness of the little-known splendors of their own country.

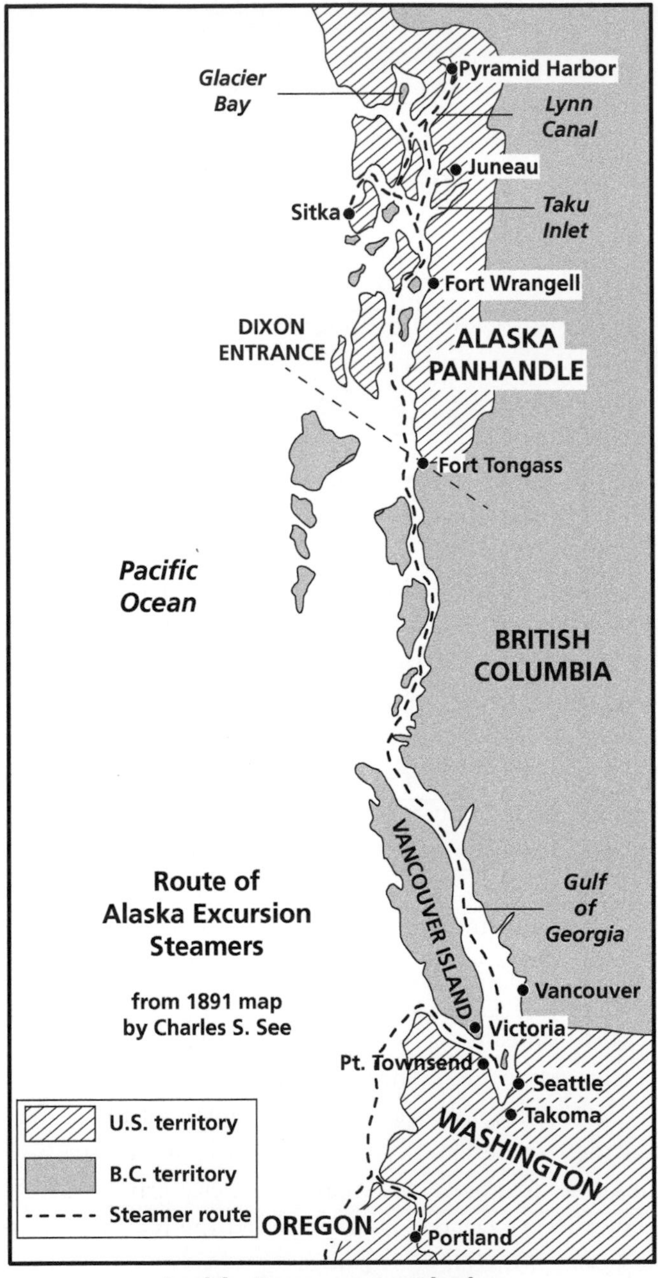

Inside Passage to Alaska

Inside Passage to Glacier Bay

"Portland to Sitka," the long purple ticket read.[1] It contained all the coupons Eliza needed for a seamless journey to Alaska and back. In the summer of 1883, as the final days of June were drifting into July, she arrived in Portland, Oregon, to meet her ship, after a long swing through California that included visits to Monterey and Yosemite, and a stagecoach journey by moonlight through the Sacramento Valley.

Anyone going to or from Alaska had to travel by way of the mail steamers. Leaving monthly from San Francisco, they touched at other ports in the Pacific Northwest to take on final passengers and cargo before starting the long push north. They sailed to Alaska along the Inside Passage, a thousand-mile lane of deep-sea channels off the coast of British Columbia. The route offered a smooth journey most of the way, thanks to an almost-continuous strip of islands that provided a buffer from the rough open waters of the Pacific. The steamers' monthly stops at Fort Wrangell, Juneau, Sitka, and other wilderness settlements gave those in the remote Alaskan territory their only link to the outside world.[2]

To an enterprising reporter like Eliza looking for the next big story, Alaska had the smell of opportunity. At a deeper, more personal level, the lure was visceral. Heading off to an area little touched by the hand of modern society stirred the restless and inquiring part of her soul that made her yearn to travel. The rawness and mystery of Alaska promised the real thing: a true journey of exploration.

Traveling as both a sightseer and a correspondent for the *St. Louis Globe–Democrat*, Eliza knew readers shared her curiosity. Alaska had been part of the United States only sixteen years. To most Americans, the territory still loomed as strange and foreign as the Congo.[3] The United States had acquired "Russian America" in 1867, changing the name to Alaska, from an Aleut word meaning "a great land." William H. Seward, who negotiated the deal as secretary of state in the Andrew Johnson administration, saw the vast region as a key piece of real estate in his "Manifest Destiny" vision of building an American empire that would cover the whole of North America and reach across the Pacific to new markets in Asia.

The purchase added almost 600,000 square miles to the United States—an area twice the size of Texas and three times as large as California. At a cost of $7.2 million, or about two cents an acre, the deal was a huge bargain. But critics saw the land as worthless at any price. Editorial cartoons lampooned "Seward's Icebox" and "Walrussia"—and the image stuck. In American parlors, references to Alaska brought to mind an inhospitable wasteland covered year-round in ice and snow, inhabited mostly by polar bears.[4]

It annoyed Eliza that jokes about the purchase "fastened themselves upon the public mind, and by constant repetition have been accepted as fact."[5] As a close watcher of national affairs in Washington, she knew the story of Alaska had yet to be fully told. She agreed with people like the scientist William Healey Dall who argued the need for greater knowledge of Alaska and its resources. Dall became the country's leading expert on Alaska after his own overland journey across Russian America in the 1860s, as part of a team sent to scout out a possible telegraph route to Europe by way of Siberia.[6] The mission ended after two years when undersea cables established a communications link across the Atlantic. But the expedition became important for American science when Dall and his colleagues carried home thousands of specimens documenting Alaska's flora, fauna, and geological record. Since then, Dall had spent years squirreled away in the Smithsonian Institution's red-sandstone castle on the National Mall, cataloguing the material and describing his findings. Eliza came to know him personally, and she relied heavily on his published work in her own reporting on Alaska.[7]

In a theme she would return to time and again in her reporting on Alaska, Eliza was critical of what she saw as Congress's virtual neglect of the territory. A few expeditions were sent to survey parts of the region and chart its coastal waters, and the U.S. Treasury benefited handsomely from a lucrative seal-fur industry off the tiny Pribilof Islands, in the Bering Sea. But amid more pressing national demands after the Civil War, there had been little support for sinking money into the sparsely inhabited area whose 32,000 residents were mostly native Alaskans.[8] Over the years, the lack of civil laws and self-government in the territory hindered settlement, land claims, and investment.

Now, stirrings in the north were prompting greater interest in the region. Prospectors were rushing north at reports of gold discoveries, and corporate mining and salmon canneries were mushrooming. Most intriguing to the average American, curiosity-seekers who filled empty berths on the mail steamers were returning from Alaska full of glowing accounts of its scenery and other attractions.[9]

***

For Eliza and a significant segment of the reading public, no one made a more compelling case for seeing Alaska than the naturalist John Muir. Though not yet the national conservation figurehead he would become, Muir had a devoted following for his eloquent articles in national magazines extolling the country's wilderness areas. After visiting Alaska for two summers in a row, he became an evangelist for the territory. "To the lover of pure wildness," he wrote, "Alaska is one of the most wonderful countries in the world."[10]

Muir, a lanky and intense Scottish-born man, first went to Alaska in the summer of 1879, after a decade of wandering in Yosemite Valley and the High Sierra. Alaska beckoned largely for its glaciers, one of Muir's longtime interests. At Fort Wrangell he met a young preacher, Samuel Hall Young, who shared Muir's interest in nature. The two men set off with four Tlingit Indian guides in a thirty-five-foot dugout canoe on what became an 800-mile journey through southeastern Alaska.

It was late October when they reached the mouth of the large bay at the foot of Mount Fairweather. *Sit-a-day-kay*, the local Hoonah Tlingit natives called it: "icy bay."[11] During the Pleistocene period the area had been covered by a thick ice sheet. When the ice melted thousands of years later, it created the large body of water surrounded by a horseshoe rim of mountains on which thick forests sprouted. With the arrival of another ice age, excess snow from the mountains funneled into the bay and hardened into glaciers. When Europeans first explored the area in the seventeenth century, the ice extended almost to the mouth of the bay. Since then, it had receded more than thirty miles north.[12]

Despite the onset of a winter storm, Muir insisted on camping for several days to explore the area. He roamed the glaciers amid sleet and howling winds, making sketches and taking notes. Finally, the bad weather forced the men to depart for Fort Wrangell. As they hurried away, Muir caught a glimpse of a magnificent glacier in the distance, the great "icy mountain" that seal hunters had reported seeing. Muir vowed to return, and did so the following summer, when he first studied the massive glacier that would soon bear his name. Its cluster of more than a dozen tributaries spilling ice into the bay formed a frozen prairie that extended, in Muir's estimation, for up to a thousand square miles.[13]

Muir described his Alaskan travels in a series of eleven "letters" in the San Francisco *Daily Evening Bulletin*. In his writings and frequent talks, Muir expressed many of the ideas that would make him a national voice for a conservation ethic. Experiences in the wilderness were not just an adventure, he proclaimed, but a means of cultivating deep spiritual connection with the natural world. He urged Americans to visit Alaska while it was still untamed. "Go," he preached, "go and see."[14]

Eliza took Muir's message to heart. Now she was following in his footsteps.

***

Awaiting the mail steamer's arrival in Portland, Eliza enjoyed the town—a vibrant, friendly place full of upbeat residents who seemed full of civic pride. From a rustic wooded settlement at the end of the Oregon Trail, Portland had grown into one of the region's most important ports, thanks to its deep-water harbor 120 miles up the Columbia River. A bustling waterfront, mercantile houses, and modern cast-iron buildings epitomized the economic boom that was transforming the Pacific Northwest with the coming of the railroads.[15]

Eliza's published accounts suggest she was traveling with another woman and her young son. They had planned to board the Alaska steamer in Portland, but changed their minds, possibly to avoid the sometimes-rough ocean passage up the long Pacific coastline. They left Portland instead on the morning of July 2, 1883, and traveled overland to meet their ship on one of its stops further north.

In Puget Sound, Eliza found the docks in a state of confusion. It was forest-fire season in the area. Clouds of dark smoke curled above the shores, smudging the blue summer sky. Telegraph lines were down in some places, agents informed her, so the shipping news was sporadic. The Pacific Coast Steamship Company had assigned the *Idaho* to make the July run to Alaska, but no one could say exactly where it was.

"For two restless and uncertain days," Eliza reported, "we seesawed from British to American soil, going back and forth from Victoria to Port Townsend as we were in turn assured that the ship lay at anchor in one place, would not go to the other, and that we ran the risk of losing the whole trip if we did not immediately embark for the opposite shore." Amid the hubbub of ships coming and going, she made inquiries while overseeing the transfer of baggage. Despite the anxiety-provoking moments, Eliza played up the drama for readers. As she wrote:

> The dock hands came to know us, the pilots touched their hats to us, the agents fled from their ticket-offices at the sight of us, and I think even the custom-house officers must have watched suspiciously, when the same two women and one small boy paced impatiently up and down the various wharves at that end of Puget Sound.[16]

As the second day passed, the travelers found themselves in Port Townsend, Washington, a busy U.S. customs port built at the base of a cliff. Fourth of July revelry charged the air. Eliza slept fitfully that night, as firecrackers exploded and rockets hissed. The off-key singing of drunken men floated up through the windows of her hotel.

The next morning, as the town awakened on a more subdued note, Eliza and her companions hurried down to the wharf. She felt a heavy weight give way to relief when a wisp of smoke appeared in the distance. It was the *Idaho*, which announced its approach with the firing of a small cannon. As the steamer slipped into the wharf, Eliza got her first view of the ship that would be her home for the next three weeks.

***

The *Idaho* was a trim and handsome wooden steamship, 215 feet long, with a shiny black hull and a row of white cabins lining the deck (Figure 5.1). Built just after the Civil War, it had been put into service in 1867, the same year the United States acquired Alaska. Many of the steamers that carried railroad workers and supplies to wilderness camps across the Pacific Northwest were old sternwheelers, with shallow hulls and flat bottoms that allowed them to nose close to shores and riverbanks. The *Idaho* was a modern ship, driven by a screw propeller.[17]

Figure 5.1. Steamship *Idaho* at dock in Juneau, 1887.
(Photo by William H. Partridge, Alaska State Library)

In July 1883, the *Idaho* was under the command of Captain James C. Carroll, one of the most experienced navigators in the Pacific Northwest. The son of Irish immigrants, he had gone to sea in the 1850s and traveled the world before joining the Pacific Coast Steamship Company as a regular on its Alaska route. In 1880, he was in command of the *California* when it carried two prospectors—Richard Harris of Cleveland, Ohio, and Joe Juneau, a French-Canadian from Quebec—up the coast to check out rumors of gold. Their discovery of high-grade deposits set off Alaska's first gold rush and gave James Carroll a stake in the earliest mining claims. He had a financial foothold in the new town of Juneau, where the wharf was named for him.[18]

Eliza took her place at the rails as the *Idaho* began its journey through the Gulf of Georgia, between Vancouver Island and mainland British Columbia. Around her, the San Juan Islands slowly floated past, their meadows dotted with sheep and lime kilns. For the next several hours, the low hum of the engines and the soothing rhythm of the ship's motion cast a blanket of contentment across the deck. Everyone ambled about admiring the scenery and poring over pink maps of the British Columbia coast.

At nine o'clock that night, the steamer pulled into the sleepy port of Nanaimo, on the north side of Vancouver Island, to dock for several hours while it took on coal.

In the days that followed, Eliza marveled at the astonishing beauty of the landscape, which unfolded in one stunning scene after another. The steamer threaded its way through straits and channels surrounded by pristine wilderness. Cliff-like mountains spouted foaming white cataracts. In some stretches, forests covered the slopes, while the emerald-green water along the shoreline mirrored every tree and twig. Eagles soared overhead in long, lazy sweeps, and hundreds of young ducks fluttered about the ship's bow. One sunny afternoon in a broad channel where the Fraser River spilled into the sea, cries of "Whales! Whales!" drew everyone to the railing. Eliza stood with the others watching for an hour as the giant creatures surfaced and dived, performing their slow acrobatics.

After several days, the *Idaho* crossed into Alaskan waters south of Prince of Wales Island, at the Dixon Entrance. During a stop at Fort Tongass, near present-day Ketchikan, Eliza and the others toured a busy salmon-canning factory, where the work was being done by local Haida natives and a few Chinese employees. Further up the channel, at Naha Bay, the salmon ran so thick in mid-July that "the turn of the tide of their splashing was like falling rain."[19]

Like all travelers, the *Idaho*'s passengers were keen to buy souvenirs. But with steady income from the salmon trade, Eliza told readers, the Haida felt no need to part with their old, well-crafted bracelets and other fine curios for bargain prices. Another local specialty, intricately carved spoons made from the horns of mountain goats, proved hard to get because a local trader had bought up all the stock. He strolled the deck of the ship offering to sell them for the inflated price of two dollars each.[20]

A stop at Fort Wrangell, at the mouth of the Stikine River, lived up to John Muir's reports of a dreary place—as he said of the town, "the roughest I ever saw."[21] Formerly a Russian stockade and U.S. Army garrison, Fort Wrangell grew wild and rough during an 1870s gold rush to the Cassiar mining district of British Columbia. But the gold deposits proved anemic, and the town's glory days had faded by the time of Eliza's visit. She and her companions found the streets quiet, with a Presbyterian mission occupying the grounds of the old fort. One element worthy of exploration were the giant carved *totem* poles that stood outside many of the modest Tlingit dwellings.

Eliza and her fellow passengers got their first thrilling experience of glaciers further up the coast at the Taku Inlet, southeast of the new mining town of Juneau. After the ship dropped anchor, she and the others climbed into boats and were rowed ashore. In single file, they marched across the moraine and up the slopes of the glacier, their footsteps crunching on the rough, gray ice. At one point they came to a wide rift in the rocks. Everyone moved in closer and huddled to listen as a priest in the group picked up a small boulder and hurled it into the crack. The air was absolutely still as they waited for the splash—until the moment was spoiled by a young man who plopped onto the ground and started munching soda crackers that he pulled from a brown paper sack. During the adventure, some in the group grew so giddy they raced across the glacier playing "snap the whip." When the steamer's whistle called them all back, they returned bedraggled but elated. Later, crew members passed by the cabins collecting the muddied clothing and boots, which they carried off to the engine room to dry.

The stop at Juneau, located in a curve of the Gastineau Channel against a stunning mountain backdrop, was marred by heavy rain. Eliza and the others bobbed their way to town on planks laid atop the ankle-deep mud. Several hundred Tlingit natives lived in Juneau along with a couple hundred white inhabitants, though many residents were off for the summer in mining fields or salmon fishing. The Tlingit women who stayed behind crouched on the wharf and wandered the streets—in calico dresses, blue blankets, and lip and nose rings—peddling baskets and other hand-crafted items. Eliza noted with amusement a sign in a barber's shop offering "Russian Baths" every Saturday for fifty cents each. As she told readers, it showed "that the luxuries of civilization are creeping in."[22] At half past five the next morning, a few hours before the ship's scheduled departure, she led a group off through a wooded trail to visit the Silver Bow mining camp.

***

The *Idaho* reached the upper end of its route ninety miles north of Juneau, in the long, narrow Lynn Canal. A thin, jutting nose of land divided the waters into two inlets. To the left lay Davidson Glacier—named for the American astronomer and Alaska explorer George Davidson—and north of that, the native lands of the Chilkat branch of the Tlingit.[23] The eastern inlet provided access to the traditional trade route of the Chilkoot Tlingit, which later became a popular highway for hordes of miners during the Klondike gold rush in the late 1890s.[24]

Before reversing course for the homeward journey, the *Idaho* docked overnight at Pyramid Harbor, a tiny settlement on the shores of the Chilkat inlet. As the steamer neared the banks, Eliza puzzled over what looked to be strips of red flannel. They turned out to be salmon, hung out to dry. "A Chilkat salmon," she informed readers, "is as bright a color, when caught, as a lobster after it has been boiled."[25] At midsummer, the sun did not set until half past nine. She sat on deck until midnight, reading from the residual glow of light reflected off the rocks. Some passengers lounged at the rails scanning the area for bear cubs, while those with tripod cameras fussed for hours trying to capture the faint, ethereal crescent moon in the eastern sky.

The next day, the *Idaho* made a detour from its usual route. Instead of heading straight south out of Lynn Canal, the ship veered west into the Icy Strait. Among its cargo, the steamer was carrying supplies for a new trading post, located somewhere just inside the mouth of Glacier Bay. At the southern entrance to the bay, the *Idaho* turned north and made its way through bobbing floes of ice.

The destination was Bartlett Cove, where a doughty settler named Dick Willoughby had recently built a log house and neighboring store. One of Alaska's earliest white pioneers, Willoughby had mining claims and trade ventures throughout the territory, and was popular in Juneau's saloon district for his fiddle-playing and dance-calling skills.[26] As the *Idaho* cruised the southern waters of the bay, Eliza and the others stood on deck pointing their field glasses toward the wooded shores, looking for signs of the settlement. Around noon, a white man and two Indians out hunting sea otters paddled up to the *Idaho* and offered to guide the steamer to Willoughby's place.

On shore, while the cargo was being unloaded, Eliza hurried off with several other women to roam the beach and explore the settlement. Inside Willoughby's store, which he had built next to a small village of Hoonah natives, the women pawed through the jumble of merchandise—lumber, salt, fishing nets, and barrel staves—looking for baskets and other Native American crafts. Outside the log cabin, Eliza complimented Willoughby on the fine appearance of his small garden, which had a large crop of strawberries coming on.

Touring the nearby Hoonah village, or "rancherie," Eliza and her companions pushed their way inside some of the dwellings for a closer look. In one tent, an Indian hunter, his face stripped of flesh, lay dying of wounds from a bear attack. In another, the women came upon an old man stooped over a cooking pot. "We peered into the family kettle," Eliza said, "and saw the black flippers waving in the simmering waters like human hands." The old man gestured his delight at the local delicacy. "Seal," he told them, tasted "same as hog."[27]

Back at the store, Eliza hovered near Captain Carroll and Dick Willoughby as they sat talking about the massive glacier thirty miles to the north—the great icy wall that native Alaskans and John Muir had described. The thunder from its falling ice was so powerful at times, Willoughby said, that it rattled the teacups in his cabin.

Willoughby had traveled with Captain Lester Beardslee of the U.S. Navy when he explored the bay in 1880, around the time of John Muir's second journey to Alaska. Beardslee and his men charted some of the waters of the lower bay, but heavy fog and strong tides of floating ice prevented them from going further north. Later, drawing on information from various accounts, Beardslee produced the first rough sketch of the bay. The name he gave the body of water—Glacier Bay—became official.[28]

Captain Carroll and his pilot, W. E. George, had copies of Beardslee's charts on board. Now, in the summer of 1883, in a period when the weather was mild and the waters calm, James Carroll made a momentous decision: he would take the *Idaho* north into the unknown waters to go in search of the mighty glacier he had been hearing about for years. No ship had ever sailed before into the upper reaches of the bay.

***

It was a bright, clear morning on the day in mid-July when the *Idaho* left Bartlett Cove and began its journey north. Standing on deck in a spot near the wheelhouse, breathing in the sharp, stringent air, Eliza felt giddy at the sense of adventure. "Away we went," she wrote later, "coursing up Glacier Bay, a fleet of one hundred and twelve little icebergs gayly sailing out to meet us, as we left our anchorage."[29] As the steamer inched its way ahead, the crew cast lead-weighted lines into the gray-green waters to take depth soundings and ensure the ship's safe passage.

Dick Willoughby had come along as a guide, and he stood on the bridge alongside Captain Carroll and the pilot as Eliza peppered the men with questions. Before long, a small green and hilly island near the mouth of the bay came into view. Eliza got one of the biggest surprises of her life when Carroll and his pilot turned to her and announced they would name the island for her. On the ship's log, they recorded the name "Scidmore Island." In her account of the incident, Eliza teasingly told readers: "Heated and suffering humanity is invited to visit that emerald spot . . . and enjoy the July temperature of 45°, the seal and salmon fishing, the fine hunting, and the sight of one of the grandest of the many great glaciers that break directly into the sea along the Alaska coast."[30]

The *Idaho* advanced cautiously up the bay, veering left and right to avoid ice floes—some as big as houses—that floated in on the tides. Near one grounded iceberg, crew members cast the anchor, then rowed off with a few photographers to get a picture of the ship amid the Arctic scenery. Eliza was relaxing on deck with some of the other women when Dick Willoughby approached them. "You ladies are very brave to venture up in such a place," he said in a voice of grave concern. "If you only knew the risks you are running—the dangers you are in!" As Eliza told readers, "We received this with some laughter, and expressed entire confidence in the captain and pilot, who had penetrated . . . unknown waters before."[31]

In the middle of the bay, the ship sailed slowly past the large Willoughby Island, which Lester Beardslee had named for Dick Willoughby.[32] Just past the island, in the eastern arm of the bay, came the first signs of the icy plain that had fueled the day's journey of exploration. Silence gripped the air as everyone stared. There, up ahead at the end of a five-mile inlet, lay the full front of the mighty glacier where it dipped down and broke into the sea. "Of all the scenes and natural objects," Eliza wrote, "nothing could be grander and more impressive than the first view up the inlet, with the front of the great glacier, the slope of the glacial field, and the background of lofty mountains united in one picture."[33]

As the *Idaho* sailed closer, Eliza could hear the subterranean rumblings of the icy cliff, which rose several hundred feet above the water. Close up, from the spot where the ship dropped anchor an eighth of a mile away, she was struck by "the dazzling effects of prismatic light." The sharp peaks and craggy surfaces of the ice shimmered in subtle gradations from the purest white to a range of blues.[34] The internal cracking of the ice emitted shots that sounded like artillery fire. As everyone on board stood watching the strange phenomenon, large chunks of snow and ice broke away and tumbled, like an avalanche, into the sea.

The glacier's beauty and power moved Eliza to poetic expression. As she wrote:

> The vast, desolate stretch of grey ice . . . that faced us had a strange fascination, and the crack of the rending ice, the crash of the falling fragments, and a steady undertone like the boom of the great Yosemite Fall, added to the inspiration and excitement. There was something, too, in the consciousness that so few had ever gazed upon the scene before us.[35]

During the *Idaho*'s voyage that summer, Captain Carroll decided to name the mighty glacier and its inlet for John Muir. Eliza would reveal years later that it was she who conveyed the news to Muir during a visit with him and his wife at their home in Northern California. His reply was laconic, she noted, as he asked her only: "Which one of the glaciers do they call mine?"[36]

The travelers ended their day at Glacier Bay by going ashore at a ravine on the north side of the glacier (Figure 5.2). As a

Figure 5.2. Tourists on the eastern shoreline of Muir Inlet in Glacier Bay, c. 1890s.
(Photo by G. D. Hazard, Glacier Photograph Collection, National Snow and Ice Data Center, Boulder, CO)

sharp wind blew, they hiked for two miles across the moraine and icy slopes, their rubber boots and shoes sinking into the glacial mud. While the photographers climbed higher for the best views, Eliza joined those who wandered around gathering interesting-looking pebbles and shreds of ancient cedar trees. They pocketed the treasures to carry home—souvenirs of a historic and most unusual experience.

***

The *Idaho*'s homeward journey included a stop of several days at Sitka, the territorial capital of Alaska. Once a noble seat of government for governors and high-ranking military officers of Russian America, the town had since fallen into "picturesque decay."[37] Still, Eliza found its rich history and local attractions so fascinating that she devoted several chapters to Sitka in her later book on Alaska.

Sailing south, the *Idaho* called at other settlements in the wilderness, including native villages, canneries, and Presbyterian missions. During long stretches along the Inside Passage, two unusual "passengers" entertained everyone on board. At Chilkat and Fort Wrangell, the crew had taken on two young black bear cubs, four or five months old, which they planned to donate to a park in Puget Sound. The younger cub learned to climb the rigging, and both grew so tame they trotted freely about the ship begging the ladies for cakes and lumps of sugar. Eliza described how Captain Carroll's toy terrier, Toots—"an aristocratic little mite of a dog," who lived on an afghan in the captain's cabin—seemed to delight in terrorizing the cubs. One evening near the end of the voyage, while the *Idaho* was passing through the Sea of Georgia, a stormy crimson sunset drew everyone onto the deck. After a while they began singing. As they did so, the two cubs came pattering across the ship. They lay down amid the group and folded their forepaws before them in a "most human attitude," Eliza wrote, as the passengers' voices floated out to sea.[38]

Eliza found the Alaska voyage that summer so sublime that she would return a year later to repeat it. She combined her newspaper dispatches from the two trips into what is now widely regarded as the first book-length travelogue on Alaska. Published in 1885 as *Alaska: Its Southern Coast and the Sitkan Archipelago*, the book shows Eliza's strengths as both writer and reporter. The book provides a great deal of information about the region's history and geography, its coastal settlements, fish canneries, fledgling mining ventures, and native inhabitants. Apart from its prosaic details, however, *Alaska: Its Southern Coast* also has a magical, almost mesmerizing quality that makes it the most lyrical of all the books Eliza would write.

In an era before the advent of movies and color photography, her captivating voice and evocative descriptions transported readers to a unique and exciting place. "The weeks of continuous travel over deep, placid waters in the midst of magnificent scenery," she wrote, "might be a journey of exploration on a new continent, so different it is from anything else in American travel."[39] From on deck, while enjoying a ride as smooth as gliding on a river or lake, one could take in scenery that embodied the nature paintings popular in galleries back East; the gigantic canvases, by artists such as Frederic Church and Albert Bierstadt, that celebrated the rugged character and scenic majesty of the North American continent. Eliza herself described areas of the Pacific Northwest coast and Inside Passage with a painterly eye. "The scenery," she wrote of one stretch, "gains everything from being translated through the medium of a soft, pearly atmosphere, where the light is as gray and evenly diffused as in Old England itself."[40]

Readers of travel writing, no matter the era, look to be moved and inspired. If traveling themselves, they want a practical view of things as well. In her Alaska narrative, Eliza described the awe-inspiring wonders along the Inside Passage, but made herself an immediate companion at the same time by conveying the leisurely, day-to-day pleasures of the voyage itself. At one point, she recounted the *Idaho*'s passage though Finlayson Channel, which she judged one of the most beautiful of all the fjords on the British Columbia coast. That morning, she was up and on deck by four o'clock, along with some of the other passengers, to witness the panoramic newness of the day. John Muir had talked of the spiritual connection with nature that he felt in Alaska. The scene that morning filled Eliza with a similar sense of rapture. "The clear, soft light, the pure air, and the stillness of sky, and shore, and water made it seem," she wrote, "like the dawn of creation in some new paradise."[41]

The Potomac Flats

Eliza adored California. After her 1883 trip to Alaska, she lingered on the West Coast into late autumn, basking in the golden perfection of its Indian summer. "California landscapes were never so beautiful as on the day I left," she said wistfully of her farewell among friends in San Francisco.[1] It was hard to break away. But she needed to be in Washington by the first week in December, when Congress convened, so she boarded a train home.

Everyone in the East had been expecting a mild winter, Eliza told readers as she resumed her *Globe–Democrat* reporting from Washington. She had barely settled in with her mother, however, when an unseasonable cold snap gripped the city in mid-November. Temperatures plunged overnight, freezing everything stiff for three days. Winds blustered through the avenues with the force of a March gale, keeping residents at home by the fire and halting construction projects all around town. Workers had to abandon their posts on the Washington Monument, now in the final stage of construction.[2]

Life was getting back to normal a few days later when, on November 16, an article in the *Washington Star* apparently caught Eliza's eye. The reporter had interviewed Major Peter Hains, of the Army Corps of Engineers, about the reclamation work being done down along the Potomac River, close to the Washington Monument.[3] The area, known as the Potomac flats, had been a wasteland for decades—muddy, smelly, prone to flooding, and littered with debris such as derelict sheds and the rotting hulks of boats.[4] Now the area was getting a major makeover. Meanwhile, residents welcomed the prospect of seeing the blighted Washington Monument completed at last, after nearly forty years.

Eliza had covered the monument's progress for years. This time around she wanted to find out more about the Potomac flats project as a subject for her *Globe–Democrat* column. To do so, she knew she had to act soon. A chance came on the morning of November 21, when she woke to find the weather favorable for what she had in mind. The sky was partly cloudy but dry, with only a faint ripple of winds.[5] It was Wednesday. On Friday, work on the Washington Monument would shut down for the season. The hole at the top would be boarded over, and the stonecutters would spend the winter preparing more blocks to install when the work resumed in the spring. Before that happened, Eliza hoped to get a good view of the Potomac flats work from the top of the Washington Monument.

As she left the house that day and set out for the Mall, Eliza had deadlines on her mind. Never did she dream of the major role she herself would one day play in making that area of town—site of the future Potomac Park, the Lincoln and Jefferson Memorials, and a four-leaf clover-shaped reservoir known as the Tidal Basin—one of the most popular tourist spots in Washington.

***

For as long as Eliza and other residents could remember, the Potomac flats had been a scourge of Washington. Not just a stinky eyesore but a public health nuisance. The flats had formed over the years as silt flowed downriver in the current and built up along the jagged shoreline west and south of the Washington Monument. The deposits created a wide arc of shoals that threatened navigation in a channel of the Potomac.

Twice a day at high tide, the river submerged the flats. When the water receded, however, it left behind a mudbank of several hundred acres, extending from below the White House southeasterly to the Long Bridge, which crossed the Potomac into Virginia.[6] Most residents avoided the rank area, but its stench at times was hard to escape. Effluent from the old Washington Canal—before it was paved over in the early 1870s—had made the flats a dumping ground for all manner of nasty waste: human sewage, the carcasses of dead animals, rotting fish and produce from Center Market stalls on Pennsylvania Avenue.

Everyone in town blamed the smelly "pestilential swamp" for breeding a "deadly miasma" that caused epidemics of malaria and yellow fever.[7] Eliza and the city's other correspondents had followed the issue for years on Capitol Hill. Every new session of Congress brought renewed calls to "do something" about the flats. But politicians who spent half the year living elsewhere felt little inclined to spend the money for a cleanup.

Then, nature forced their hand. The area flooded periodically. In 1881 the river rose so high it inundated the business district on Pennsylvania Avenue and covered the National Mall nearly to the base of the U.S. Capitol. Congress responded by appropriating $400,000 for a project to raise the level of the flats.[8] The plan devised by the Corps of Engineers entailed filling in the low-lying area with mud and silt dredged from the bottom of the Potomac, a move that would simultaneously improve navigation in the Potomac. The Corp of Engineers assigned the job to Hains, a West Point graduate and Civil War veteran with experience building lighthouses in shoaly areas along the Atlantic seaboard.[9]

Now, on this November day in 1883 when Eliza went to view the project, the Army engineers and their contractors were hard at work along the Potomac, filling in the flats using methods similar to levee building along the Mississippi. A drumbeat of thumping came from dredging machines lined up in the middle of the river. Mud scows scuttled the river-bottom dirt to shore, where it was transferred onto hopper cars, carried across a makeshift tramway, and dumped in piles for workmen to spread across the flats.[10]

In his interview with the *Star*, Hains declared himself "very well satisfied" with the progress made in the first year, considering the "countless difficulties" that had arisen.[11] First, the excavation had snagged when the dredgers struck boulders on the river bottom, embedded in hard clay. Then, the dumped soil didn't settle properly, so the contractor had to bring in steam pumps and hoses to saturate the muddy piles with water. The latest headache occurred during the recent storm, when high winds ripped several of the mud scows loose from their moorings and slammed them into the embankment of the Long Bridge causeway. One of the vessels sank, taking with it a hundred tons of coal meant to power the engines.

Despite the setbacks, Hains felt optimistic. As Eliza and other residents learned from his remarks, he expected to have the first 150 acres of the flats filled in by the end of December. All told, the project would eventually give Washington several hundred acres of "reclaimed" land. The intent, Hains indicated, was to earmark the area for use as a public park.

***

Making her way through the streets, Eliza felt invigorated by the crisp pewter chill of late November. Fall was always an exciting time in Washington, as the city once again came alive. Horse-drawn carriages passed by at a brisk trot. The air crackled with the self-importance of politicians and their entourages pouring back into town for the opening of Congress. Residents who had fled to their country estates or favorite resorts for the summer were reopening their homes—airing out the parlors, drawing up guest lists, planning menus, ordering new gowns and a season's supply of kid gloves.[12] President Chester Arthur was feeling rejuvenated in the White House after his own late-summer vacation, an expedition to Yellowstone Park under Secretary of War Robert Todd Lincoln and a seventy-five-man military escort.[13]

After six months away, Eliza marveled at the sight of so much construction. "Since last winter Washington has enjoyed as much of a building boom as any young Western town," she informed readers. Apartment buildings, mansions, whole blocks of new residences were mushrooming "like magic."[14] The housing demand came in part from people who were moving to the capital to work for the rapidly growing federal government, but also from transient residents who came to spend the winter. As Eliza's colleague Frank Carpenter reported in the *Cleveland Leader*, for many of the country's richest families, it was now "as fashionable to have a winter house in Washington as it is to have a summer one at Newport or Saratoga."[15]

As in any city, development was uneven. There were still pockets of poverty and neglect. Yet no one could deny, as Eliza observed often in her newspaper column, that Washington was growing "more beautiful every year."[16] In a gushing article, *The Century Illustrated Magazine* hailed what it called "The New Washington." In a mere decade, the magazine noted, the nation's capital had "ceased to be a village" and was now so beguiling that "during the winter all the world and his wife goes there for a visit."[17]

Nothing better epitomized the leap of progress than the steadily rising Washington Monument. After leaving her home and turning the corner onto Fifteenth Street, Eliza headed south toward the monument, past the long white colonnaded front of the Treasury building, past the sloping green lawn at the rear of the White House. Because the red and yellow leaves of autumn had already fallen to the ground, the trees lining the sidewalk formed a lattice of dark limbs. Reaching the end of the block, Eliza stopped and stared. There, across the street on the grounds of the National Mall, the massive column of granite and marble blocks soared upward like the walls of a mighty fortress. The progress during her absence was clear. In six months, the structure had grown seventy feet taller, pushing it to almost 400 feet, or nearly three-quarters of its final height. A safety net encircled the flat-topped peak, where the arms of stone-setting cranes protruded from the corners.[18]

Like everyone else in town, Eliza was thrilled to know the thing would soon be done. It had disfigured the Mall for far too long. Since the laying of the cornerstone in 1848, construction had progressed in fits and starts, delayed at first by funding problems, political bickering, and the Civil War, later by concerns that the sandy shoreline soil along the Potomac could not adequately support the weighty structure. As the country celebrated its centennial in 1876, Congress had vowed to complete the monument once and for all. This time around, if everything remained on track, Eliza told readers in the West, the towering landmark would finally be finished in another year. "Unless"—she could not help adding in a partisan barb—"the Democratic Congress makes trouble with the annual appropriations for the work."[19]

Approaching the entrance, Eliza had to weave her way past building materials and tourists ambling about the grounds. People went down to the site every day to gaze in wonder at what would soon be, at 555 feet and five and one-eighth inches, the tallest man-made structure in the world.[20] Eliza did not share their awe. She agreed the project was an impressive feat of engineering. But to her eye, the design—based on an Egyptian obelisk—was a huge failure of human creativity, "the ugliest thing for the money human hands could design." In its final effect, the monument conveyed no lesson or symbolism; it had no character of its own. The thousands of blocks of stone, she suggested to readers, could have been put to much better use in a lighthouse, a bell tower, even a levee along the banks of the Mississippi River. Imagine George Washington's despair, she wrote acidly, at seeing so costly a structure built in his honor "without any compensating beauty to palliate the crime of bad taste."[21]

Eliza mounted the wide steps cut into the earthen berm at the base of the monument. The entrance had been placed on the eastern side of the building, to face the rising sun. The interior, with stony walls fifteen feet thick at the base, felt as cold and dim as a tomb. A faint sliver of sky blinked at the top. One day the building would have an interior staircase of 896 steps, but it was still under construction. For now, Eliza and other visitors who wanted to go to the top had no other option but to ride the open-sided elevator that ferried the workmen and materials to the upper levels. Designed with a double cast-iron framework, the lift had a platform at the center to carry the giant blocks of stone, and a built-in system of pulleys to hoist the stones into place. Steam carried through underground pipes powered the elevator, while its wire cables unspooled from a deep well below the floor.[22]

The ride looked daunting, but Eliza routinely faced nervy adventures for the sake of a story. Steeling her will, she lifted her skirts and stepped onto the platform, taking her place among others going to the top. With no railings to cling to for safety, she planted herself close to two blocks of stone and held on tight. The elevator began to rise, and for "eight mortal minutes" it inched upward. "The ascent is not a pleasant experience," she reported later in the *Globe–Democrat*. "In the darkness of the shaft a toe-hold on the moving platform seem[s] very insecure."[23] She fought dizziness and a queasy feeling in her stomach.

At last, with a surge of relief, she heard the *tap-tap-tapping* of chisels and saw the burst of daylight indicating they had reached the top. She stepped out onto the large wooden platform, about twenty feet square. Laborers walked deftly about, hoisting the stone blocks with the heavy chains and swinging them into place along the wall while others mixed and spread the mortar (Figure 6.1).

Eliza gazed over the edge of the platform. The landscape below fanned out for miles in every direction like a giant relief map. On the far western horizon she could see the Blue Ridge Mountains, sixty miles away in central Virginia. Closer in lay the rolling and wooded hills of Virginia and Maryland, and the broad and winding Potomac River and its branches. On the ground below, Washington resembled "a toy world," scattered with red-brick houses, green stretches of park, and the white marble and granite buildings of the federal government.

Turning her attention to the Army's reclamation work, Eliza studied the operations: the dredging machines and river tugs, the tramway, and the tiny shapes of men spreading soil across the muddy field. A long retaining wall of rocks stretched along the shoreline. "The reclaiming of these malarial flats is progressing rapidly," she told readers in her reporting from that day,

Figure 6.1. Platform elevator and pulley apparatus atop the Washington Monument in midconstruction, c. 1883, with women sightseers at the rear.
(Courtesy of U.S. Army Corps of Engineers)

though the project would, in fact, drag out to the turn of the century. In the end, it would extend the National Mall by a mile to the west of the Washington Monument and give the city 700 acres of new, usable land—today's East and West Potomac Parks, separated by the Tidal Basin.[24]

"All of this reclaimed land is to be converted into a public park," Eliza wrote in her *Globe–Democrat* column. "When laid out with trees and flower-beds, drives and walks," she added, it "will be the largest park in the city, and one of the most attractive pleasure grounds." Its roadways, she predicted, were bound to be "the fashionable resort for afternoon rides and the display of magnificence in future administrations."

On this day, as Eliza took in the scene below, the raw, torn-up field hardly looked inspiring. But an image of that vast, empty stretch of ground would linger in the recesses of her mind. The new park taking shape on the flats was a blank canvas, ripe with possibilities. Before long, the sight of all that unused land would spring to mind when her travels took Eliza halfway around the world, to Japan.

Cherry blossoms, it would occur to her. That's what the city needed to make its new park something extraordinary.

***

On her return to Alaska the following summer, Eliza again chronicled her journey for the *Globe–Democrat*, while also filing a couple of long articles for the *New York Times*.[25] This time, she traveled with several friends. They were in Victoria, British Columbia, the first port of call outside the United States, when they learned that the *Idaho* had to be sent to San Francisco for repairs. During the week they spent in Victoria, the women hired heavy London-built carriages and went sightseeing on well-paved roads that took them through lush woods and along the shores. In the local climate, vegetation thrived. Ferns along the roadside grew ten to twelve feet high. Acres of wild rosebushes filled open fields, and honeysuckle vines smothered cottages and trellises.

Though lovely, the town struck Eliza as quaint and dull, laid back in its everyday affairs. Businessmen seemed to drift into their offices around ten in the morning and head for home by four o'clock, "as if the fever and activity of American trade and competition were far away and unheard of."[26] A clerk at the post office greeted Eliza with a yawn, then sent her down the street to buy postage stamps at the stationer's shop.

Finally, on a sunny morning "when the whole Olympic range stood like a sapphire wall across the Straits," the Alaska steamer puffed its way into port.[27] Watching it draw near, Eliza felt disappointed: the weather-beaten vessel was not the handsome *Idaho* but the *Ancon,* an old converted coal carrier with a giant paddlewheel churning at its side. Happily, she found Captain James Carroll on board, along with many of the crew members she had come to know a year earlier.

The journey re-created many of her memorable experiences from a year earlier, though unfortunately, the return to Glacier Bay was marred by rain. In the heavy mist, the crew member on watch managed nonetheless to catch sight of Scidmore Island in the distance, and he pointed it out to Eliza as the ship passed nearby.

The homeward journey late in the voyage brought a "fairyland" moment one night while the *Ancon* was sailing through the Gulf of Georgia. Eliza's description of the incident offers another example of what made her Alaska book so evocatively compelling. That evening, around midnight, the captain raced along the deck rapping on cabin doors. "Wake up!" he cried out. "The whole sea is on fire." Eliza and the others roused from their beds and flung open their stateroom doors and windows. Around them, everywhere they looked, the water shone like a sheet of liquid silver, an effect caused by tiny light-emitting plankton that filled the sea.

The millions of points of light, Eliza wrote, mingled in what seemed like "a solid stretch of miles of pale, unearthly flame." In the strange, ghostly light that filled the sky, the shores of Cape Lazlo, twenty miles away, seemed near at hand. As the *Ancon* glided through the sea, she continued:

> A broad pathway of pale-green, luminous water trailed after us, and the paddle-wheels threw off dazzling cascades. Under the bows the foaming spray washed high on the black hull, and cast

> long lines of unearthly, greenish white flame, that illuminated the row of faces standing over the guards as sharply as calcium rays.[28]

In the strange, eerie darkness aboard the ship, no one could think of sleep. Everyone gathered close to watch as one of the ship's officers lowered a bucket into the sea, then pulled it up, and placed it on the deck. Each time he stirred the water, the shimmering glow of the sea occurred again in miniature.

***

The winter of 1884–5 was brutal for Eliza as she performed her usual social reporting duties in Washington while also struggling to produce a book manuscript on Alaska for the Boston publisher D. Lathrop and Company. She started by combining her newspaper dispatches into a seamless narrative. To supplement her field reporting and the conversations she had with many people she had met along the way—traders, teachers, miners, and government officials—she raced around Washington acquiring additional information from maps, surveys, U.S. Census data, and official reports. She reached out to *Harper's Weekly* and an amateur photographer in San Francisco for illustrations. William Dall gave his permission to use a new map of Alaska he had just made for the U.S. Coast and Geodetic Survey.[29]

On March 15, 1885, Eliza signed off on the final pages and mailed them to Boston. Published a few weeks later, *Alaska: Its Southern Coast and the Sitkan Archipelago* made her a published author for the first time, at the age of twenty-eight.[30]

Producing the book had been a "nightmare," Eliza told a friend several weeks later, once the deadline was well behind her. The whole awful experience convinced her that "book making and authorship are not half the fun things" people made them out to be. She was happy, she added, to stick with newspaper work, which offered greater "peace of mind."[31]

Despite her anxiety about the book, she need not have worried: most critics were enchanted by it, both for the "great deal of information that is new to most people" and the author's "bright and entertaining" style.[32] *The Literary World* told its readers that "no book yet published" compared with Miss Scidmore's in relating "the history, geography, topography, climate, natural scenery, inhabitants and rich resources of the wonderful *terra incognita*."[33]

The positive reception was gratifying. But by then, Eliza was exhausted; all she could think of was getting away. Early in the summer, she and her mother wrapped up their affairs, pulled their trunks from storage, and shopped for last-minute necessities before boarding a train for the West Coast. They were off on vacation—one that would also be an emotional reunion. They were going to Japan, where Eliza's brother George had been working for the U.S. Consular Service since 1880. The two women had not seen him in years. They planned to spend the summer with George at his posting in Kobe, a busy commercial port at the head of Japan's Inland Sea. During their six-month stay, they would visit some of Japan's major cities and cultural attractions before returning home at the end of the year.

Eliza insisted, in a letter to a friend in Boston, that she had no intention of working that summer. She wanted, for once, to travel like a tourist, purely enjoying the sights. "I am rather resting and feasting my soul this summer," she wrote. Besides, she added, "too much has already been written on Japan and I could only fall farther short of it than the rest of them."[34]

It was a hollow vow. Within days of arriving in Japan, Eliza would resume her travel letters for the *St. Louis Globe–Democrat*. Meanwhile, as they set off for the long journey ahead, mother and daughter had no idea how profoundly the trip to Japan would change both their lives. Nor could they imagine the near tragedy that would leave them rattled before they ever set foot in the country.

PART 2
FAR AND WIDE

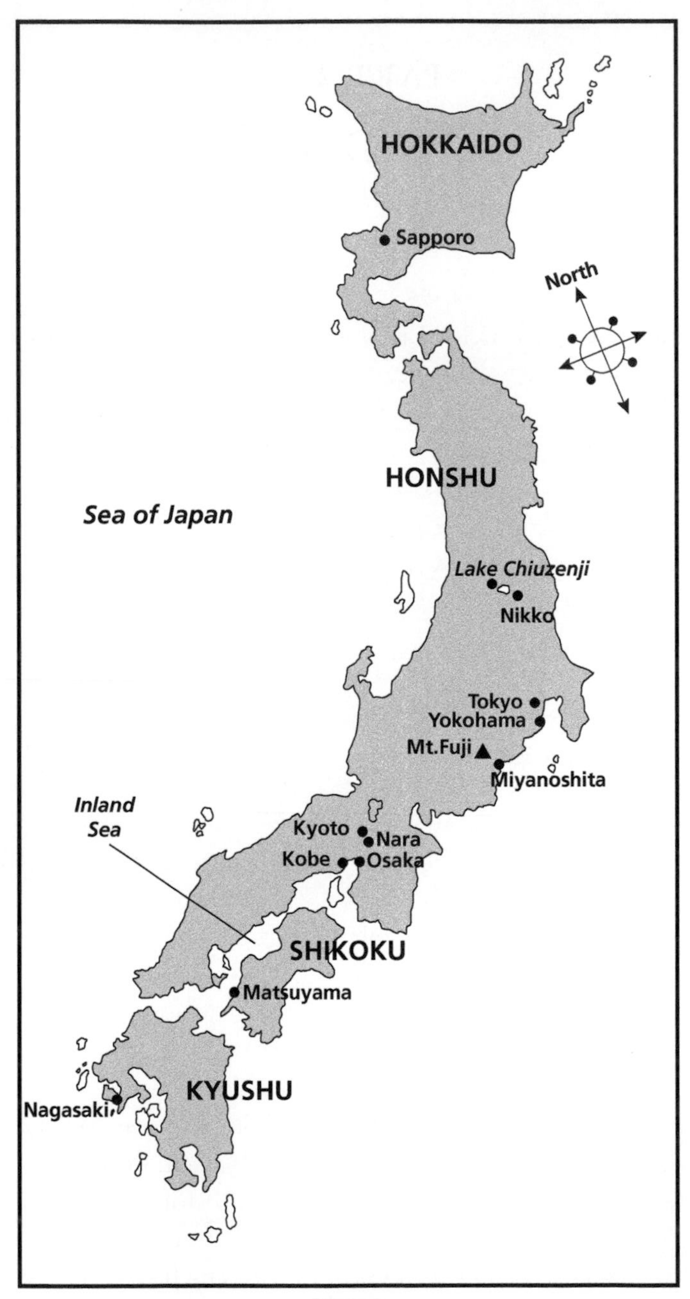
HOKKAIDO
Sapporo
North
HONSHU
Sea of Japan
Lake Chiuzenji
Nikko
Tokyo
Yokohama
Mt.Fuji
Miyanoshita
Inland Sea
Kyoto
Nara
Kobe
Osaka
SHIKOKU
Matsuyama
KYUSHU
Nagasaki

Japan

Jinrikishas in Japan

Squawking seagulls circled the blue sky as Eliza and her mother bumped along the San Francisco wharf in a horse-drawn taxi. Shortly after noon on June 2, 1885, they joined the stream of passengers arriving for their sailing at the Pacific Mail Steamship Company's departure sheds, at First and Brannan Streets. Shouting porters pushed through the crowd, maneuvering past dray horses and carts that stood unloading cargo. A thunderous din roared from the far end of the dock, where coal was being transferred from hopper cars into chutes that fed the giant ship.[1]

The two women had booked passage on the *City of Tokio*, the largest and finest steamer on the Pacific. When launched a decade earlier, the 420-foot-long behemoth had created such a stir that President Ulysses S. Grant and thousands of other spectators traveled to the shipyard in Chester, Pennsylvania, for the christening.[2] Commissioned specially for the Pacific Mail's Far East route, the *City of Tokio* and its sister ship, the *City of Peking*, plied the waters between San Francisco, Yokohama, and Hong Kong, carrying up to 2,000 passengers and huge stores of trade goods like silk, tea, and hemp. The two ships had transported thousands of the Chinese workers who went to America after the Civil War to help build the transcontinental railroad, but their numbers in steerage had dropped significantly since the passage of an 1882 law restricting Chinese immigration. President Grant himself, after leaving office, sailed home aboard the *City of Tokio* in 1879 following a two-year world tour with his family that included a long stay in Japan. More recently, just a few months earlier, the ship had carried the first group of Japanese laborers who went to Hawaii to work on sugar plantations.[3]

Eliza and her mother were standing on the after deck, saying goodbye to friends who had come to see them off, when a Chinese worker raced by beating a gong and crying out that it was time for all nonpassengers to go ashore. Eliza watched in fascination as people hurried away in a mad dash, "slipping down the gangplank like beads off a string."[4] Minutes later, she and her mother were astonished to find themselves among a small knot of people; in a bizarre development, the entire contingent of cabin-class passengers totaled only fifteen. They would be swallowed up in the cavernous ship.

At 2:45 the *City of Tokio* slowly pulled away from the wharf. In the bright and breezy afternoon, the mammoth ship glided past the city front, then turned its prow westward toward the Golden Gate Bridge and Point Lobos. Forty-five minutes after the departure, the powerful engines churned the blue water as the ship headed out to sea.

***

The Pacific Mail Company prided itself on the comfort of its ships, which boasted gilt trimmings in the dining rooms and lounges and parlors as cozy as drawing rooms back home. Despite the plush conditions, Eliza chafed at the dull monotony of the three-week journey across the Pacific. From San Francisco to Yokohama, a distance of 4,500 miles, hardly another passing ship was seen. The captain's route through the northern latitudes brought heavy fog and cold winds that put the deck off limits much of the time.[5] Stuck indoors, Eliza and the others—all Americans except for the Scottish wife of a missionary in Tokyo and a Japanese man with limited English—passed the hours playing cards and practicing their Japanese. The idle time gave her and her mother time to get to know the incoming U.S. minister, Richard Hubbard, who was traveling out to Japan with his family. Many office seekers who lobbied for diplomatic posts considered Japan a backwater. But Hubbard, a highly educated former congressman and Texas governor, liked and admired the Japanese, and was sympathetic to their frustration with unfair trade treaties that had been pushed on them by Western countries. As Hubbard would tell his superiors in Washington only a few months into his new job, "I know of no people among whom it could be more pleasant to dwell than among these people."[6]

On the tenth day at sea, the ship crossed the 180-degree meridian, marking the passage from west to east longitude—from the Occident to the Orient ("the East"). Eliza and Mrs. Scidmore celebrated over a private supper in their cabin. Because of the time difference, they went to bed on the night of June 14 and woke up on June 16.

Five days later, the cold and damp weather turned balmy as the *City of Tokio* headed south into the equatorial climate of the *Kuro-siwo*, or Japan gulf stream. When the coastline of Japan came into view, Eliza and the others pushed against the rails, straining for a glimpse of the fabled Mount Fuji. To the Japanese, Fuji—a perfect volcanic cone rising 12,000 feet—was a sacred place and a source of artistic inspiration for centuries. To people in the West, Fuji was the very symbol of Japan. No image of Japan was better known in America and Europe than "The Great Wave," the artist Hokusai's woodblock print showing small boats caught in a high claw of waves, with the snowy peak of Fuji in the background.[7] As Eliza was to discover, the mountain was fickle: a violet silhouette one day, pale and ghostly the next. Often it disappeared altogether in a mist. During her six months in Japan, she later reported, "the matchless mountain refused to show herself from any point of view."[8]

Late in the day on June 23, the *City of Tokio* cut west to enter Tokyo Bay. Yokohama and its deep harbor lay at the upper end of a long, jagged peninsula. Twenty miles further up the bay was Tokyo, and the waters where Commodore Perry had sailed thirty-two years earlier seeking negotiations with Japan.

That night, a lighthouse at the entrance to the bay cast its beam across the ship as Eliza and Mrs. Scidmore strolled the deck before retiring for the night. They went to bed excited at the prospect of seeing George the next morning and arriving in the enchanted land of Japan.

At midnight, a cracking jolt shook them awake. They leaped from their beds and clutched one another in fright as the ship trembled. A grinding of metal shrieked as the engines were thrust into reverse. In heavy fog and a strong current, the *City of Tokio* had gone aground twenty-three miles from Yokohama, its hull smashed on the rocks. The captain sent the purser off in a boat to Yokohama for help, and as the ship began taking on water, the crew tossed cargo into the sea to lighten the weight. Most of the goods were insured, but the greater loss was huge: the *City of Tokio* was later declared unsalvageable.[9]

The fog had lifted by morning when a line of rescue vessels pulled up to the steamer. George was there, in a small launch of the Imperial Japan Post Office, its interior handsomely outfitted in blue velvet upholstery and gold lacquer. The family's reunion was all the more poignant for the brush with danger (Figure 7.1). After joyful embraces, they settled into the snug cabin, and for the next two hours, they chattered about their lives as the boat steamed up the bay. "We had a merry time of it," Eliza reported in the *Globe–Democrat*, turning the mishap into an adventure, "and for 'survivors' and 'shipwrecked souls' were in high glee at being 'rescued' by such a pretty little life-boat."[10]

***

The small vessel puttered to shore through a harbor filled with ships of all sizes and nations: black merchant steamers and white men-of-war; hotel omnibuses; rustic brigs and barks, and pink and red "canal" steamers that sailed from Suez to the Orient. Low, flat-bottomed *sampans* drifted about, some with entire families aboard. Southeast winds stirred a drizzle across the harbor that made going ashore a wet affair.[11]

After the picturesqueness of the harbor, Eliza felt disappointed by Yokohama. The clubs, hotels, and residences lining the wide waterfront avenue, or Bund, made the city seem like a piece of the West grafted onto Japan; "too European to be Japanese, and

Figure 7.1. Eliza Scidmore with her mother and brother George, c. 1890s.
(Wisconsin Historical Society, WHI-91609)

too Japanese to be European."[12] Her mood changed quickly when they walked to the end of the stone pier. There, "one dream of the Orient was realized," she wrote. Awaiting them was a long row of *jinrikishas*, the "big, two-wheeled baby-carriage" that offered the chief means of getting around in Japan.[13] As the drivers—in dark blue tights, loose blouses, mushroom-shaped hats, and straw sandals—took up their positions, Eliza climbed into one of the vehicles and settled atop a red cushion. A small, slender man lifted the poles of the shaft, and they set off. Traveling in a line with her mother and George, Eliza found herself pulled through the streets on the way to their hotel.

Yokohama had been established chiefly as a port for Japan's trade with the United States and other Western nations, and it remained the country's main gateway for foreigners. Merchants and traders, government agents, sailors, missionaries, and tourists shared the streets with the native Japanese and other Asians.

The first wave of Americans had arrived a decade earlier, many from New England, where transcendentalist thinkers drew inspiration from the Asian religions of Hinduism and Buddhism; Eastern philosophy, with its emphasis on asceticism and unity with nature, offered a sharp contrast to the vulgar excesses of the Gilded Age and the dislocations of rapid industrialization in America.[14] Japan also became a favorite stop on the round-the-world "globetrotter" tours pioneered by the travel organizer Thomas Cook, and a popular destination for Victorian collectors on the hunt for Asian art.[15] Two years before Eliza and her mother went to Japan, the wealthy Boston socialite Isabella Stewart Gardner developed such an interest in Asian cultures during a year-long trip with her husband that she would later build a palace-like museum in Boston in which to display the many treasures she acquired during extensive collecting trips across the region.[16]

Though everyone in cosmopolitan Yokohama mingled freely during the day, they retreated at dusk to racially separate neighborhoods. Most Western residents of the city lived atop the Bluff, in gingerbread-trim bungalows set amid well-tended gardens. Official government business was conducted in Tokyo, a fifty-minute train ride, but many diplomats preferred living in Yokohama because of its Western-style amenities, which included English-language newspapers, men's clubs, a cricket field, and even a racecourse.

Before going on to Kobe with George, Eliza and Mrs. Scidmore spent several days shopping and sightseeing. At the base of the cliff in Yokohama lay the crowded Chinatown, and the large Japanese district full of shops that opened directly onto the street to display their wares. In front of the Japanese teahouses and restaurants, slender young women sat playing the three-stringed *samisen*. With her weakness for beautiful textiles, Eliza spent hours with her mother in the silk shops, where clerks carried in bolts of fabric by the basket and unrolled them to display the jewel-like tones of rippling crepes and brocades.

Eliza studied the kimono-clad Japanese women with fascination. Walking by on their wooden clogs, they seem to have stepped out of the fans, prints, porcelains, and other Japanese objects that were all the rage in parlors back home. Most captivating of all were the little children, their tiny bald heads bobbing, and their black eyes darting here and there as they went everywhere tied to the backs of their mothers or siblings (Figure 7.2).[17]

Within days, Eliza's infatuation with Japan was complete. "I am bewitched and bewildered," she wrote in a letter to America, and "afraid that I am going to wake up from a dream."[18]

***

Kobe, a pleasant city of 80,000 set at the base of a mountain slope, was a busy commercial port of Osaka and Kyoto that produced

Figure 7.2. Japanese woman and babies, from National Geographic Society's Eliza Scidmore Collection.
(Courtesy of National Geographic Society)

most of Japan's mass-produced goods for foreign markets. It was also a great tea port. All summer, Eliza breathed the fragrant air of roasting tea leaves.[19]

The small expatriate population in Kobe made for a quiet social life, and George's household of four servants left Eliza and Mrs. Scidmore with much time on their hands. Eliza channeled her energy into a detailed study of the tea trade, which she described in the *Globe–Democrat* and later published as a chapter in her book on Japan.[20] In her rudimentary Japanese, she prodded the staff about everything from dietary habits to the price of laundry for a series of dispatches on daily life in Japan. She and her mother visited Osaka, where they saw the famous castle at which *samurai* troops made their last stand in the battle that led to the downfall of the *shoguns* and the restoration of power to the emperor.

The Scidmores and everyone else in Kobe crowded the docks that summer to greet the young emperor, Mutsuhito, during his overnight stop on the way to Hiroshima and Yamaguchi. The one hundred and twenty-second emperor in a family of sovereigns that dated back twenty-five centuries, he had come to the throne at age fifteen. Though the town feted him with pageantry and fireworks, the emperor himself, Eliza reported, was a slender, modest young man who carried himself with quiet dignity and lived simply, as an example to his thirty-three million people. In contrast to the opulence of most courts, his palaces were as plain as Shinto shrines. The Mikado, as his people called him, didn't even own a royal yacht but arrived in Kobe on a steamer of the Mitsu Bishi line.[21]

The stay in Kobe was cut short when George got word of a transfer. He was being sent temporarily to Shanghai in the fall, a move that came with a promotion to vice consul general.[22] An angular man of thirty-one, with dark sideburns and a slender face, George had done the family proud since joining the Consular Service a decade earlier. Thanks to his legal training, he had already distinguished himself in Japan by compiling a digest of leading cases decided under the U.S. consular court system.[23]

August was not a good time to visit Kyoto, fifty miles from Kobe, as the weather then was brutally hot. But Eliza didn't want to miss seeing the city that had been the seat of Japan's emperors for a thousand years, before the capital was moved to Edo (Tokyo). She and her mother went to Kyoto expecting to stay a week or ten days; they were there more than a month. To avoid the worst of the heat, they made early morning excursions—to potters' studios, silk factories, temples, and old imperial residences—then lounged the rest of the day at their hotel, two miles above the town, which a pair of brothers named Yaamis had converted from an old temple.

By chance, the two women had arrived on the final day of a festival. That evening, they watched a scene "like fairyland on Midsummer's Night" as residents carrying white paper lanterns suspended from long sticks paraded through a temple courtyard.[24] The stay in Kyoto ended with an incident that was memorable for other reasons.

The occasion was a banquet that local merchants held for some Japanese officials visiting from Tokyo. Upon learning of the ladies' keen interest in Japanese culture, the men invited Eliza, Mrs. Scidmore, and two other Americans to join them. When the guests arrived that evening, at the dining hall on the upper floor of a temple teahouse, they found their hosts—forty-four men, all in dark silk kimonos—waiting to greet them in a deep bow.

The elaborate meal the men arranged lasted several hours. Gathered around the room at the low banquet table, the diners made their way through an endless array of courses: soups, omelets, and jellies; lily bulbs and bamboo shoots; raw fish, abalone, eels, and fowl. Everything was served with rice and washed down with gallons of sake drunk from tiny porcelain cups. In a gesture of esteem for their guests, the merchants included a Japanese culinary delicacy that proved unnerving for Eliza and the other uninitiated foreigners.

The city's famous *maiko* dancing girls provided the evening's entertainment, and at one point, two of the young women entered the room carrying a large tray. On it lay a large carp, its green and silvery scales shining as though it had just been caught. Later, Eliza described the nauseating horror of watching as thin slivers of flesh were cut from the back of the live fish and passed around. "We tried to think that the fish before us had not been touched with a knife," she told readers, "but that bits from the back of another fish had been cunningly concealed behind it and passed to us." Fighting revulsion, she lifted the shreds of pinkish yellow flesh with her chopsticks, dipped them in soy sauce, and swallowed them. The fish's eye, she reported, seemed fixed in reproach.[25]

Eliza stayed only briefly that fall in Shanghai, where the gaudily lit Bund and fast life of clubs and racetracks held little appeal. She went on to Yokohama to spend the final weeks of her trip. One day, a stiffly formal card arrived for her at her hotel. Unable to read Japanese, Eliza sought out Tatsu, a servant who had impressed her with his command of English and a dignified manner "quite out of keeping with his broom and dust-pan." He was, she discovered during her stay, a former *samurai* who proved knowledgeable on everything from embroideries and blue and white pottery to ancient relations between Japan and Korea. When the samurai lost their privileged status after the Meiji restoration, some became teachers, diplomats, and bureaucrats; others, like Tatsu, had to enter household service. When Eliza approached Tatsu with her letter, he examined it, sucked in his breath, and delivered the translation with a deep bow: "Mikado want to see Missy, Tuesday, at three o'clock."[26] It was an invitation to the annual chrysanthemum party at the imperial palace.

Not long after, when Eliza packed for the return to America, she did so alone. Eliza Catherine Scidmore had decided to stay on in the Far East to live with her son for a while. For nearly three decades, mother and daughter had been nearly inseparable. Now, at the age of thirty, Eliza was truly on her own for the first time.

***

Back in Washington, gossip in the spring of 1886 included rumors of wedding plans in the White House. The stories turned out to be true. A year into his presidency, Grover Cleveland planned to wed—and his choice of bride made tongues wag. She was Frances Folsom, the twenty-two-year-old daughter of his late law partner, and his legal ward. A bachelor in his forties when he took office, the president had grown smitten upon meeting the young woman when she was a student at Wells College. They were married in the Blue Room on June 2, 1886, the first presidential wedding ever held at the White House.

The social event of the year apparently put Eliza in demand. "The old, tried newspaper correspondents had to give in to the great demand for every detail of wedding gossip, and they summoned to their staff the full corps of lady writers who center upon the capital," the city's *Hatchet* newspaper noted on July 2.

> Many of them put much good money in their purses. Miss E. R. Scidmore skimmed the cream of the milk and honey season. She did the work of ten of the best papers in the country, and in a week took in $1,000, enough to take her on a summer and autumn trip through the old world. She sailed Saturday for England in the company of several lady friends.

The item was likely an exaggeration, as the weekly *Hatchet* modeled itself on humor magazines like *Puck*, with "amusing anecdotes" and fictionalized "reporting."[27] Earning a thousand dollars in a week from freelance newspaper dispatches, at a time when the salary of a congressman was about $6,000 a year, seems improbable. If the report was true, the sum may have included royalties from Eliza's Alaska book, which had been published a year earlier. At the least, the account in the *Hatchett* demonstrated the attention Eliza was attracting in Washington for her reporting success.

She did spend the fall in Europe. In late June, Eliza and her friends sailed to Liverpool on the *Gallia*. They stayed several weeks in London, where Eliza visited the House of Commons, attended an exposition on the cultures of India and other British colonies, and reported on the heavy presence of Americans in England, including celebrities such as the New England preacher Henry Ward Beecher and Lady Randolph Churchill, mother of the future prime minister. The women spent the Fourth of July holiday at the country estate of an Englishman Eliza had met while hiking in Yosemite.[28]

From London, the women went on to Norway, where they cruised the fiords and visited a Lapland community, then to Stockholm and Copenhagen. They ended their European jaunt with several weeks in Paris, where the aroma of roasting chestnuts filled the chilly autumn air. Eliza scheduled dress fittings for herself and rushed around buying a trunk full of new clothes for her mother.

In the year since their parting, Mrs. Scidmore had lived through the harrowing experience of typhoons and an earthquake.[29] She had suffered—twice—the indignity of being tipped out of a jinrikisha. In the latest fright, she had lost all her possessions in a fire that broke out at Yokohama's Windsor Hotel at 4 o'clock in the morning and burned so fast the building was leveled in half an hour. She and George, along with the other guests, barely escaped with their lives.[30] The incidents left her feeling as helpless and bewildered as "a young robin," she told relatives back in Wisconsin. She complained, as she did periodically, that they didn't write often enough. But in case she gave the impression that she was homesick, she hastened to reassure them: "I ain't."[31]

Now in her early sixties, Eliza Catherine Scidmore was, like her daughter, a stalwart traveler. After Eliza returned to the Far East in the summer of 1887, the two women spent much of the year traveling. In June and July they went to Korea and China. The trip included visits to Chefoo, a popular resort in northern China with a clean, sandy beach, and Tientsin, a foreign settlement and commercial port on the Peiho River. Eliza bristled at the attitude of many Americans and other Westerners they met. "Residents in foreign settlements along the coast have only scorn or at best pity for the zealous traveler who wants to see Chinese by themselves, in their own homes, and to experience some of the ancient customs."[32] Tapping her consular contacts, she arranged a trip to Chang Sha, an old city in the interior of China, for a daylong visit with a Chinese mandarin and his family at their country castle, or *yaamen*. The host sent a fleet of several houseboats to escort their party of three through the long waterway of rivers, creeks, and canals. Most of the people they met at farmhouses along the way had never seen foreigners.[33]

In late summer, the Scidmores and several companions made the annual exodus to Nikko, a temple district in the mountains of Japan, so popular with vacationing diplomats that foreign legations rented entire hotels there for the season.[34] The trip included a steep ascent by pack horse and *kago* (basket chairs hanging from poles) to Lake Chiuzenji, "a vast sapphire" three miles wide and eight miles long.[35] They walked from Chiuzenji to the white sulfur springs at Yumoto along a pine-forested path overlooking the lake and waterfalls. At the hot springs, Eliza followed her reporter's eye to the public bathhouses, where Japanese men, women, and children shared the water. At her approach, the grown bathers plunged deep into the pool. "They know, though they cannot understand," Eliza wrote, "that the European finds something objectionable, and even wrong, in so insignificant a trifle as being seen without clothes."[36]

***

A year later, Eliza organized a trip that was a defining experience for many travelers in Japan: climbing Mount Fuji. Every summer, about 30,000 Japanese made the spiritual journey, dressed in the baggy white clothing of pilgrims. Most traveled in large groups, their bells tinkling and walking staffs tapping the ground as they crowded the paths. At each temple along the route, they had their clothing stamped with a red seal.

Eliza and her party—three women; four men, including two veteran climbers of Pike's Peak; and two young Japanese valets—started off in late July from Miyanoshita, a mountain resort near Hakone Lake. With guides and porters, they departed the next morning at 6 o'clock, the women riding in kagos, the four foreign men on horseback, and the Japanese attendants on foot. As they crossed through a narrow pass in the mountain range, Fuji was visible across the green plain, its head in the clouds. After transferring to jinrikishas, they rode six miles more to Subashiri, the village where pilgrims prayed before beginning the ascent. Hundreds of their flags and towels fluttered from the teahouses.

The next morning, when the group breakfasted quickly at 4 o'clock, Fuji shone pink and lilac in the first light of day. They were at 4,000 feet, and the summit looming above them seemed impossibly high up. To Eliza, the mountain "looked twice its twelve thousand feet above the sea, and the thought of toiling on foot up the great slope was depressing."[37] A stone staircase arched by a tall *torii* marked the actual beginning of the ascent. From there, the ground was holy and straw sandals were required of all hikers. The porters carried dozens of pairs, a crucial accessory for walking on coarse cinders of lava that cut quickly through foreign boots.

Ten rest houses, or pilgrim stations, stood at intervals along the slope. Eliza and her group hoped to reach the eighth one by late afternoon, either to sleep there or push on to the summit. Three porters were assigned to each of the women, "with orders to take her to the top if they had to carry her pickaback," Eliza reported. Roped together, "we were literally hauled and boosted up the mountain, with only the responsibility of lifting our feet out of the ashes."[38]

The path led for several miles through a flower- and moss-studded bower, then opened onto sloping ground. The path ahead was wide, with no boulders to navigate and no dangerous precipices. "It was apparent that the walk would be merely a matter of perseverance," Eliza wrote. The whole cone of Fuji was visible, surrounded by thin white clouds.

They were still well below the eighth station when a heavy mist moved in. Gusts of wind followed, bringing showers that turned to pouring rain. The climbers pushed on for two hours, chilled and exhausted by the time they reached the hut. The stone-covered log cabin consisted of a single room, about twelve by thirty feet. Inside, it was crammed with stranded travelers. Eliza counted forty-two people. The proprietor barred the doors with planks to keep out the storm, and the cabin filled with blinding smoke from the fire. With no room to move about, Eliza and her companions gobbled down a sandwich, collected their allotted pair of futons, and retired early on the cold and hard floor, as blasts of wind seeped through the walls. Eliza used a basket lid as a pillow. Before closing her eyes, she scanned the room. The single hanging lamp cast a faint glow over a sea of heads tucked beneath blue futons.

From Saturday night to Tuesday morning, the howling tempest kept everyone prisoner in the dark, smoke-filled hut. Eliza marveled at the cook's ingenuity in feeding everyone. She watched him shave thin slices of dried fish with a plane, then stew the pieces with mushrooms and a seasoning of soy and sake. A huge copper cauldron of rice seemed to replenish steadily, like the loaves and fishes in the Bible.

On the third morning, the clouds lifted. In the crystalline air, the view extended for miles across a broad, green plain sparkling with flashing diamonds of lakes. The panorama extended a sheer ten thousand feet to the level of the sea. Eliza's party set off again, the porters trailing behind like a flock of canaries in their bright-yellow oil-paper capes and hats.

After climbing another thousand feet, they reached the edge of the summit, passed beneath a torii, and arrived at the last station, where they were greeted with cups of hot sake. They had barely rested when the storm took up again. This time, the rain fell in whirling sheets. Instead of making the obligatory circuit of the rim and its shrines, Eliza and the others quickly collected their certificates of ascent, then hurried to begin the descent. The rain-soaked cinder path was so slippery that the return journey to station eight was a steady downhill plunge. Weighted down by her sodden skirt, Eliza slipped discreetly out of it and wrapped herself in a red Navajo blanket. A yellow-clad porter fastened his rope, and together they skidded down the slope. "Sinking ankle-deep," Eliza wrote, "we travelled as if on runners through the wet ashes, sliding down in minutes stretches that had taken us many hours to ascend."[39]

Late that day, they reached the teahouse where they had stored a change of clothing before the climb. Despite their exhaustion, they decided to press on through the night to Mayanoshita. Riding in the cramped kago, Eliza listened in the dark to the silence of the countryside. The only sound came from the roaring streams swollen after days of heavy rain. Fireflies drifted through the rushes and stands of bamboo, and the night air over the rice fields was warm and heavy.

The light of morning was just coming up when they reached the summit of the pass, where they stopped for a breakfast of bread and chocolate. "The sacred mountain was clear and exquisite in the grey light of dawn," Eliza wrote. But the moment didn't last. "While we waited to see the sunrise on Fuji," she wrote, "a dirty-brown fog scudded in from the sea, crossed the high moon, and instantly the plain faded from view."[40]

A Singular Vision

Eliza sailed home to America late in 1888, her head filled with sensations of Japan. During the long crossing, alone in her cabin and circling the deck in the sea-salted air, she had a chance to slow down and take stock of her life. At thirty-two, she was in transition, unencumbered for the moment by pressing deadlines. In eleven years, she had published around 600 columns and travel letters in the *St. Louis Globe–Democrat.*[1] The work had made her a seasoned journalist and provided steady income. Yet the time had come to move on.

After nearly three years abroad, she was a different person. She had first gone to Japan as a freelance newspaper reporter: hardworking, proficient, but with little status. Now she was returning a world traveler and successful author, thanks to the strong sales of her Alaska book. Restless for change, she welcomed new opportunities, and felt sure her knowledge of Asia would pay off. Her homecoming that winter marked the start of a new phase in Eliza's career, one that would bring her greater authority and prestige.

Back in Washington, she wasted little time reaching out to editors. Assignments poured in. Harper and Brothers commissioned a book on Japan. She agreed to write a monthly Washington column for the San Francisco *Daily Call.* Magazine work alone would keep her busy for months. A revolution was occurring in magazine publishing, fueled by technological advances and an urge for self-improvement among middle-class Americans. Readers hungered for the information and culture that magazines provided in their absorbing mix of content: serialized novels, in-depth reporting, seductive advertising, and beautiful illustrations. Thousands of titles were in circulation, with more cropping up all the time.[2]

As Eliza had expected, editors looking for good content welcomed the kind of fresh and vivid material she had to offer. Americans had not outgrown their fascination with the country that first enchanted them a generation earlier at the Centennial Exhibition; Japanese objects were now in vogue, an object of pride and a badge of sophistication in many homes.[3] For years Eliza had written anonymously as "Ruhumah." As she started publishing under her full name, with national bylines, she soon established herself as one of the country's best-informed observers of Japan and other areas of Asia.

She wrote about the silkworm industry for *American Farmer*. Oriental teapots and Japanese theater for *The Cosmopolitan*, a smart new magazine of arts and letters. A handful of articles for *Harper's Bazaar* included one on home life in Korea, based on her long visit in 1887 as the guest of Judge Owen Denny, an American adviser to Korea's foreign office.[4] Eliza lectured on the subject that winter to the Women's Anthropological Society of America, an organization she joined before her now better-known affiliation with the National Geographic Society.[5]

Eliza even started writing for children. In the highbrow children's magazine *St. Nicholas*, which carried stories by authors such as Louisa May Alcott and Mark Twain, Eliza instructed American boys and girls on how to use chopsticks. Another article introduced them to Japan's nine-year-old crown prince, Haru, who had "the rank of a colonel in the Japanese army" and went everywhere in a miniature military uniform and a cap with a gold star.[6]

A lengthy piece for *Frank Leslie's Popular Monthly* illustrated Eliza's ability to go behind the scenes to report on subjects generally off limits to outsiders—such as Japan's royalty, and its women. In her article on Empress Haruko and her attendants, Eliza expressed a personal lament on the passing of some of Japan's ancient customs under the press of modernization. At the imperial chrysanthemum party in the fall of 1885, she had been dazzled by the empress strolling the grounds in her traditional dress: scarlet-red *hakama* (divided skirt), a heliotrope-blue *kimono* embroidered in wisteria, blue-black hair tied with bits of silky white rice paper and topped by a gold crest with three sun-like rays. The following year, the ladies of the court, following the emperor's example, adopted Western fashions for public functions. At a later garden party, Eliza observed that "all the soft, pink, electric-light reflections from the clouds of cherry blossoms . . . could not make picturesque the dreary little groups of women in dark, ugly Parisian dresses, with bonnet-strings, bangs, bustles, and all the abominations of that kind." Except for the cherry blossoms, the scene could have been "in any other civilized capital of the globe."[7]

Eliza agreed with those who admired Japan's devotion to progress. Yet many elements of "Old Japan" that made the country so distinctive were fast disappearing. How far would it go?

***

The lush flowering of springtime Washington always reminded Eliza of how much she had come to love her adopted hometown. She felt proprietary pride at having watched its transformation from a battered war-torn town into one of America's most pleasant cities, and a popular destination for travelers. Everywhere she looked Eliza saw tourists. Whole families beamed with excitement as they strode energetically along the avenues and traipsed through the halls of Congress and other institutions, the women in bustles and trim little hats, the youngsters in pigtails, straw boaters, and dark stockings. Bicycles, invented in England and showcased at the 1876 Centennial Exhibition, were starting to appear in the streets. The National Mall beckoned as an oasis, now that thousands of shade trees planted during the Grant administration had reached maturity. Public receptions at the White House were a picture of democratic harmony, with residents and out-of-towners of all ages and social classes mingling on the south lawn.[8]

And of course, the Washington Monument had become the hottest place in town; every month 10,000 visitors came to see the now-world-famous landmark.[9] For several years, going to the top had required visitors to climb the iron staircase of 897 steps. Now they had an option. The open-sided workmen's lift Eliza had ridden nervously during her earlier reporting had since been converted to a passenger elevator, complete with padded benches, carpeting, and a full-time attendant. At the monument's official opening on October 9, 1888, thirty-two people crammed into the elevator—packed "like sardines in a box"—for the maiden ten-minute ride to the top.[10]

Meanwhile, Eliza noted with satisfaction, the Army Corps of Engineers' reclamation work on the Potomac flats was coming along. The bleak and barren area would benefit greatly from landscaping and other improvements. But why settle for more of the same old thing—the poplars, elms, willows, and other common trees that the city's landscapers turned to time and again? Since her time in Japan, Eliza had started seeing the Potomac riverbank in a new way. She could picture it: filmy clusters of pink and white cherry blossoms floating above the water's edge, anchored by a tangle of dark, graceful limbs; above, a canopy of blue sky and white clouds; in the background, the towering stone spike of the Washington Monument.

No original idea is born in a vacuum. Any act of imagination, whether an artistic work or a new invention, takes shape in a mind poised to receive it by virtue of an individual's unique combination of life experiences, interests, and sensibilities. So it was that Eliza's keen knowledge of Washington's beautification efforts and her travels in Japan inspired her to see novel possibilities for that desolate corner of the nation's capital. As Eliza said years later to explain her thinking at the time, she realized that a grove of flowering cherry trees along the Potomac "would afford an annual flower-show at the season when all Washington is at home and the city receives its greatest number of visitors and sight-seers."[11]

Of all the impressions Eliza carried home from Japan, none had left so vivid an image in her mind as the springtime sight of cherry blossoms—*sakura*. It was magical. When the buds of Japan's cherry trees burst forth, they transformed the landscape almost overnight. Masses of cherry trees dotted the hillsides, lined the roads and waterways. Every major locality had special parks and temple grounds to which the Japanese flocked to stroll beneath the blossoms in an ancient tradition known as *hanami* (from "*hana*" for flower and "*mi*" for seeing). In the sixteenth century, a legendary feudal warlord, Toyotomi Hideyoshi, staged lavish sakura parties for thousands of guests. Over time, the ritual filtered down to every level of society. As Eliza explained to friends back in America, "Every one goes to see the cherry-blossoms as a matter of course."[12] In her eyes, nothing better epitomized the essence of Japan than the sight of men, women, and children promenading beneath the heavenly blossoms in their colorful kimonos, a rich tapestry of humanity gathered in kinship to celebrate nature's glory (Figure 8.1).

As with everything she grew passionate about, Eliza would later make a thorough study of flowering cherry trees, which were not unique to Japan nor totally unfamiliar in the United States by the middle of the nineteenth century. The trees were native to several countries of East Asia, though the nature-loving Japanese had elevated cherry blossoms to new heights of splendor over more than a thousand years, thanks to their talent for horticulture.

"Necromancers," Eliza called the Japanese: wizards who teased beauty from nature.[13] Japan's national calendar unfolded month to month with festivals celebrating every kind of flower—camellia, plum, azalea, lotus, chrysanthemum, and maple.[14] Each plant brought unique pleasure, but cherry blossoms reigned supreme, in part for a powerful symbolism that made them a favorite motif of Japanese art and literature for centuries. In the

Figure 8.1. "Rambles Under the Cherry Trees," by Kazumasa Ogawa in *The Hanami* (Flower-picnic), A. B. Takashima, 1899.
(Special Collections and Archives, The Claremont Colleges Library)

duality of their spectacular but brief blooming, sakura offered a reminder of the impermanence of life and the ephemeral nature of joyous experiences. Buddhism used the term *mono no aware*—"the pathos of things"—to convey the feeling that cherry blossoms evoked in viewers.[15] Eliza herself wrote that a cherry tree in bloom was "the most ideally, wonderfully beautiful tree that nature has to show, and its short-lived glory makes the enjoyment the keener and more poignant."[16]

By the time Eliza visited Japan in the mid-1880s, individual flowering cherries were growing in parts of the United States. Commodore Matthew Perry had carried cherry trees home from his voyages to Japan in the 1850s, and scientists and other travelers imported them for study or as a garden novelty. A few varieties appeared in nursery catalogs.[17] Washington itself boasted a couple of fine specimens that generated wide admiration every spring.[18] Given the unique appeal of cherry blossoms, Eliza wondered after her experiences in Japan, why had the trees not caught on more widely in America, especially as a striking addition to public parks?[19] To most Americans, cherry trees brought to mind a juicy red fruit. The idea of ornamental cherries prized strictly for their flowers was a strange concept. Most people in the United States had never seen one.[20]

As Eliza started thinking about possibilities for a cherry blossom park in Washington, two places in particular came to mind, both of them popular spots in Tokyo for cherry blossom viewing. One was the formal Uyeno Park, surrounding the burial temples of the Tokugawa clan of shoguns who had governed Japan for two centuries until power reverted to the emperor in 1868. The park was so famously picturesque during *hanami* season, when handsomely dressed Japanese came to admire the cherry blossoms from the comfort of jinrikishas, that the scene became a favorite subject of woodblock prints and hand-colored photographs.[21] Eliza loved the park's weeping, pendulant varieties of cherry trees whose vivid pink blossoms, framed against dark green pines, looked like "pillars of ruby fire."[22]

A week or so after peak blooming at Uyeno Park, a more boisterous gathering occurred along the Sumida River at a place known as Mukojima. When its mile-long avenue of cherry trees came ablaze, Eliza reported, "all the million inhabitants of Tokio seem to stream in unceasing procession day and night."[23] Mukojima was a "people's park" during cherry blossom season. Acrobats, jugglers, orators, and vendors gave the scene a carnival air. Crowds picnicked beneath the trees, while houseboats full of revelers cruised the river. Sake, Japan's mild rice wine, flowed freely, a tradition long linked with *hanami*. One prudish Western visitor expressed his disgust at the "dissipation."[24] But Eliza found the scene entertaining, even charming. Though alcohol often led to drunkenness and brawling in Westerners, she noted, sake seemed to cast a spell of playfulness over the Japanese. The laughter at Mukojima was so infectious, she wrote, "that even sober people seem to have tasted of the insane cup."[25]

The jollity and fellowship Eliza witnessed during *hanami* in Japan made a lasting impression. In the years that followed, she would say, time and again, that she hoped to create a "veritable Mukojima along the river's bank" in Washington.[26]

***

The story behind Washington's earliest Japanese cherry trees has been muddied over the years by misinformation and a scarcity of official records. Some of the confusion stems from Eliza's own account of the events. During the 1920s, in the final years of her life, she published two lengthy articles in the *Washington Star* describing how Washington came to acquire the trees. She did so, she said, to set the record straight in the face of erroneous explanations. Her own two articles contain a few discrepancies and minor inaccuracies but are consistent in major details, most of which are supported by other records.[27]

Eliza wrote that she began proposing the idea of cherry trees in Washington after seeing sakura during her first trip to Japan in 1885. That seems implausible, as she arrived that year in July and left in November, a period outside cherry blossom season. She may have misremembered the timing when recounting the events forty years later. The most likely scenario, based on her travel schedule and newspaper datelines, is that Eliza first actually experienced cherry blossom season in Japan in the spring of 1888, during her second trip. The images would have been fresh in her mind when she sailed home late that year.

Sometime in the months following her return to Washington, Eliza sought out a senior official to make her pitch. That path likely led through the White House—as it would more than two decades later when she finally found a way to make her vision a reality with the help of First Lady Helen Taft.

***

Americans were just getting to know their new president when Eliza went to the White House late in the spring of 1889 to report on the first family for *Harper's Bazaar*. The election of Benjamin

Harrison, a dour and religious man, had not garnered much national excitement, in part because he campaigned mostly from the front porch of his home in Indianapolis. One of the greatest assets he brought to the White House was his attractive family. His wife, Caroline, was a warm and gracious woman whose vivacious personality made up for her husband's chilly manner. Four generations of the family moved into the Executive Mansion. Press photographers and gawkers congregated outside the gates, hoping to see the president's small grandchildren as they rode around the grounds on a small cart pulled by a pet goat.

When Eliza arrived for the interview, she was led upstairs for a tour of the family's private quarters. With twelve people sharing the space, they were cramped. Over the years, presidential offices had encroached into the living area, including the second-floor oval library. The situation, Eliza reported, forced the Harrisons to use the wide upstairs corridor as a sitting area and reception room. Showing Miss Scidmore around, the first lady pointed out the desk and piano she had wedged among the cloth-covered tables, tufted-velvet slipper chairs, large landscape paintings, and potted palms.

Her biggest challenge, Mrs. Harrison confided, was keeping up with the huge volume of mail. Lacking any government-provided clerical help, she had to answer all the letters herself with the help of her married daughter, Mary, who served as her mother's social secretary. Americans, Eliza told readers, wrote about "all sorts of possible and impossible things": pensioners to ask about their papers; temperance activists to insist the White House stop serving liquor; housewives to request recipes for church bazaars and patches of the president's old ties for "crazy" quilts. Some people expressed a wish for an extravagant gift, like a typewriter or sewing machine, and "one-third of the letter-writers asked for money outright," Eliza wrote. "There is much that is pathetic and affecting in such a correspondence," she noted, "but even pathos loses by several hundred repetitions a week."[28]

Eliza's interview with Mrs. Harrison in *Harper's Bazaar*, followed a year later by a piece in the *New York Herald*, introduced readers to an energetic and civic-minded first lady who made it her mission to leave the Executive Mansion in better shape than she had found it.[29] Nearly a century old, the building had peeling wallpaper, moldy floors, an outdated kitchen and fixtures—and rats. The attic held a jumble of abandoned furnishings, with nothing labeled to distinguish the treasures from the trash. A woman who cared deeply about historic preservation, "Carrie" Harrison rolled up her shirtsleeves and got to work, directing a cleanup from top to bottom. She also lobbied to secure the funds for major renovations, which Eliza described in later reporting.[30]

In her interactions with the first family, Eliza developed a lasting friendship with the Harrisons' daughter, Mary, known by her married name, Mrs. Robert McKee. They would stay in touch for many years as part of a small circle of women who got together now and then in Washington.[31] The Harrisons invited Miss Scidmore to join them occasionally at the White House for private entertainment. Eliza also became a regular among the select group of women who assisted the first lady at public receptions.

On a chilly January afternoon in 1890, Eliza was standing by when a Marine Band struck up the music at 3 o'clock to announce the start of Mrs. Harrison's first open house of the season. As a long line of well-wishers waited to pay their respects, the first lady entered the flower-filled Blue Room on the arm of Col. Oswald Ernst, who made the introductions. Afterward, "among those invited to linger in the East Room was Miss Eliza Scidmore," the press noted.[32] That singling out, as society-watchers knew, was an honor reserved for friends and special guests of the president and his family.

***

Colonel Ernst became a familiar figure to Eliza at White House events.[33] The president had named Ernst in September 1889 to head the Office of Public Buildings and Grounds, a Corps of Engineers unit that oversaw federal lands and property in the nation's capital. A highly experienced Army engineer and former instructor at West Point, he assumed duties that included managing official functions at the Executive Mansion and serving as a de facto aide to the president and his family.[34]

Colonel Ernst was just the man Eliza needed to see. As superintendent of the office that oversaw the city's public parks and open spaces, he had the clout it would take to approve her idea of planting some Japanese cherry trees on the land being reclaimed along the Potomac. One day Eliza sought him out. In their meeting, she presented him with a packet of photographs from Japan to show what she had in mind. Since his men "had to plant something in the great stretch of raw, unclaimed ground by the river bank," she argued with enthusiasm, "they might as well plant that most beautiful thing in the world—the Japanese cherry tree." In a dignified manner befitting his position as a senior Army official, Ernst maintained his formal bearing as he allowed Miss Scidmore to have her say. But in the end, her appeal fell on deaf ears. "He listened patiently and seriously to my fairy tales," Eliza wrote. Then, "nothing more," she said. "Nothing happened."[35]

Many years later, Eliza would express—repeatedly—her frustration with Ernst and other park overseers in the Army Corps of Engineers who refused to see the merit of her idea.[36] One of those men, Col. John M. Wilson, both preceded and succeeded Ernst as the head of the Office of Public Buildings and Grounds in President Grover Cleveland's two nonconsecutive administrations. Like Ernst, Wilson was a West Point–trained engineer who went on to serve as superintendent of the U.S. Military Academy. A Medal of Honor recipient for his bravery in the Civil War, Wilson was a tall and handsome bearded man whose stern, buttoned-up appearance could intimidate strangers. Soldiers who had served with him, however, knew his softer side; he commanded great loyalty and respect for his sharp mind and the "genuine affection" he showed toward those who worked for him.[37]

In her own meeting with Wilson, Eliza apparently saw through the gruff demeanor. Later, she had a bit of fun—at his expense—describing his reaction to her proposal for the cherry trees.

> That delightful person had ideas of his own, and he swept the pretty pink pictures aside and turned his grim official countenance, his practical engineer's eye on me and said:
>
> "Yes! And when the cherries are ripe we would have to keep the park full of police day and night. The boys would climb the trees to get the cherries and break all the branches."
>
> "But these cherry trees do not bear any cherries. Only blossoms," I ventured.
>
> "What!! No cherries! No cherries! Huff! What good is that sort of a cherry tree?"[38]

In time, once she eventually got her way on the cherry trees through other avenues, Eliza would publicly decry the "indifferent and obdurate" men who treated her and her idea dismissively. She respected their professionalism, as Army engineers, competent administrators, and conscientious public servants. But their biggest sin, in her eyes, was a lack of imagination. They were all, to a man, conventional in their thinking, Eliza charged, with rigid ideas about what constituted a proper park in the nation's capital. "So fixed was the type, so standardized their minds on park planting, so much they represented the unknown and the untried" that she grew frustrated by "the whole lot" of them, she wrote.[39]

One historian has noted that the Army Corps of Engineers felt very protective of its authority over the city's public parks, including the one taking shape along the Potomac. "Officers of the corps insisted that they would be the ones to develop the filled land, just as they had created it."[40] They put their faith in sturdy, all-American trees, like elms. Apart from resenting the intrusion of an outsider—a bothersome, presumptuous woman, no less—they may have seen the "Oriental" trees Miss Scidmore favored as highly ill-suited for the symbolic heart of America's capital. They were "foreign" trees, after all, and too delicate perhaps for a city with grand visions of itself.

By character and temperament, Eliza was little disposed to self doubts and half-baked convictions. All her life, she stuck fast to her principles, confident about the soundness of her own judgment. Over the years, she would never lose sight of her vision of Japanese cherry trees in Washington. Despite the wall of opposition, she must have reasoned that sooner or later, she could bring others around to her way of thinking. It would take nearly three decades, but she was right.

***

Steamy weather could make Washington unbearable in the summer, and August 1889 was no exception. Eliza fled the city to visit relatives in Wisconsin. Twenty-seven years after first leaving Madison, the native daughter returned wearing the laurels of success as a world traveler of literary renown. One day Eliza and her cousin Mary Atwood joined a rail excursion that Mary's father—still the publisher of the *Wisconsin State Journal*—organized to carry a party of dignitaries to a celebration at the famous Bethesda hot springs in Waukesha.[41] As they dined at the elegant Fountain Spring House and stayed on for the fireworks and grand ball, Eliza felt no need to shrink in feminine deference among the powerful men in their group, who included the governor and secretary of state, several state commissioners, a judge, and the mayor of Madison. She could hold her own among them.

Eliza treasured most that summer the time she spent with her Oakley relatives at their farm, "The Evergreens." Her cousin Mary Oakley was the Scidmores' most faithful correspondent, and Eliza adored her Aunt Jane like a second mother.[42] As a young woman, Jane had been the family beauty, with a cameo profile and dark glossy hair; several years at a female academy in Canton, Ohio, honed her social graces and domestic skills. Admiring her aunt's artistic needlework, Eliza beseeched her to "piece me a quilt" like those Jane had made for others in the family.[43] Unfortunately, their time together was cut short. One day they all went into town and found a letter waiting for Eliza. *The Cosmopolitan* had written requesting the article on Oriental teapots. Eliza retreated to her room to work on the piece.

Before returning to Washington, Eliza made a trip to Boston. *The Cosmopolitan* planned to run her teapot article with ink drawings based on pieces in the ceramics collection at the Museum of Fine Arts. While in town, Eliza stayed at the Brunswick, a fashionable hotel in the Back Bay area. There she socialized with Lilian Whiting, a widely syndicated columnist who had rooms at the Brunswick. Whiting, who had favorably reviewed Eliza's book on Alaska, listened with enchantment as Eliza described her various adventures. "Miss Scidmore," Whiting told her readers, "has the genius for travel." She was also, the columnist added, "as good a conversationalist as she is a writer."[44]

The Cosmopolitan shared Whiting's praise. In December 1889, the magazine included Eliza in its profiles of two dozen men and women of letters—historians, scientists, journalists, and authors—who made up "Literary Washington." Eliza shared the spotlight with such notable figures as George Bancroft, author of a multivolume history of the United States; Lincoln biographers John Nicolay and John Hay; Ainsworth Spofford, head of the Library of Congress; and the novelist Frances Hodgson Burnett, whose internationally popular novel *Little Lord Fauntleroy* sparked a trend of dressing young boys in velvet suits with lace collars. *The Cosmopolitan*'s writer ticked off Eliza's travels in awe: "She has been twice to Europe, twice to China and Japan, spent two summers in Alaska before the tourists' travel and mining seriously began, and 'did' Colorado and California pretty thoroughly."[45]

That winter, George and Mrs. Scidmore arrived back in the States. At thirty-six, George was now a deputy consul-general and an instructor at Tokyo's new English Law School.[46] He endeared himself to the expatriate community in Japan by his devotion to his mother; they had grown close since Eliza's departure for America and went everywhere together.[47] George had returned on home leave amid a hero's welcome at the State Department. He was due to receive a medal from the emperor of Japan, a reward for his bravery after he jumped into the Yokohama harbor the previous summer to save a Japanese man from drowning.[48]

George and Mrs. Scidmore found Eliza hunkered down at the Elsmere, a boarding-style hotel in a brownstone on H Street NW, two blocks from the White House. Papers littered the room: research materials, half-completed articles, bits of correspondence. She was juggling deadlines with a heavy calendar of social obligations. One of her oldest and dearest friends, Ida Thompson—a fellow stalwart in the "great company of independent spinsters" in Washington whom Eliza would recall fondly many years later—lived just around the corner, in a home furnished with Oriental treasures paid for by Ida's wealthy banker father.[49] Eliza's growing celebrity brought invitations to dine at some of the best homes around town. She was still a relatively young woman, yet few people had seen as much of the world as she had. It made her seem as "exotic" as many of the places she visited.

Eliza Catherine Scidmore was tickled by her own social success that winter. Years earlier she had enjoyed status among the "Wisconsin people" in Washington for her longevity in the nation's capital. Now she turned up regularly at official events alongside an old friend from Wisconsin, Mrs. Jeremiah Rusk, whose husband President Harrison had named his agriculture secretary. "Mrs. Scidmore and Miss Scidmore": the paired names appeared often on the society pages. Three decades after moving to Washington to make a new home, the Scidmores had "arrived."

Now that she expected to be in America for a while, Eliza looked forward to resuming her reporting on Alaska. She planned to return that summer. She acquired a commission to write a chapter on Alaska for the U.S. Census and would also travel as a correspondent for *Harper's Bazaar*.[50] To help pay expenses, she arranged to report on the trip for the *Milwaukee Sentinel*.

With Eliza buried in deadlines, it fell to George to escort their mother to Madison to visit relatives. He stayed only briefly, then left to meet Eliza in St. Paul.[51] They traveled together by train and parted on the West Coast—he to sail on to the Far East, she to head north to Alaska. The scene that greeted her in the Pacific Northwest was much changed since her last trip to Alaska, six years earlier.

Among the Scientists

Having gotten a late start on the travel season, Eliza found the Pacific Northwest overrun with tourists on their way to Alaska. In the years since her last trip, a cruise industry had sprung up. With more than 5,000 people booked that summer for a voyage, the steamers were packed.[1] Large groups arrived in special trains from the East, after the major tour companies, Thomas Cook and Raymond & Whitcomb, added Alaska to their lists. Many titans of industry organized private excursions. As Eliza told readers in the *Milwaukee Sentinel*, "The celebrities passing through the Puget Sound towns have given the local papers columns of interviews and items." Though still in its infancy, Alaska tourism, she reported, seemed "destined to reach greater proportions."[2]

The Pacific Coast Steamship Company helped spur the boom with special excursion steamers it added to its routes in 1887.[3] Ships now left for Alaska every two weeks from May to September. With all-inclusive packages available for $130—about a month's salary for a government clerk—"it is cheaper to travel than to stay at home," one tour promoter argued. Touting Alaska as the "American Switzerland," slick brochures of railroad and steamship companies played up the "must-see" experience of a stop in Glacier Bay, where Eliza's old friend Captain James Carroll, now one of the territory's biggest boosters, built a walkway to assist passengers going ashore. Travel advisories included packing tips that made an Alaska cruise sound rustically romantic. Gentlemen had no need for their "swallow tail or court dress, or Sunday-go-to-meeting clothes in Alaska." Ladies were advised to raise the hem of their skirts to keep them from dragging on wet steamer decks and mossy riverbanks.[4]

As word of its splendors spread, Alaska was becoming "a new place to dream over," as one historian notes.[5] Summer visitors published scores of accounts about their experiences, but no one besides John Muir had done more to popularize the region than Eliza had.[6] Her 300-page book, *Alaska: Its Southern Coast and the Sitkan Archipelago*, was now an essential reference work for tourists, and a vicarious pleasure for armchair travelers back home who didn't make the trip but wished they could. "Spent a greater part of the day in reading Miss Scidmore's book on Alaska," one woman diarist wrote.[7]

Eliza's reporting on Alaska coincided with a new age of mass tourism. Americans with rising incomes and more leisure time were hitting the road as never before; destinations "out West" were especially popular.[8] In an era of keen interest in science and nature, many people saw an Alaska trip as educational. Photography hobbyists went for the dramatic scenery. For those with "nervous prostration," a common diagnosis blamed on "the much greater demand which the conditions of modern life make upon the human brain," the bracing northern air and gentle cruising offered just what the doctor ordered.[9]

***

Eliza traveled that summer on the *Queen*, a sleek two-masted beauty that boasted mahogany trim, brass fixtures, and stained-glass accents. Passengers boarded amid a perfume of fresh flowers, settling into the 250 cabins on three decks that blazed at night with electric lights. The *Queen*'s patrons that summer, Eliza told readers, included Mr. Armour of Chicago meat-packing fame, traveling with family and friends.[10]

As on her first trip to Alaska, Eliza was to meet the Alaska steamer in Port Townsend, the busy customs port where she had been stranded overnight in the summer of 1883. This time, after checking into her hotel, she quickly got to work. She wanted to mail off a query letter before her ship sailed.

The story idea came to her after reading a pair of articles by John Muir in the latest issues of *The Century* magazine (previously *Scribner's Monthly*, renamed after a change of ownership). Like many who had read his writings in national magazines and newspapers on the West Coast, Eliza idolized Muir as a modern-day prophet. No one spoke more passionately and eloquently than he did about the natural world and the need to protect its wonders for the benefit of present and future generations. When she herself had started reporting on Alaska, she sought out Muir at his home in Northern California.[11] On trips to the West Coast, she and her mother got to know some of the Muirs' friends, including Ezra Carr, Muir's former geology professor at the University of Wisconsin; his wife, Jeanne, a longtime mentor to Muir; and Captain Calvin L. Hooper, commander of a revenue cutter in the Pacific Northwest.[12]

Pressed by the demands of a growing family and a busy orchard business he ran with his in-laws, Muir had published little in recent years. Now, on her way to Alaska, it delighted Eliza to see his byline again. The August *Century* carried a lengthy piece by Muir extolling the scenic wonders of Yosemite, his spiritual home in Northern California. In a related article that followed, he made the case for putting a fragile area of Yosemite under federal protection, along the lines of Yellowstone, the country's first and only national park.

Eliza had learned that Muir was in Glacier Bay for the summer. She would have a chance to see him when her steamer made a routine stop in the bay. Sitting at the desk in her hotel room in Port Townsend, she composed a letter to *The Century*, proposing to write a profile of Muir for the magazine. To those who had read Muir's articles in *The Century*, she wrote, a sketch of him "could not fail to be an interesting subject." Few people recognized "what a remarkable man Prof. Muir is, or how full of unusual incident his life has been."[13]

She had written many query letters over the years. This one carried much greater weight, as she was writing to Richard

Watson Gilder, whom many people regarded as the most important literary figure in New York. A slight man of unassuming appearance, Gilder would not have turned heads in the city's streets. Yet he put a large stamp on public taste and opinion as editor of *The Century*, America's leading "thought" magazine and a trend-setter in mass-market magazines.[14] Every month, 200,000 of the country's top people in every field—from presidents on down—read *The Century's* mix of literature and reporting. Millions read its long retrospective series on the Civil War, and the magazine paid an enormous sum to excerpt the biography of Abraham Lincoln by his secretaries John Nicolay and John Hay. Apart from its editorial excellence, *The Century* had a crusading spirit that distinguished it from competitors like *Harper's* and *Atlantic Monthly*.[15] *The Century* supported progressive issues like tenement housing improvements, free kindergarten, civil-service reform, and international copyright protection. In its latest pet project, the magazine had taken up advocacy of wilderness protection, a concept still new to Americans. The editors saw John Muir as the perfect figurehead for the cause.

She was well qualified to profile Muir, Eliza told Gilder, by virtue of knowing "this 'poet scientist' and 'glacial prospecter,' his family and many of his friends." To strengthen her case further, she promised to get "picturesque material for illustrations." Though *The Century* did not print photographs, its art staff enjoyed a reputation as the best in the business at making sophisticated woodcut illustrations from photos and famous works of art.[16]

Now that she herself had joined the "Kodaking" craze sweeping the country, Eliza saw a chance to photograph Muir in Glacier Bay, framed against the glacial backdrop. Two years earlier, George Eastman had revolutionized photography by introducing a simple box camera that came preloaded with a roll of black-and-white film, enough for a hundred snapshots. The baggage at Eliza's feet contained one of the new cameras: a boxy two-pound device covered in dark Morocco leather, with nickel and lacquered-brass fittings.[17]

Signing off on her letter to Gilder, Eliza rushed off to mail it. How it would be received in New York, she had no idea. But there was a good chance Gilder would recognize her byline from her articles in *St. Nicholas*, the company's children's magazine. Breaking into the esteemed parent magazine itself would be quite a feather in her cap, putting her in the top tier of American writers and journalists.[18] The letter to Gilder was her first query to the magazine, but it would hardly be the last. Her letter opened the door to what would become the most important professional relationship of Eliza's career.

***

John Muir had been in Glacier Bay for two months when Eliza met up with him on a day in mid-August. That morning, she rose early and was on deck at 5 o'clock as the *Queen* entered the bay. The temperature was chilly, the air thin and sharp. Snuggled under a lap robe, Eliza sat with other passengers nibbling rolls and sipping coffee as three tall peaks came into view up a broad inlet to the west. Each stood in sharp silhouette, bathed in radiant yellow light and surrounded by the white of glaciers and snow fields. A "water-color country": that was Alaska, Eliza told an interviewer later that year.[19] As the steamer headed north in Glacier Bay, it swung back and forth, veering to avoid icebergs. Writing in the *Milwaukee Sentinel*, Eliza likened the scene to sailing through a treasure chest of gems:

> Each piece of ice took the early sunlight and became a vast jewel, snow white as opal, blue as turquoise and sapphire and softly green as aquamarine and peridot. One huge transparent mass floated colorless until it crossed the sun's rays and then flamed suddenly and became a huge golden and smoky topaz, flashing prismatic lights from every splintered edge.[20]

The *Queen* reached the northern end of Muir Inlet later that morning and dropped anchor a quarter-mile offshore from the eastern moraine of Muir Glacier. The icy cliff of its massive front wall loomed just ahead. Eliza and the other passengers climbed into boats and were rowed ashore, free to wander for several hours. Clutching her camera and notebook, Eliza hurried down the beach to the area where Muir was camped.

Muir had returned to Alaska that summer for the first time in ten years. He was relishing not just the solace of nature but freedom from the tyranny of deadlines. Writing for *The Century* had been an ordeal. A year earlier, in San Francisco, Muir had met Robert Underwood Johnson, *The Century*'s associate editor, when Johnson went west to recruit writers. The two men hit it off, and Muir invited Johnson to go camping with him in Yosemite. They made an unlikely pair: the bespectacled editor from the East and the gangly, bramble-bearded naturalist with the agility of a mountain goat. They slept under the stars, and as they talked into the night, Johnson, a published poet, realized that Muir viewed nature itself as poetry. "To some, beauty seems but an accident of creation: to Muir it was the very smile of God," the editor recalled. The sight of his companion scrambling about in the wild conjured an image of "John the Baptist as portrayed in bronze by Donatello."[21]

Surrounded by the beauty of Yosemite, it troubled the men to see how badly the area was managed under a state-appointed commission. Acres of trees had been cut to stumps. Sheep—"hoofed locusts," Muir called them—had trampled meadows of wildflowers. Livestock pens and seedy tourist ventures marred the landscape. Agreeing that something needed to be done, the two men came up with the idea of a campaign to make Yosemite a national park. Muir could inspire public support by writing about Yosemite in *The Century*, Johnson suggested, while he himself would urge his influential friends to lobby Congress.[22]

Muir agreed to the plan, then struggled for months with the writing. Elegant thoughts that formed so fully in his head sprawled into incoherence when he tried putting words to paper. Johnson prodded him along, in letters that grew ever more frantic as the deadlines neared. Finally, when their mutual patience had been nearly exhausted, Muir mailed the two pieces off in the spring of 1890. Craving the sanctuary of the wilderness, he secured his wife's blessing, boarded a steamer, and fled north to Alaska.

In Glacier Bay, Eliza found Muir, still lanky and fit at fifty-two, in his element among a group of companions. He and his friend Henry Loomis had arrived in late June. They got company a week later when a steamer dropped off another party, along with tons of equipment and supplies. Harry Fielding Reid, a geology professor at the Case School of Applied Science in Cleveland, had come with a handful of students to do scientific studies in the area for ten weeks.[23] The two groups agreed to collaborate on field work, and Muir arranged the delivery of lumber to build a cabin as a base camp (Figure 9.1). Perched at the water's edge, the small hut

Figure 9.1. John Muir, far left, and geologists in 1890 at the cabin they built in Glacier Bay.

(Photo by J. F. Morse, courtesy University Archives, Sheridan Libraries, Johns Hopkins University)

had two windows and a chimney of boulders cemented together by glacial mud.

Though Eliza's profile of Muir in *The Century* never transpired, she described her encounter with him in the *Milwaukee Sentinel*. "The summer colony at the glacier has been a very sociable and congenial one," she reported.[24] The Reid party spent their time taking measurements and staking markers for studies to determine the flow and movement of Muir Glacier. Reid was also compiling one of the earliest maps of Glacier Bay, which Eliza would borrow years later for an updated edition of her book on Alaska (Figure 9.2).[25]

In contrast to his companions, Muir had arrived with only field glasses and sketchbooks, Eliza reported. He fashioned a primitive sled and went off on his own for ten days, climbing to the upper

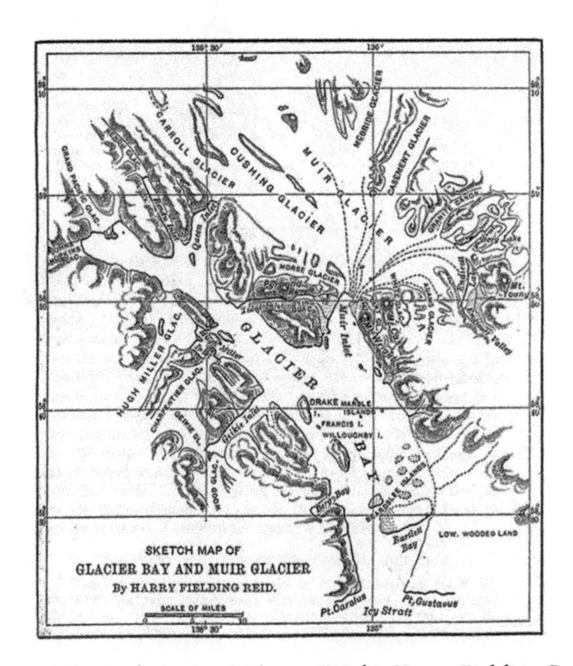

Figure 9.2. Early map of Glacier Bay by Harry Fielding Reid.
(From Eliza Scidmore's 1893 *Appletons' Guide-Book to Alaska*)

heights of the glacier, exploring its tributary streams and neighboring forests. Muir showed Eliza his field notebooks. The pages were covered with tiny sketches, "each as clear, fine, sharp and perfect as if done on the largest and most permanent scale," she wrote. Weather conditions in Alaska that summer had produced vivid displays of the aurora borealis, Muir informed her. He and the other men had sat up many nights in a row watching, as Eliza described it, "the whole sky aflame, banded, quivering and flashing with strange lights."

During the visit, Eliza dashed about taking photographs. Weeks later, back in Washington, she sent the Muirs copies of her photos, including some of the cabin at Glacier Bay. "Mr. Muir's summer chalet by the ice," she called it.[26] She had also gotten good snapshots of Vancouver, Juneau, and other places on the steamer's stops. Overall, she was quite pleased with the results. Her Kodak negatives had all "come out so well," she wrote, "that I am provoked that I did not take more."[27]

***

After the Alaska trip, Eliza spent some time in San Francisco before heading home. While in town, she went to see the work of the landscape painter William Keith. Scottish born, like Muir, the burly, bearded artist was Muir's best friend and frequent hiking companion. Keith was not there when Eliza visited his studio on Montgomery Street, but she took an immediate fancy to one of his works. It was a painting of the Taku Glacier, which Eliza considered the most beautiful of all the tidewater glaciers along the coast.[28]

She came away infatuated by the painting. It really ought to go on display back East, she decided. She already had a place in mind: the art gallery of the Cosmos Club, a private men's enclave at Lafayette Square in Washington, across the street from the White House. The National Geographic Society held its twice-a-month Friday-night meetings in the club's public rooms.[29] In 1890 Eliza herself became a member of the society, which had been founded two years earlier "for the increase and diffusion of geographical knowledge."[30] Its organizers—three dozen eminent men from science, business, and government—hoped to build a broad organization of people interested in learning more about the physical world. Unlike other professional societies that were springing up as science became more specialized, the National Geographic did not restrict its membership to experts in their fields—nor to men. A few women had been admitted from the start, possibly because many of the early male members had strong, accomplished women in their own lives.[31]

Eliza's reporting on Alaska made her a natural fit. No less an authority than William Dall, the "dean" of Alaska exploration and one of the National Geographic's cofounders, had reviewed her book favorably. Her writings also impressed another man with a strong interest in Alaska exploration, one whose influence mattered greatly: the society's chief founder and patron, Gardiner Greene Hubbard (Figure 9.3). A wealthy lawyer and financier with roots in Boston, Hubbard—a tall, patrician-looking man with a vine-like beard—viewed science as an engine of American

Figure 9.3. National Geographic Society co-founder Gardiner Greene Hubbard and his wife, Gertrude, at their summer home of Twin Oaks in Washington.
(Bain Collection, Library of Congress)

progress. Fortuitously, among the many scientific ventures he backed were Alexander Graham Bell's experiments that led to invention of the telephone. When Bell, a young teacher of the deaf, fell in love with Hubbard's deaf daughter, Mabel, and married her, the relationship between the two men became personal as well as professional. Eliza would develop a close friendship with both families during her long affiliation with the National Geographic Society, and Hubbard would become her greatest champion in the organization.

Like other scientific societies, the National Geographic Society published a journal for its members—but not very systematically. Bound in a dull brown cover, the lofty-sounding *National Geographic Magazine* consisted at times of a single article, often with a stuffy title like "The Classification of Geographic Forms by Genesis." The first attempt to include photography, in the July 1890 issue, was so unremarkable the fuzzy black-and-white image of Herald Island, in Russia's Chukchi Sea, resembled a distant humpback whale rising from the sea.[32] It would take years for the magazine to find the right editorial formula.

To help meet its educational mission, the society also organized outings for its members and sponsored a public lecture series that often drew several hundred people, especially when the speakers showed lantern slides. In the spring of 1890, not long before Eliza returned to Alaska, the society had decided to undertake its first scientific expedition, in partnership with the U.S. Geological Survey. "After much discussion," Eliza reported, "it was decided to stay by our own continent and country, and go to the whitest peak of Alaska, rather than to darkest Africa, to New Guinea or the Antarctic."[33]

The society fielded a team to explore Mount St. Elias, in a mountain range straddling Alaska and Canada. It was thought to be the tallest peak on North America, but several attempts to reach the summit and measure the elevation had failed. For the National Geographic expedition, Major John Wesley Powell, head of the U.S. Geological Survey, offered the services of two of his men: to lead the group, Israel Cook Russell, a thirty-seven-year-old veteran of Alaska exploration, and Mark Kerr of the bureau's survey corps as the topographer. A few weeks before Eliza arrived in Alaska that summer, the team made its way to Yakutat Bay, 150 miles northwest of Glacier Bay, to begin the climb of Mount St. Elias.

Eliza followed the developments with proprietary pride. The society would announce the scientific results of the expedition that fall in Washington, and it occurred to her that with Alaska in the spotlight, Mr. Keith's Taku Glacier painting would be a nice touch. She decided to inquire whether he might loan it out. It was a "splendid picture," she told John Muir in a letter. And yet—never short of strong opinions—she couldn't help thinking that one element seemed a bit off. "The water tones are perfect, but the ice ought to be more dazzlingly white and blue," she told Muir. Perhaps he could take up the matter up with Mr. Keith, she suggested. "If you are in the city soon will you not go in and look at it and tell me what you think of it for color."[34]

***

The Muirs invited Eliza to visit them before she returned home in the fall of 1890. Muir's wife, Louisa Wanda—"Louie" to family and friends—was a well-educated woman and talented pianist with a deep interest in current events and issues such as women's rights. But she was also shy and stayed close to home, helping to manage the family's fruit-growing ranch near Martinez, California. She welcomed the company of worldly visitors like Miss Scidmore, who shared Louie's conviction that her husband had been put on the earth to defend God's creation from the travesties of greed and destruction.[35] Louie regretted that Miss Scidmore could not join them that fall when a mutual friend, the writer Alice Rollins, and her husband spent a day with the Muirs in Martinez. It was harvest time, and they were "up to the scalp" in grapes, as John Muir told his editor; tokay, Muscat, and zinfandel, they were shipping 2,000 boxes a day.[36] As the Rollinses left to return to the city, Louie

Muir sent a box of grapes with them for Eliza at her hotel. When Alice tried to deliver them, however, the clerk reported that Miss Scidmore had already checked out.[37]

Eliza had hurried down the coast to visit a wealthy friend in Santa Monica, before returning East by way of the southern route.[38] After a short stay in Washington, she went on to New York to meet with her editors about her book on Japan, which Harper and Brothers planned to publish in the spring.

One day in the first week of October, Eliza marched off to *The Century*'s offices in Union Square to pitch a story idea. She found the staff buoyant at the rousing success of the magazine's Yosemite campaign. The results had been swift and stunning. Muir's pleas in *The Century* to protect a large area of the Yosemite wilderness had struck a chord with Americans. Letters poured into Congress. A bill was introduced; it quickly passed both the House and Senate, and on October 1, 1890, President Benjamin Harrison signed the law creating Yosemite National Park.[39]

With that issue happily settled, Eliza hoped to persuade *The Century*'s editors to "take up the cause of Alaska" and its welfare as the magazine's next crusade.[40] Before she could effectively make her pitch, the matter got sidetracked. Local newspapers were reporting news from San Francisco on the Mount St. Elias expedition.

Eliza was furious. The National Geographic Society had gotten "scooped" on its own story. Even more galling, she discovered, the breach had been committed by one of the expedition's own men.

The team of nine had reached Yakutat Bay in late June and started up the mountain. Bad weather plagued them from the start. Heavy rain and blinding snow hampered their progress, and avalanches threatened to crush the men in their tents. When Russell and Kerr forged ahead in an attempt to reach the summit, they got trapped in a snow slide and barely escaped with their lives. Finally, the team had to abandon the climb, though they managed to make observations for a topographical map of the area.[41]

After the group disbanded at Port Townsend that fall, Israel Russell departed for Washington to brief the sponsors. Mark Kerr went on to San Francisco. While there, the young man briefed members of the Geographical Society of the Pacific on the expedition. Reporters picked up his comments and rushed to interview him. Now he was being quoted in all the papers. Eliza found Kerr's conduct despicable. It was not just a breach of scientific protocol, but also an affront to his superior. Israel Russell was a modest man and an expert in his field, widely respected by his colleagues as a "scientists' scientist."[42] He deserved better.

Determined to give Mr. Russell his proper due, Eliza took it upon herself to act. While Russell was heading eastward by train, she contacted the Associated Press's manager in Washington—a man she knew, named McKee—to give him "a few facts and reasons" about the situation, as she put it.[43] Perhaps he could send out something about Mr. Russell. McKee agreed, and on Monday, October 20, newspapers across the country carried the AP's column-and-a-half interview with Israel Russell on the Mount St. Elias expedition.[44]

Eliza was still fuming about the situation days later in a letter to Muir. "Mark Kerr's Great Feat. Mark Kerr alone on Mount St. Elias . . . [with] no whisper about the Russell Expedition or anyone but M. K. the young dude, who went as Mr. Russell's assistant," she wrote in fury. "All Mr. Russell's work and effort . . . to go to the glory of the insignificant Kerr."[45]

Muir met Kerr that fall in California and told a friend that he found the young man "all right."[46] Eliza was not so quick to forgive and forget. Her monthly column for the San Francisco *Morning Call* offered a chance to have the last word. On December 7, she chided the scientists on the West Coast for their "nice little trick . . . in stealing the first thunder of our Mount St. Elias expedition." The morality of the situation, she pointed out, seemed clear: "There should be honor even between geographic societies." She couldn't resist adding a jab at Mark Kerr by mocking his inconsistent reports on the height of Mount St. Elias; he had given different numbers, from 13,000 feet to 15,330 feet, to several audiences. "Now," Eliza wrote sarcastically, "what can a country do with a mountain that elongates and shortens itself at that rate?"

***

As the year 1890 drew to a close, scientists flocked to the nation's capital for annual meetings. "Alaska is having quite a popularity and prominence in Washington just now," Eliza informed Louie Muir in a New Year's Eve letter. Harry Fielding Reid had dazzled geologists with lantern slides of his photographs from Glacier Bay, which included pictures of John Muir and his cabin. Her old friend Captain James Carroll was in town, "as full of original salt and sin as ever," and amusing everyone on Capitol Hill with his colorful stories and opinions.[47] The biggest scientific event of the season occurred the evening of November 26, when 800 people streamed into Lincoln Music Hall to hear Israel Russell report on the National Geographic's Mount St. Elias adventure. Dignitaries in the audience included the British and Russian ministers, the veteran Alaska explorer A. W. Greely, and faculty from Johns Hopkins University, who took the train down from Baltimore. Bursts of applause filled the auditorium when Russell flashed images of snowy landscapes and other scenes from the expedition.[48]

Two nights later, Alaska was again the focus when ninety members of the National Geographic Society gathered for a meeting. During the program, Israel Russell unveiled the Taku Glacier painting Eliza had admired in San Francisco.[49] Mr. Keith had agreed to her request to send it for display at the Cosmos Club. And of course, no one could stop talking about the successful Yosemite park campaign. "Everyone has been so delighted to find Mr. Muir again in print and have so enjoyed his Century articles this year," Eliza told Louie Muir. "I hope that you will keep him to it and not let him rest until he has given Alaska its fair share of his attention."[50]

Meanwhile, Eliza and her mother were immersed in domestic tasks. They had windows to measure for curtains, crates to unpack after Eliza ordered them sent from a storage warehouse across town. That fall, she bought a house in Washington. The handsome brick townhouse at 1501 21st Street NW, which she purchased for $11,000, sat in a row of new residences a block south of mansion-lined Massachusetts Avenue, in the fashionable neighborhood of Dupont Circle.[51] Besides giving the women a permanent home in Washington, the new house offered space in which to display the many treasures Eliza had begun to acquire in Asia. Future guests would enter a parlor lined with Chinese cabinets and Japanese tapestries, faience and lacquer ware, idols, and paintings from old Buddhist temples. Eliza would serve tea in quaint Chinese cups as a haze of incense wafted through the room.[52]

At Christmastime, a large, bulky package arrived in the mail, stamped with a Wisconsin postmark. Eliza opened it to find the quilt she had requested a year earlier from her Aunt Jane. Stitched into the cloth were threads tying Eliza to the place of her early childhood. Three decades had passed since she left Madison with her mother and brother during the Civil War. Home in the intervening years had been as much a state of mind as a physical place. As her travels spiraled ever wider, the world was becoming her home. With the new house in Washington and her brother's residence in Japan, Eliza now had a firm foothold on two continents.

For a woman forced to make her own living, however, security would never be a sure thing. For the time being, at least, she and her mother would need to live frugally. Jane Oakley explained the circumstances to her son and his wife when Eliza decided not to send out Christmas presents that year: "Their new Home she says absorbs everything."[53]

10

A Voice for Conservation

The year 1891 brought several milestones in Eliza's career, starting with the publication of her book on Japan and her first article in *The Century* magazine.[1] Of all her writings, *Jinrikisha Days in Japan* would most define her literary legacy, eventually becoming a classic of travel literature. Critics loved it. Miss Scidmore's new book, *The Literary World* told readers after the book's release that spring, "may safely be recommended as the best general guide for travelers in Japan."[2] The account benefited, reviewers noted, from the author's lengthy stay in the country and her access to parts of Japan that most tourists never saw. The straightforward reporting style was an added plus, making the book more useful than idiosyncratic works like the British traveler Isabella Bird's *Unbeaten Tracks in Japan* and the American scientist Percival Lowell's philosophical *Soul of the Far East.*[3]

Eliza's assignment for *The Century* flowed from her tendency to chew over issues she cared about like a dog with a bone. In this case, she fretted over Alaska's welfare. From the time she started reporting on the region she was critical of what she saw as the federal government's lax oversight of the territory.[4] In one example she cited often, there was still, a quarter-century after America's purchase of Alaska, no systematic plan for surveying and exploring the region.

Now she was anxious about a dispute with Canada involving a boundary line between southeastern Alaska and the coast of British Columbia. The exact line of demarcation was contentious because of ambiguous language in an Anglo-Russian treaty that predated the U.S. acquisition of Alaska. Discussions had gone on for years with no resolution. But the matter was growing more urgent because of gold discoveries and other interests in the region. Heightening the tension, Canada had recently issued maps showing a different boundary line from the one long recognized by the United States.

More talks on the issue were due in the fall of 1891, and Eliza wanted to make Americans understand just how much was at stake. During her meeting with *The Century*'s editors the previous fall in New York, she had urged them to promote awareness by running a pair of articles, as they had done in their successful Yosemite campaign.[5] John Muir could write on Alaska's natural wonders, and William Dall, America's foremost Alaska expert, on the boundary issue. Eliza suggested in a letter to Louie Muir that she might use her wifely persuasion to get her husband on board: "Dr. Dall on the Boundary Line and Mr. Muir on the glaciers and scenery would help the territory on amazingly. *The Century* has such a power."[6]

But both men had other priorities, so early in 1891, Robert Underwood Johnson arranged for Eliza herself to draft a thousand-word piece for the magazine's "Open Letters" section. She tackled the subject with her usual vigor, only to discover—as John Muir had before her—just how rigorous *The Century*'s editing process could be. Johnson returned the piece, requesting a rewrite. Chastened, Eliza tried again. "I return the mss. subdued to a more pacific tone," she replied, "and am greatly obliged to you for the privilege of withdrawing and altering."[7]

In her article, published that summer, Eliza explained how Canada's alternative boundary—known as the Cameron Line, for the army officer who drafted it—leaped bays and inlets, breaking that part of the Alaska coast into alternating patches of U.S. and British Columbia territory. If the alternative boundary was allowed to stand, she warned, the United States stood to lose mining fields and other valuable resources in the area. Most alarming, the "gerrymandering of scenic Alaska," as she called it, would put a major part of Glacier Bay and several of the area's most spectacular glaciers under the British flag. No longer would U.S. steamers be able to land passengers at Muir Glacier.[8]

As the meetings neared that fall, Eliza brooded about the situation in a letter to Muir. She feared that George Dawson, of the Geological Survey of Canada, "will make a grab for everything worth having on the mainland coast." He needed to be "muzzled," she insisted. "I do not care very much what he does about the Pribylof Islands," she added jocularly, "but Revillagigedo and Scidmore Islands cannot be spared him, and you must make a stand for your own glacier and the Taku."[9] Despite her sense of urgency, the matter would not be settled until after the turn of the century, when an international tribunal decided in favor of America's position.

***

By the time *The Century* article appeared, Eliza was back in Alaska. On Saturday, July 4, *The Alaskan* reported her arrival in Sitka: "Our little town is honored by the presence of Miss E. Ruhamah Scidmore, whose book on Alaska is so well and favorably known as the first volume which described graphically the wonderful scenery of the Inside Passage, the habits and characteristics of the natives, and a good deal of the legendary lore which pertains to Sitka and its environments." Eliza had come to meet friends for a month-long camping trip to Glacier Bay, where they planned to use John Muir's cabin. She later detailed their adventure for *California Illustrated Magazine* and *Harper's Weekly*.[10]

Her party included local judge Charles Sumner Johnson and his wife, Mary, and their young son and the family's setter; the Johnsons' cook; a local Russian named Koster, going along as their guide; and Theodore Richardson, an artist who taught drawing in Minneapolis and spent his summers painting in Alaska.[11] They all stood waiting amid piles of boxes and barrels on the morning when the *Queen* pulled into the docks at Sitka. Eliza's old friend Captain James Carroll grumbled a greeting: "Do you people think of spending the winter there?" Charles Johnson was taking tents, guns, and ammunition; the women had packed up food and linens, even sash curtains for the cabin windows; the Johnson boy swung a camera; and Koster sported rifle, ax, and bucksaw. The *Queen*'s passengers stared in befuddlement as the campers boarded, Eliza wrote later in her published account of the trip. So did a group of local Hoonah Indians who had gathered to watch the departure. Camping on a cold and barren glacier for geological exploration, as Muir and his companions had done, was one thing; doing it for sport seemed bizarre.

With every cabin booked at high season, Eliza and her group had to make do in the *Queen*'s steerage space. In the middle of the night, she was jolted awake by the loud clanging of the anchor chain running through the hawse holes. The steamer had to sit anchored at Bartlett Bay until dawn, before navigating the thirty miles of ice floes up ahead. At daybreak, Eliza lay in the stillness listening to the *drip-drip-drip* from the deck awnings. It meant, she knew, that the bay would be shrouded in mist and fog.

The *Queen* reached Muir Glacier at midday, amid leaden skies and drizzling rain. The landscape that usually stirred Eliza to feelings of wonder now looked desolate and barren. She and the others climbed into rowboats and were taken ashore where the cabin sat perched on a flat stretch of the rocky moraine. After everything had been delivered, their party lingered at the water's edge to wave goodbye. Finally, the *Queen* raised anchor, dipped its flag, and disappeared into the mist. In the silence of the departure, Eliza felt the force of the moment: for the next several weeks, they would live isolated from the world. Their closest neighbors were a group of Hoonah in a fishing camp at the mouth of the bay. Juneau and Sitka were both 160 miles away. "Seven souls and a setter constituted our world," she wrote.[12]

Inside the dry cabin they found mattresses, chairs, tables, and a gasoline stove that Muir and his colleagues had left behind the previous summer. Outside the front door, an eighth of a mile away, the icy wall of Muir Glacier stretched across the inlet. In the weeks that followed, its moans and booms established the rhythm of the days. The roar of crashing ice in the middle of the night woke the campers and caused the dog to rise up howling. Every morning when Koster headed off to collect water from a stream, the others hurried out to the beach to see what changes had occurred overnight. "We spent hours and whole half-days on watch for spectacular displays, and rushed from the house at every rumble," Eliza reported.

Charles Johnson spent his time wandering across the moraine with his gun and the dog, in search of grouse and other wildlife. Theodore Richardson set up a sheltered spot on the beach and painted from dawn to dusk. Eliza stood by, studying his technique. In nature, an iceberg looked like a mass of crystallized light. To achieve the clear, cerulean effect on canvas, she observed, Richardson first laid in a series of washes and undertints in tones of rose-madder, emerald-green, violet, and black. Because he used large quantities of blue pigment, Eliza teasingly called him "Cobalt" in her later account of the trip in *Californian Illustrated Magazine*.[13]

In Sitka, a reporter who interviewed Eliza quoted her as saying she went to Glacier Bay "to make scientific observations."[14] A group of Hoonah told her once, she said, that only a few generations earlier, *sitt-gha-ee*—"the great cold lake"—had been a vast plain of ice extending nearly to the entrance of the bay; in her own five trips since 1883, Muir Glacier had receded by a mile.[15] She had heard a theory that the sun and tides played a major role in glacial dynamics, so she kept a daily log during her stay, but her results did not support the idea.[16]

As a long spell of warm days passed, Eliza expected that at any moment, the heat of the sun might cause a great crack-up. The group set up the camera every day on the beach, ready to snap the drama when it occurred. Then, one morning at 4 o'clock, the big bang came. They all rushed outside, and "in the cold, clear light," Eliza wrote, "we saw a half-mile of the fantastic ice-wall topple into the sea and a second range of icy mountains crash upon the ruins."[17] Two hours later it happened again.

The campers made the most of every hour, until one day in late July the Alaska steamer came plowing through the icy bay to pick them up and take them back to Sitka. The "glacial picnic," as Eliza called it, had come to an end.

Weeks later, writing up her trip, Eliza penned a short note to John Muir. He knew botany. Could he tell her the name of the small, fragrant white wax plant she had seen in the woods around Sitka? Also, the "exquisite little tulip-y sort of flower" that grew in patches of heather above Muir Glacier. "It is the purest and truest blue that I have ever seen in a flower's petals, and Mr. Richardson, our artist, was quite excited over it," she wrote. "It was shameful," she added, "for us to be there in the midst of all that rich flora and not even know the names of the things."[18]

***

Her literary skill made Eliza an obvious choice to serve as corresponding secretary of the National Geographic Society. Her unanimous election at the society's end-of-year meeting on December 23, 1891, attracted notice in the press: "This is the first time a woman has held a position of such honor in a geographic or scientific society of so much importance."[19]

Her position on the board made Eliza a peer of Washington's most distinguished scientists and a member of the city's elite. One summer afternoon, she and her mother were among the 300 people who mingled at a reception hosted by the society's president, Gardiner Hubbard, and his wife, Gertrude, at Twin Oaks, their country home on Woodley Lane, north of downtown Washington. The other guests included cabinet officials and congressmen; scientists like Major John Wesley Powell, explorer of the Grand Canyon; and Crosby Noyes, editor of the *Washington Star*, there with his wife and daughter.[20] Devotees of Japanese art and culture, the Noyes family became good friends of Eliza and strong advocates of her efforts to have flowering cherries planted in downtown Washington.

Eliza assumed her new role in the National Geographic Society when its membership of 400 people included only a dozen women.[21] Besides the professional accomplishments that qualified her for the position, she possessed other traits that made her an asset to a young organization that had little money and depended heavily on its members to keep things running. Those who knew her understood that Miss Scidmore was someone who got things done. That she was also fiercely loyal to the National Geographic Society, and quick on her feet, she had demonstrated during the press debacle following the Mount St. Elias expedition. In gratitude for her support, Israel Russell had named a mountain for her when the society sent him back to Alaska in the summer of 1891. Mount Ruhamah is a 5,620-foot peak in the St. Elias range.[22]

The National Geographic Society and others have called Eliza a geographer.[23] Though she had no formal training in the discipline, that was true as well of many men in the nineteenth century now widely regarded as leading geographers of their day. With the field still evolving as an academic discipline in America, most geographers were an eclectic mix of scientists—"made-over geologists" and others, such as hydrologists, cartographers, meteorologists, and naturalists.[24] Some studied at universities, but many developed their expertise through self-education, mentoring, and field work, including a large concentration of scientists in Washington who had taken part in the four Great Surveys of the United States sponsored by the federal government after the Civil War.[25]

Though best known as a travel journalist, Eliza exhibited hallmarks of good science writing in her clear, precise language and keen powers of observation, whether describing the icy surface of a glacier or the sumptuous texture of ancient kimonos. Visiting remote places and producing new knowledge about them made her an explorer of sorts—a "discoverer." Scientists in Washington respected her work for its strong grounding in firsthand reporting and diligent research. One Alaska historian wrote that Eliza's descriptions of Glacier Bay greatly aided scientists who did early field work in the region.[26] In a study of *Alaska: Its Southern*

Coast and the Sitkan Archipelago, another scholar wrote that Eliza's work conveyed authority in part because she avoided the sentimentality, self-deprecatory references, and "overly moralizing tone" that marked the writings of many other women travelers in the nineteenth century. "She never apologizes for herself," the scholar added; nor did she generally call attention to herself as a woman.[27]

***

Eliza's dual association with *The Century* and the National Geographic Society put her in the forefront of the country's burgeoning conservation movement. In the wake of the successful Yosemite campaign, the United States now had four national parks: Yellowstone and three in California. In the winter of 1893, Eliza started advocating openly for what she hoped would become the next one. On February 21, she wrote to Johnson at *The Century*, suggesting an article on "the next new National Park—the Rainier Reserve in the State of Washington."[28]

She pitched the idea a day after news came from the White House saying that President Benjamin Harrison had just signed a proclamation that put nearly a million acres in the Mount Rainier area under federal protection. His action was the latest in a series of moves in which he created a system of "national forest reserves" during his term, under the guidance of his interior secretary, John W. Noble. The new policy was meant to keep large tracts of forested federal land from being gobbled up by lumber companies and other commercial interests.[29]

There was no guarantee that the new forest reserves would become national parks. But in the case of Mount Rainier, Eliza was hopeful. In her eyes, the volcanic, snow-capped cone rising 14,410 feet out of the Cascade range was as magnificent and iconic as Japan's Fuji.[30] The most glaciated U.S. mountain south of Alaska, Rainier spawned a half-dozen rivers and supported a great diversity of flora and fauna; subalpine meadows of wildflowers ringed the upper zone, and thick stands of ancient forest cloaked its slopes.

Going to and from Alaska and the Far East, Eliza had gotten wind of a movement in the Pacific Northwest to seek permanent federal protection for Mount Rainier. She was all in. As Eliza told Johnson in her query letter, "those Washington [state] people mean to have this Reserve made a national park with wagon roads, trails, etc." She herself planned to do her part, she added, by appealing to the newly created Sierra Club to build huts and other amenities for hikers.[31]

Johnson answered immediately (Figure 10.1). Yes, he wanted an article. But he wanted a broader piece, of 2,500 words or so, that explained the entire new forest reserve system to Americans. The unusually quick response took Eliza by surprise. She was swamped at the moment. Now in the final throes of her latest book, a comprehensive guide to Alaska, she was writing frantically and sending pages off to the printer every day.[32] She also had

Figure 10.1. Robert Underwood Johnson, editor at *The Century Illustrated* magazine for both John Muir and Eliza Scidmore.

(Photo by Albert Bigelow Paine, from *The Critic*, c. 1903; via Wikimedia Commons.)

slides to prepare for a lecture on Japan she was set to deliver for the National Geographic Society in late March.

After negotiating a slight delay, Eliza got to work on the article for *The Century*. Johnson arranged for her to meet with Secretary Noble. He also wanted a good map illustration showing the location of the new forest reserves. Unfortunately, Eliza found the Land Office staff too busy with other projects. Determined to deliver, she tracked down a U.S. Geological Survey map and worked with topographers to draw the new reserves from Land Office information.[33]

After sending the article around for several experts to review, Eliza mailed the piece to Johnson on March 28, at more than a thousand words over the assigned length. Published in the September issue of *The Century*, her article explained the origin of the new policy and how it would help ensure, for the good of the country, the many scenic, recreational, and environmental benefits that forests provided. The United States, she told readers, lagged behind other countries when it came to sensible management of its forests. "The average American, living only for the present day and the dollars of the moment, in this extravagant age of wood does not consider the lumberless condition of the next century."[34]

As Eliza had hoped, the president's decree provided impetus for stepped-up efforts to make the Rainier area a national park. A coalition of scientific societies partnered with nature activists and other advocates to lobby Congress. Eliza served on the National Geographic Society's Rainier committee, chaired by Gardiner Hubbard.[35] John Muir headed a parallel committee formed by the Sierra Club. When Muir and Johnson went to Washington in the fall of 1893 for talks, Muir regretted, in a letter to his wife, that the jam-packed schedule prevented him from seeing Miss Scidmore; every evening he returned to his hotel room exhausted.[36] Later, when bills for the Rainier park were finally introduced in Congress, Eliza briefed the Muirs on Gardiner Hubbard's swank evening of wining and dining a group of senators at his home to win their support. "As part of the scheme of 'whooping it up,'" she wrote, the evening featured lantern slides and a lecture by Ernest C. Smith, a veteran climber of Mount Rainier.[37]

Though the Mount Rainier campaign later stalled over different visions for the park, Eliza would finally see the efforts pay off in 1899, when President William McKinley signed an act establishing Mount Rainier National Park. Eliza's devotion to the issue made her a heroine to some people in the Pacific Northwest. "Miss Scidmore, One of the Workers for the National Park, Visits Tacoma," the local *Daily News* announced when she spent a few days there in 1897 after one of her trips to Japan. Calling her "one of the most enthusiastic promoters of the government park project," the reporter noted that she had visited the city a number of times and was "a great admirer of Puget Sound country."[38]

***

In a break from her writing, Eliza delivered her Japan lecture on the evening of March 24, 1893. Arctic explorer and National Geographic vice president General Adolphus Greely introduced her to the audience of 600 assembled in Washington's Builders' Exchange Hall.[39] The lady author who appeared before them was a tall and slender woman of thirty-six, with "a graceful carriage and an intelligent face," as one reporter described her around that time.[40] Dressed in a high-collared dress with the pillowy leg-of-mutton sleeves that were the height of fashion, Eliza addressed her listeners with a firm and steady gaze, and began speaking with passion and authority.

Her talk on "Japan and Its Inhabitants" gave those seated before her the evocative travelogue they had come to see. The lecture featured ninety lantern slides, showing temples and teahouses, cherry blossom celebrations, and other rituals. But Eliza also touched on more serious matters, including tensions that simmered beneath the outwardly friendly relations between Japan and the United States. America had been one of Japan's earliest and staunchest Western allies. Yet the United States had acted as shamefully as European countries, Eliza charged, in taking advantage of Japan in unfair treaties that posed a great tax burden on the still heavily agrarian population. "All the vultures of Western civilization preyed upon the last little new-comer in the family of nations, and it learned the lessons of progress at a fearful cost."[41]

More darkly, Eliza raised the issue of racial stereotypes that colored Western perceptions of the Japanese and other natives of the Far East. Cultural differences made it difficult to comprehend the "yellow brain," she said, using a common phrase of the time.[42] Despite that gap of understanding, she continued, it was wrong to patronize the Japanese as likable but meek and inferior. Take, for example, Japan's impressive social and technological strides under its thirty-year-old Meiji government. In a point she would make time and again over the years, Eliza stressed to her listeners: "There is little in our Western world that is not paralleled in Japan, and much there that we never dreamed of—arts and crafts we cannot approach nor cease admiring, a refinement impossible for others to acquire."

Like many other Japan enthusiasts, Eliza was becoming an avid collector of Oriental art. Her brother George's diplomatic privileges eased the shipping process, and she already had the start of what would grow into a fine collection of several hundred pieces. Some of her Japanese pottery, mostly blue and white porcelains, were displayed at the National Museum in Washington, and she lent Asian textiles to the Boston Museum of Fine Arts and other institutions.[43]

One winter's night, Eliza attracted notice when she turned up at a Dupont Circle mansion for a fancy-dress ball hosted by a senator's daughter. Murmurs of admiration surrounded her appearance in a traditional festival costume of Kyoto's famous *maikos*, or dancing girls. A *Washington Post* society writer gushed over the ensemble: a green crepe under-robe, topped by a black silk kimono embroidered across the back with silver characters and multicolored flowers. The silver stitching on the garment turned out to be the words of a Japanese song, the reporter noted, "so that when Miss Scidmore wears it she may be said to be a veritable poem." The writer's judgment of the "handsome young woman" wearing it was no less effusive: "Her blue-gray eyes are full of varying expression and humor, which, with a superb physique, denotes the superabundant vitality and health which have made her one of the most noted women travelers and writers in the world."[44]

***

In the summer of 1893, Eliza joined the crowds flowing west to Chicago for the World's Columbian Exposition, taking place on the shores of Lake Michigan to mark—a year late—the four-hundredth anniversary of Christopher Columbus's voyage to the New World. First, she made a detour to western New York for a speaking engagement at the Chautauqua assembly. Started as a camp for Sunday school teachers, the now-public event drew loyal Chautauquans from far and wide every summer for a packed program of lectures and entertainment, offered in a wooded setting along a seventeen-mile lake. Eliza got prominent billing as one of the opening-day speakers. That night, on July 1, she stood bathed in the spotlight, looking out onto a sea of faces in the assembly's new amphitheater to deliver the first of four illustrated talks she gave that week on Japan and Korea.[45]

Train companies ran special excursion lines from Lake Chautauqua to the Columbian Exposition. At the end of the week, Eliza paid the ten-dollar fare and rode west to Chicago. The strangers sharing her car couldn't stop talking about the attractions ahead. For many, Chicago itself was a curiosity. After burning to the ground in the Great Fire of 1871, the city had risen again out of the wide, flat prairie to become America's second-largest city, a dynamic metropolis of a million people and a major transportation hub for the continent.

The exposition itself dazzled. "Make no little plans, they have no magic to stir men's blood," Chicago architect Daniel Burnham had told his team of designers.[46] Their dream-like creation on 553 acres in Jackson Park fulfilled his wishes. The fair's centerpiece was a cityscape of neoclassical buildings, built of a durable but temporary material called staff—a mix of jute fiber, plaster of Paris, and cement. A coat of whitewash made everything gleam in the sun, leaving Eliza and other fairgoers with lasting images of the magical "White City." A system of canals and lagoons brought to mind scenes of Venice, while grand plazas, colossal statues, graceful bridges, and spouting fountains completed the picture (Figure 10.2).

Taking in the beautiful grounds would have to wait. First, Eliza had to handle last-minute details of a scientific symposium the National Geographic Society was sponsoring in late July. Back in Washington, Gardiner Hubbard had called on her, as corresponding secretary, to help him organize the event at the last minute.[47] At her hotel in Chicago, Eliza was still fielding letters only days before the meetings began. "Dear Sir," a Brazilian explorer wrote, regretting that the short notice prevented him from delivering the requested talk on the Amazon River.[48]

The first international meeting of geographers ever held in the United States, the two-day event included hundreds of delegates from Europe and South America.[49] In keeping with the

Figure 10.2. World's Columbian Exposition in 1893, with the Japanese pavilion on the wooded island at left.
(From *Official Views of the World's Columbian Exposition*, Arnold and Higinbotham,1893)

exposition's tribute to Christopher Columbus, many of the presentations focused on the theme of discovery. Eliza herself presented a paper and served as one of the recording secretaries. She opened the second day of the sessions, at eleven o'clock on July 28, with a talk on "Recent Explorations in Alaska." Once again, this time in an international forum, she castigated the federal government for its slack attention to Alaska, which she framed as a shameful dereliction of the nation's duty. "Un-American," she called it in a letter to one of her editors.[50] The National Geographic Society, she was happy to report in her talk—published later as her first article in *National Geographic*—showed a far more enlightened attitude in its scientific exploration of Mount St. Elias.[51]

Eliza's patriotic pride in America's physical splendors matched the mood of a nation that was coming to see itself as destined for greatness. Nothing offered a better testimony to that than the ambitiously conceived Columbian Exposition. A few days after Eliza arrived in Chicago, a young professor from Wisconsin named Frederick Jackson Turner presented a paper in which he reported 1890 census data indicating that America's long-held view of itself as a "frontier" nation was becoming outdated, now that settlement was so widespread across the West.[52] As though to underscore the point, the showman "Wild Bill" Cody was in Chicago that summer with his band of Rough Riders, portraying the "taming of the West" as a conquest of native tribes who contested the advance of white "civilization."[53] America was entering a new phase of national development. Having expanded to the far western reaches of the continent, the nation would soon look toward new frontiers across the Pacific.

Eliza was also in Chicago to report on the exposition for *Harper's Bazaar*. Like millions of other fairgoers, she strolled the "midway," a new entertainment concept featuring cafés, theaters, street scenes, and human sideshows sponsored by forty nations. The mile-long strip offered "authentic" experiences such as a Turkish bazaar, a street in Old Vienna, and a Lapland village; encounters with Bedouin camel drivers, Africans of Dahomey, South Seas Islanders, and a hootchy-koochy dancer named "Little Egypt." The polyglot mix of humanity mingled in the shadows of the world's first revolving Ferris Wheel, which carried passengers 264 feet above the ground in thirty-six cars, each the size of a bus.[54]

In her reporting for the magazine, Eliza focused heavily on Japan's exhibits. Seeing it all required miles of walking, as the country had organized fourteen different displays.[55] The Japanese minister, Gozo Tateno, a dinner guest of Eliza's the previous winter in Washington, explained the approach as a bid by his country to not just win expanded markets for its goods, but to give the world a "more accurate and comprehensive knowledge" of Japan to prove its worthiness "of fellowship in the family of nations."[56]

Eliza walked readers through Japan's attractions: women silk weavers, a noblewoman's boudoir, a forestry exhibit of native woods, and a teahouse where paying fairgoers could sit on crepe cushions to watch a *cha no yu* master whisk green tea into a froth. Though other visitors would not have noticed, Eliza felt the porcelains, bronzes, cloisonné, and other wares on display suffered in comparison to Japan's offerings in Philadelphia two decades earlier. "The commercial progress of the country is more apparent than any artistic progress," she observed, bemoaning the turn toward mass production. One of the most popular exhibits at the fair was Ho-o-Den, or "Phoenix House." On a wooded island in the middle of the fairgrounds, Japanese carpenters had constructed a traditional dwelling that featured home interiors from three different centuries. "A jewel-box in three compartments," Eliza called it.[57]

In her limited free time, Eliza congregated with thousands of other women in the highly popular Women's Building, designed in an Italian Renaissance style by a recent female graduate of MIT. Inside, the rooms buzzed with a spirit of sisterhood. American women had come a long way since their campaign, two decades earlier, to be taken seriously at the Centennial. Now, they had much to be proud of in their accomplishments across society—in industry and the professions, science and the arts, education, and social reform. Eliza had a chance to mingle with other female journalists, whom newspapers and magazines were hiring in record numbers.

On the second floor, she found her work represented in a special 8,000-volume library set up to showcase the work of female authors from around the world.[58] One cabinet held forty-two translations of *Uncle Tom's Cabin* and the inkwell of its author, Harriet Beecher Stowe. Other displays featured the writings and autographs of women such as Mary Queen of Scots, Catherine de Medici, George Eliot, and Elizabeth Barrett Browning. The collection of works by sixty women authors from the District of Columbia contained three by Eliza: *Alaska: Its Southern Coast and the Sitkan Archipelago*; *Jinrikisha Days in Japan*; and *Westward to the Far East*, a slender guidebook she wrote for the Canadian Pacific Railway. Her newest book, *Appletons' Guide-Book to Alaska and the Northwest Coast*, was published that summer, too late to be included.

Before leaving Chicago, Eliza added the title of "lecturer" to her résumé. She had caught the eye of Major James B. Pond, who ran the best-known speaker's bureau in the country. His roster of clients over the years included celebrities such as Mark Twain, the stage actress Ellen Terry, P. T. Barnum, and Booker T. Washington.[59] In need of a slick publicity photo, Eliza sat for a studio portrait that has remained for a century the best-known image of her (Figure 10.3).

***

In the years that followed, Eliza continued reporting on Alaska. She revisited the issue of the Alaska boundary dispute for *The Century*. In *National Geographic* she defended John Muir's claim of being the "discoverer" of Glacier Bay, arguing that he seemed justly entitled to the honor as the person most responsible for "bringing its wonders to the knowledge of the world."[60]

National Geographic considered Eliza one of its specialists on Alaska and made her an associate editor. Her knowledge of the area proved timely when gold was discovered in the Klondike in 1896 and prospectors flocked north to join the boom.[61] In what

Figure 10.3. Publicity photo of Eliza Scidmore from *The Book Buyer*, c. 1893.
(D.C. Public Library, The People's Archive)

may have been her last trip to Alaska, Eliza traveled on a section of the Stikine River in 1898, following the route that hordes of gold miners had used to reach the Klondike. She saw sleds, tents, and other debris abandoned along the riverbanks. Firewood lay stacked in piles, waiting to service the river steamers. But, as she told readers, the gold rush was over by then, and few people visited the area.[62]

The trip left her dispirited. "I have seen the edge of the Klondike and it was more than enough," Eliza told her editors at *The Century* in a letter from Vancouver. The rowdy packs of miners who filled the steamers spoiled the tourist experience. "I am rejoiced to be back in civilization once more," she wrote, "and Japan will be still more grateful as an antidote to Alaska."[63]

It had been fifteen years since her first trip to Alaska. Now, under an onslaught of crowds, the luster was beginning to fade. A more lasting damper on her travels in the region occurred a short time later, in the final days of the nineteenth century.

In the summer of 1899, the railroad magnate Edward Harriman organized a luxury voyage for his family to southern Alaska, the Bering Sea, and Siberia. He had the steamer *George E. Elder* outfitted for nature studies and invited two dozen scientists, conservationists, and artists—including John Muir and several founding members of the National Geographic Society—to join him. On September 10, a month after the group left Glacier Bay, a major earthquake struck southeastern Alaska. The seismic shaking filled the bay with so much ice that excursion steamers could no longer visit Muir Glacier and had to change their routes.[64]

By then, Eliza had already shifted her sights to wider exploration of the Far East.

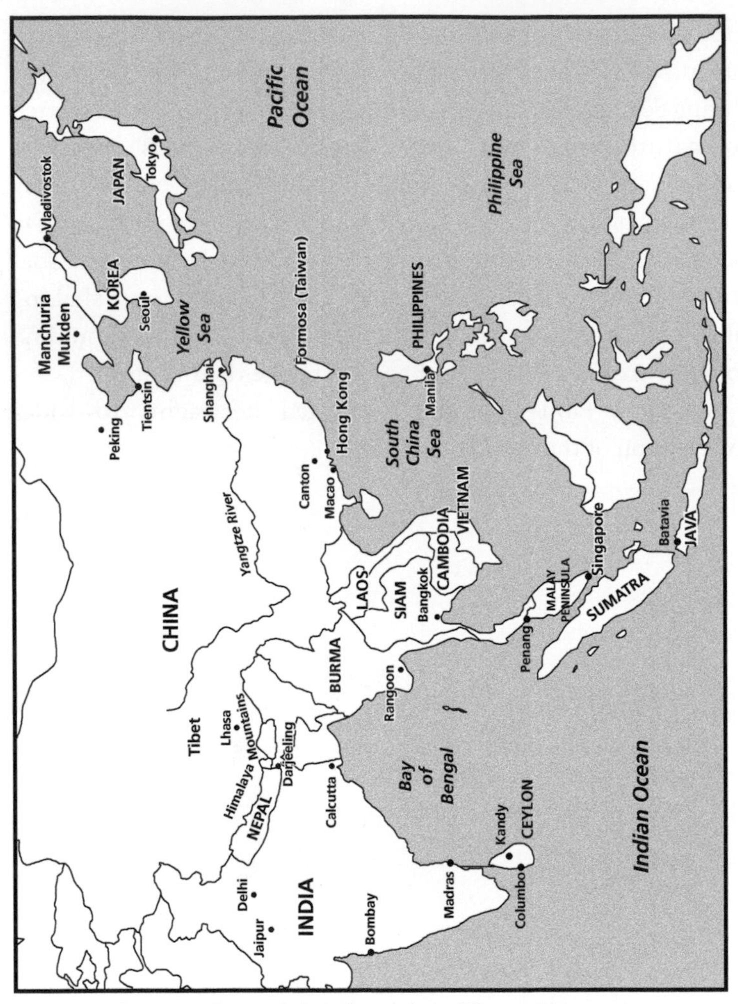

East Asia, with Victorian Place Names

11

New Highway to the East

By the 1890s, no one offered better service to the Far East than the Canadian Pacific Railway. Thanks to transoceanic connections, its passengers could circle the globe. Eliza traveled at times under the company's sponsorship after the railway hired her to write a compact guide for its Pacific route.

The initial publication of the sixty-page booklet, *Westward to the Far East*, makes it clear that Eliza had learned the freelancer's secret to success: repurposing material for multiple markets. The Canadian Pacific's ticketing office first offered the guide in the fall of 1891, while Eliza's newly published *Jinrikisha Days in Japan* was selling briskly in bookstores. *Westward to the Far East* highlighted mainly principal cities in Japan and China, though later editions expanded to include Eliza's thumbnail descriptions of places in Korea, Siam (Thailand), Java, Burma, and India (Figure 11.1).

From its founding, the Canadian Pacific had aimed to become a gateway to Asia; a "New Highway to the East," as the company touted itself.[1] Soon after the railway completed its transcontinental rail line in 1885, company officials scrambled overseas to arrange service across the Atlantic and Pacific. The job of selling the premium service to globetrotters and other travelers fell to William Van Horne, the company's former chief engineer and second president. Like many wealthy men of the Victorian period, the American-born, cigar-chomping Canadian was an avid collector of Oriental ceramics, a passion he indulged by amassing, at his baronial mansion in Montreal, a museum-quality collection of pieces that he catalogued in great detail down to the tiniest teacup.[2]

Figure 11.1. Cover of the 1900 edition of Canadian Pacific Railway guidebook written by Eliza Scidmore.
(Chung Collection, Rare Books and Special Collections, University of British Columbia)

In his marketing efforts for the Canadian Pacific, Van Horne mounted an advertising blitz in which he hired artists, writers, and photographers to promote the railway's service to Japan and other destinations in Asia. He failed in his efforts to recruit Lafcadio Hearn, a Greek American newspaperman who went on to become one of the most famous Western writers in Japan. But the company scored a coup in attracting Miss Scidmore, whose bylines on Japan were appearing frequently in publications such as *Harper's Bazaar*, which Van Horne read.[3]

In *Westward to Far East*, Eliza rose to the task of making Japan sound irresistible. She advised those with a hankering to see the country not to put it off. The country was becoming more Europeanized every year, she wrote, and "the sooner he goes the more Japanese the tourist will find it."[4]

Her descriptions of travel via the Canadian Pacific made the experience sound like one big rolling caravan of pleasure. After the trains left Montreal for the five-and-a-half-day journey to the West Coast, additional passengers came aboard at junctions along the North American route. As the locomotives raced westward at impressive speeds of twenty-five miles an hour, travelers dined, slept, smoked, and read in posh comfort.[5] At Vancouver, B.C., those going on to Asia transferred to one of the Canadian Pacific's state-of-the-art steamers. For years the company had leased oceangoing vessels from Cunard.[6] But the release of Eliza's guide in 1891 coincided with the company's launch of its own line of luxury steamships, known as the "white empresses" for their white hulls and namesake female heads of Japan, China, and India.[7] The *Empress of Japan* and its sister ships were so grand and elegant, Eliza reported, that crowds gathered at the wharf on sailing days to cheer as the steamers glided out to sea for the 4,300-mile journey to Yokohama.[8]

The ocean voyage itself was glamorous, Eliza promised. The company treated the formal dinner hour as a more serious affair than in Old England itself. If the passengers felt like a ball, the crew decorated the promenade deck with flags, added more electric lights, brought up a piano—"and lo! a ball-room worthy of Pacific dancers."[9] Eliza romanticized the cosmopolitan mix of people aboard Canadian Pacific ships as they plied the waters of the Far East. A novice world traveler from the American Midwest might find himself rubbing elbows with tea, silk, or opium merchants; planters from Manila or Java; Anglo-Indians, British colonials, and missionaries.[10]

Because the Canadian Pacific was looking to expand tourism in its own backyard, Eliza also reported on leisure destinations closer to home. In *Westward*, she described places such as Banff, in the Canadian Rockies, where the company built a chateaux-style resort hotel to lure travelers who wished to "dwell awhile in the wilderness."[11] The railway's local attractions included small-steamer excursions to picturesque sites along the British Columbia coast. In September 1892, soon after her fifth trip to Alaska, Eliza signed on for a cruise up Gardner Canal to see the famed Price's Cannery, built in an Edenic setting of fern-covered cliffs and a waterfall.[12] On the morning of the *Islander*'s departure from Vancouver, she showed up wearing a splint and bandages.

Out riding with a friend from Washington, she had tumbled from a horse and broken her collarbone. The bone was set, but a few hours into the trip it slipped, and a doctor and nurse had to be taken on board.[13] Eliza refused to let the mishap interfere with her plans for going on to Asia. The day her ship sailed, she boarded on schedule, then spent the thirteen-day crossing recuperating from her injury.

***

In the summer of 1894, Eliza set off on a fourteen-month round-the-world trip as part of her work for the Canadian Pacific Railway.[14] She went at a time when more people than ever were circling the globe. For many pleasure travelers, "a girdling of the earth is now the *grand tour* which a little round of continental Europe used to be," she wrote in *Westward to the Far East.* With time and distance "almost annihilated by modern machinery," the trip from New York to Yokohama took no longer than it once did to travel from New York to Liverpool.[15]

The American reporter Nellie Bly had helped spark excitement for round-the-world travel with her sensational 1889 record-breaking trip of seventy-two days, sponsored by Joseph Pulitzer's *New York World.* Not long after Eliza's trip, the author Mark Twain—badly in need of money to pay his debts—embarked on a year-long odyssey in which he gave readings as far away as Australia, New Zealand, India, and South Africa.[16] Eliza approached her own world tour as one of personal discovery and adventure but also serious business. Apart from her mission to update *Westward to the Far East,* she gleaned material for a half-dozen articles in *The Century*, *Harper's*, and the *Chicago Daily Tribune*, as well as a substantial part of her later books on Java and India.

Eliza's professional relationships with well-connected men in Washington's scientific community brought material support for her travels in Asia. George Brown Goode, the assistant secretary of the Smithsonian Institution in charge of the National Museum, arranged the loan of a camera from the Smithsonian's collection, along with a steady supply of film.[17] The boxy camera model Eliza borrowed, a No. 4 Junior Kodak, represented a major advance in amateur photography in allowing users to remove and reload film rolls instead of sending the camera to Kodak for film processing. Eliza sent her film and photographs to the Smithsonian, where a couple hundred of her images are archived today (Figure 11.2).[18] Meanwhile, she sent copies of many of those photos to her editors, who had them converted to drawings or other artwork to illustrate her published articles.[19]

Eliza's relationship with Goode and the National Museum was reciprocal. For years she had lent the Smithsonian some of her Asian ceramics and other objects for public display, and she occasionally donated small artifacts of interest that she acquired.[20] Goode's keen interest in her 1894–5 trip stemmed from his curatorial interest in what she might learn relevant to the museum's ethnological and other work.[21] His support for her travels included a letter of introduction she carried with her. Addressed

Figure 11.2. Woman thought to be Eliza Scidmore, at a photo shoot in Japan.

(National Anthropological Archives, Smithsonian Institution)

"to all friends of science" and stamped with a gold seal, it asked that Miss Scidmore be shown every courtesy as a representative of the Smithsonian Institution and the National Geographic Society traveling for "anthropology" and "photography" and "scientific observation and study."[22] The letter would prove especially handy in a testy encounter Eliza had with a colonial bureaucrat.

As it turned out, Eliza set off on her round-the-world trip just as Japan and China were going to war. Tensions that had simmered for years over competing interests in Korea led to a declaration of war in August 1894. The outcome a year later would shock the world. Though everyone expected China to prevail, Japan's heavily modernized military emerged victorious, leading to a major shift in the region's balance of power as Japan became the dominant nation in Asia for the first time.

Eliza's travels that fall, which she made with at least one other female companion, included passage along the 389-mile Inland Sea—Japan's most "poetic" region, as she described it.[23] The signs of military preparations were evident everywhere she looked, from Kobe down to Nagasaki. The emperor of Japan had moved the government from Tokyo to Hiroshima. At ports all along the sea, men were leaving for the warfront. "No more last look at his native country could be dearer or more lovely to the Japanese soldier than that along the Inland Sea," Eliza wrote in the *Chicago Daily Tribune*.[24] At Nagasaki, the women went for a jinrikisha ride and came upon troops drilling in the public square. Japanese officers politely made way for the American ladies to move in closer to watch.

The war also intruded while they were staying at Miyajima, a "sacred" island known for its shrines and temples, tame deer, and giant red *torii* rising from the sea. Toward the end of their weeklong visit in mid-September, word came of a Japanese victory at Heijo (Pyongyang), in northern Korea. Caught up in the local spirit of joy, Eliza made a gesture of largesse that illustrates her status as an American woman of privilege and means traveling in the tropics. Most visitors to the island offered a temple donation before their departure. She proposed something else. To celebrate the military victory, she arranged with the high priest to sponsor an "illumination," usually reserved for special occasions such as moon-viewing festivals. One evening at sunset, everyone in the village assembled on the steps of the main temple to launch a thousand oil-saucers into the sea. The sight was so spectacular it generated press coverage across the region. "Surely," Eliza said in her later magazine article on Miyajima, "never did one obtain so much pleasure and glory by an expenditure of four yen."[25] It had cost her the equivalent of two dollars.

In planning her Asia trip, she had expected to spend a month in Peking. The war now made that inadvisable, so she decided to go to Java instead, in what would be her first visit to the island. Though the trip would be rushed, it would lead to one of her best-known books.

***

The forty-eight-hour journey from Singapore was smooth, but brutally hot. As the Blue Funnel steamship glided south through the Java Sea, sailing past the east coast of Sumatra, Eliza and the two American women with her lay draped limply across their steamer chairs. The temperatures, in the mid-80s, were not that high. But it was the deceptive heat of the tropics, "that steaming, wilting quality of the sun in Asia that so soon makes jelly of the white man's brain," Eliza wrote.[26]

Their arrival at Tanjung Priok, the port of Batavia (Jakarta) on the west coast of Java, made Eliza forget all the discomfort. It thrilled her to be on what the great naturalist Alfred Russel Wallace had called, during his explorations in the Malay archipelago, "probably . . . the finest island in the world."[27] Besides its botanical wonders, Java had fascinating monuments and ancient relics of past civilizations. Yet the island got few tourists. Eliza was hoping, as she told one of her editors, that she might "have the first word to say about Java" for English-speaking leisure travelers.[28]

Wallace's description of Java as the "garden of the East" gave her the title of her later book.

Though Eliza liked to think of herself as fair and open-minded in her reporting, the book revealed the kind of biases she could bring to her work. A number of passages in *Java, the Garden of the East* are colored by her low opinion of the Dutch who governed Java. "The Dutch do not welcome tourists, or encourage one to visit their paradise of the Indies," she told readers.[29] She laid the attitude to their resentment of the world's harsh criticism of Dutch rule in the East Indies. For half a century, colonial overseers in the islands had imposed strict, and at times harsh, agricultural practices in which native Javanese farmers were required to provide revenue to Holland in the form of export crops or compulsory labor. The policy, known as the "culture system," was halted in the 1870s. But the Dutch could not seem to shake their reputation as cruel and tyrannical administrators.[30]

The train that took Eliza and her companions the nine miles to Batavia chugged past palm groves and patches of banana and cocoa trees. The Javanese homes, she reported, looked like woven baskets perched on stilts. Naked children frolicked in the road, and slim, barefooted young mothers in batik sarongs walked by carrying their babies in slings knotted at the shoulder. In Batavia, which the Dutch had established as a walled town in 1621, the women rode a cart through what was now a sprawling city of whitewashed walls and red-tiled roofs, home to a hundred thousand people. Most Dutch colonial officials lived in the suburbs, in grand homes surrounded by green lawns.

When their party reached the Hotel Nederlanden, a cluster of buildings surrounding an open courtyard, Eliza climbed down and went inside to register. Nothing, she reported, had prepared her for the scene that greeted her. "We were sure that we had gone to the wrong hotel."[31] But no, people assured her the Nederlanden was the best hotel in Batavia. Many Dutch families lived there year-round. Eliza came upon them lounging about in the lobby and on the wide verandas—barefooted men in pajamas, smoking their pipes of Sumatran tobacco, the women clad in native sarongs and white dressing jackets. It was "an undress parade that beggars description," Eliza wrote.[32] Her derisory views would come back to haunt her years later, when her Java book was finally published.

Her stay at the Nederlanden included an account of the traditional Dutch *riz tavel* ("rice table"). At the sound of the noonday dinner bell, she and her companions were swept up into the mad dash of hotel residents who rushed to claim a seat around the long table in the vaulted dining room. Taking up their spoons, everyone started by heaping their soup plates with boiled rice. As the Javanese servant girls carried a succession of dishes into the room, the diners added pieces of fish, beef, and bird; onions and strips of omelet; and spoonsful of thin curry, followed by condiments of chutney, coconut, peppers, and almonds. After the rice course came beefsteak and a salad, finished off with fruit and coffee. For "squeamish folks, unseasoned tourists, and well-starched Britons," Eliza told readers, a hotel *riz tavel* could have the perverse effect of killing one's appetite.[33] She found it as fascinating as an anthropological study.

Understandably, the hotel fell quiet in the afternoon while everyone slumbered. Then, a stir arose again at 4 o'clock, when the residents roused themselves for a bath and tea. Eliza could hardly believe her eyes at what followed. The doors of the guest rooms opened, and out of them stepped the Dutch ladies, transformed in brilliant gowns and coiffed hair. For the next several hours, until the dinner hour at 9 o'clock, they drove through the neighborhood in barouches and victorias, drawn by giant horses imported from New South Wales, making the rounds of social calls as though back home in Europe.[34]

***

With her great love of plants and flowers, Eliza had come to Java keen to see the famed botanical garden at Buitenzorg (Bogor), forty miles south of Batavia. Sir Stamford Raffles had established the foundation of the garden during his brief term as the British governor of Java in the early nineteenth century. Under a series of talented Dutch horticulturists who followed, the garden became the finest of its kind in the world, with thousands of specimens of plant species collected from across the tropics.

Eliza found the town itself like a pocket of paradise. Along the main avenue, graceful kenari-nut trees formed a cathedral arch a hundred feet high. Their tall, straight trunks hosted a universe of life: stag-horn and bird's-nest ferns, rattan and creeping palms, blooming orchids, and air plants. On her strolls, Eliza spotted nutmegs lying as thick as acorns on the ground. At one point her nose led her to the perfume of a frangipani (*kamboja*) tree. It turned out to be so thick with fragrant blossoms that within minutes, she filled her tipped umbrella with the white waxy flowers.[35]

With magazine ideas in mind, Eliza had come to Buitenzorg hoping to interview the botanical garden's world-famous Dutch director, Dr. Melchior Treub, for a profile in *The Century*. Unfortunately, when she went round to the garden, she learned he was ill and out of the office. As she told her editors later, she sought him out at home, but found him half-dozing on his verandah, too weak to talk for long.[36]

She turned her attention instead to a study of the region's many unusual fruits. At the Hotel Bellevue where she was staying, she hired a Moslem servant named Amat, and they went off together to the local market, where they heaped their baskets with produce. Back at the hotel, Eliza and her companions conducted a tasting session on the verandah, while Amat explained the names of the fruits and showed how to open them.[37] They sampled bright-red rambutans, with chestnut-like burrs and pulpy white flesh; the slightly acidic *carambola*, or "star-fruit"; papaya; apple-like *salak*; and *pomelo*, a drier variant of grapefruit. The spiky, football-size durian, which most Westerners found revolting for its rotten odor, had a custard-like pulp that was actually quite tasty. By the end of the day, however, Eliza had come to agree with travelers who declared that no fruit of the tropics was more delectable than the purple-orbed mangosteen. The taste of its tissue-like fruit dissolving on her tongue made her think of a fine Italian ice—a touch tart, a touch sweet.[38]

***

Of all the experiences Eliza packed into the month on Java, the most lasting impression she came away with was her visit to Borobudur to see the monumental ruins of a Buddhist temple dating from the seventh century. People she talked to in Singapore had warned of how difficult and expensive travel on Java was because of poor transportation and the lack of experienced guides. It came as a happy discovery, then, to learn upon her arrival that the Dutch had just opened a new rail line that ran from one end of the island to the other.[39] Still, the route did not cover the entire 300-mile distance from Batavia to Borobudur, in central Java. Getting to Borobudur posed challenges, and it was in Jogjakarta, where the women had to spend the night, that Eliza got into a huff with a local official.

She and her companions were out for an evening stroll when the hotel manager flagged them down with a message: the resident officer wanted to see them. The tall, severe man they met with got right to the point by informing them they could not stay in Jogjakarta but had to return to Buitenzorg at once. He had learned that the American ladies did not have the permits foreigners needed to travel in the interior of Java. Eliza explained that because the travel passports they applied for had not been ready by the time they left Buitenzorg, officials there had waved the women on, saying the papers would be sent after them. But the official in Jogjakarta would have none of it. He insisted the women could not remain in his district without the permits. Stiffening her spine and summoning the imperious attitude she could exhibit so well, Eliza matched him in obstinacy. In her book, she would parody the "grand burlesque" of their run-in—only to face embarrassing blowback about her reporting from an incensed and influential reader. Her unpleasant encounter with the resident officer was finally resolved, she wrote, only after she produced her letter of endorsement from the Smithsonian. Upon examining it, he backed down and agree the women could go on to Borobudur, "since you come to us so highly commended."[40]

The next morning, the hotel manager sent the ladies off in a canopied carriage pulled by four small ponies and driven by a betel-chewing old coachman in a sarong and batik turban. During the journey through the countryside, Eliza drank in the landscape. She was struck by the meticulous look of everything. "There was not a neglected acre on either side for all the twenty-five miles," she wrote. "Every field was cultivated like a tulip bed; every plant was as green and perfect as if entered in a horticultural show."[41]

When their party stopped for a rest at a travelers' pavilion, Eliza rushed off to tour a *kampong* of native dwellings. "The friendly, gentle little brown people welcomed us with amused and embarrassed smiles," she wrote.[42] Behind the condescending language, her intentions were kind, and her interest in learning more about local life and customs was sincere. The incident illustrated what one scholar has called a strength of many women travel writers in the nineteenth century, in that they were more inclined than men to slow down and engage with native people they encountered along the way.[43]

The weather in the central plains of Java was hot and humid when the women reached Borobudur at midday. The slow-spoken manner of the elderly Hollander who ran the government rest-house seemed in keeping with the torpor of the tropics. While her companions napped after check-in, Eliza rocked on the porch, gazing at the darkened ruins of the former temple. Despite its decrepit state, Borobudur seemed overpowering in its effect. "In those hazy, hypnotic hours of the afternoon," Eliza wrote, "one could believe the tradition that the temple arose in a night in a miraculous bidding, and was not built by human hands."[44]

Dating from the golden age of Buddhism on Java, Borobudur was abandoned by the time Islam took hold in the region in the fourteenth century. For centuries the temple stood "lost" amid the tangle of jungle growth. Sir Stamford Raffles sparked outside interest when he began excavating the ruins during his brief tenure in the East Indies. The Dutch were now weighing preservation options, but progress was slow.

The day after the women arrived, Eliza made a more intimate tour of Borobudur at sunrise for a closer examination. The temple rose—like a giant wedding cake—in a series of stacked square terraces topped by a large openwork dome, or stupa. Seventy-two more stupas containing statues of the seated Buddha surrounded the structure, whose outer walls featured miles of stone relief panels that offered a fascinating record of life on Java in the seventh and eighth centuries. The scenes showed royal figures on thrones accepting tribute from minions; peasants plowing the fields with water sticks and buffalo; women bearing water vessels on their heads; and elephants sporting fans and state umbrellas in their trunks.

It troubled Eliza deeply to see the monument's decay. Much of the stonework had crumbled, and many of the site's 500 individual statues of the Buddha were broken or missing altogether. The massive stone walls were bulging and leaning. Eliza still had Borobudur's sorry condition on her mind several weeks later when she expressed her concerns to George Goode at the Smithsonian, in a letter posted from Kandy, Ceylon. As she told him, Borobudur was "certainly one of the wonders of the world, a marvel of art, architecture and religious zeal; but it is heartrending to see the utter ruin that must come of all that mass of sculpture, unless something is done quickly to save the leaning walls."[45] At the very least, she suggested, iron rods might be used to prop up the precious galleries of bas relief carvings. Whatever it took, was there not some way he and his scientific colleagues could use their influence to get the Dutch to act more expeditiously to preserve the world treasure?

***

Eliza spent three months that winter in India, where she made a great sweep of the major cities. Later, while cruising the Red Sea, she wrote Goode an eleven-page letter describing the various collections she had seen at a half-dozen museums, in Calcutta, Lahore, Delhi, Lucknow, Jaipur, and Bombay.[46] She also made a long journey to see the Himalayas. Though the trip was marred by the bitter cold in northern India, the mountains themselves proved "all they are cracked up to be," Eliza informed John Muir in a letter filling him in on her travels. She saw the famed mountains from three different vantage points—at Darjeeling, Simla, and the Khyber Pass—and they were splendid. But Mount Everest left her disappointed. As she told Muir, after she and her party left in the middle of the night to climb to a high place to see the sunrise over the mountain, "what did I see as the biggest, highest thing on earth? A little white nub of a thing off on the sky line, just a size larger than my thumb. I'll back my Mt. Ruhamah against it any time when it has its snow on."[47]

In the final leg of her trip, Eliza traveled through Cairo and Constantinople on her way to London. She timed her itinerary to reach London in late July, in time for the Sixth International Geographical Congress. She was attending as one of five members—and the only woman—of the National Geographic Society's official delegation, headed by its president and cofounder, Gardiner Hubbard.[48]

Now thirty-eight, Eliza wore her intelligence and charm confidently as she mingled among the scientific greybeards at the conference, which offered what the press called "some of the most brilliant social meetings of the season."[49] The U.S. ambassador proudly squired her around, and American correspondents reported on her social success in articles such as the one in the *Chicago Tribune* headlined "Honor to an American Girl."[50] It proved a great embarrassment to the conference organizers when word got around that Miss Scidmore, wearing her badge, had turned up for a social gathering hosted by the Duke of York only to be turned away by the guards because "women were not permitted."[51] Hearing of the incident later, the duke summoned her for a private audience. She was also singled out as a special guest at a garden party held by the Baroness Angela Burdett-Coutts, a wealthy British philanthropist who had participated in the international women's program at the Chicago World's Fair.

The conference activities included a reception in honor of the U.S. delegation, sponsored by the glamorous young couple George Curzon and his American-born wife, the former Mary Leiter. Mary had been a highly successful debutante in Washington. Eliza knew her casually from their occasional encounters over the years at teas and receptions. At a time when American heiresses of great wealth were being heavily courted in London by members of the European aristocracy, Mary Leiter—a graceful and intelligent young woman nearly six feet tall—had created a sensation that spring by her marriage to Curzon, a member of Parliament and heir to a barony.[52] Eliza would interact more deeply with the Curzons years later in India, after George Curzon had been named the viceroy and Mary assumed a high official title of the British aristocracy as the vicereine. In their respective ways, Eliza and Mary Curzon represented American womanhood at its best.

After a decade of success and growing celebrity in America, Eliza was now thrust into an international spotlight as an intrepid and knowledgeable world traveler. Her dizzying 1894–5 trip points to some of the traits, apart from her intelligence and stamina, that help explain her remarkable record of achievement. Her published work reveals a humanistic outlook and an associative mind that enabled her to see stories everywhere she went. Art and archaeology, history, science, and human cultures, it was all interconnected. She never tired of wanting to know the world, and curiosity made her irrepressible. As suggested in the general observation of a scholar of Victorian women travel writers, in educating herself, Eliza educated the public.[53]

12

"Miss Scidmore, of Everywhere"

In Yokohama, the Scidmores made their home for many years along the Bund, the wide road facing the harbor. It was a street of warehouses, commercial offices, hotels, and bungalows with gardens of camellias and shade trees. A long esplanade, separated from the road by pine trees and benches, offered residents and travelers a place to sit or stroll in view of the busy harbor. A half-dozen mail steamers, men-of-war, brigs, and barks usually lay at anchor, surrounded by a flurry of sampans whose boatmen scurried for mail and arriving freight. Hotel steam-launches circled the waters cruising for guests.[1]

Jinrikishas spun up and down the street between the landing dock and the hotels. The Scidmores lived practically next door to the Club Hotel, popular with businessmen for its game room and selection of fine spirits (Figure 12.1). Globetrotters gravitated to the Grand Hotel, at the far end of the block, which boasted a long seaside piazza with a telescope that looked a long way out to sea. When Eliza needed to make travel arrangements, she could get it done quickly with a short jinrikisha ride down the street to No. 14, where the Canadian Pacific Railway had a branch office (Figure 12.2).

Traveling Americans often turned up at the Scidmores' door, a reflection of both George's consular duties and the family's reputation for hospitality. The elderly Mrs. Scidmore basked in the attention of guests. The open house she held every November on her birthday drew a large share of the diplomatic community, as well as Japanese officials who stopped by to pay their respects. Visitors relished hearing her lively stories—of life in Washington

Figure 12.1. Club Hotel on the waterfront Bund in Yokohama, with the Scidmores' residence to the right of it.

(Via Alamy)

Figure 12.2. Mrs. Scidmore, center in jinrikisha, outside the Canadian Pacific Railway's offices on the Bund in Yokohama.

(Wisconsin Historical Society, WHS-150493)

during the Civil War, and how she had known President Lincoln "personally."[2]

Besides serving as a deputy U.S. consul general in the 1890s, George Scidmore was in demand for his legal expertise. He served for a while as a barrister and solicitor for the Imperial Government of Japan in the British Courts, and spent time in Fiji as a special agent investigating Americans' land claims there.[3] A devoted bureaucrat and model of rectitude, George had discovered that the bachelor life suited him. When not working, he rowed in the harbor for exercise, leaving his mother to run the house with the help of several Japanese servants.

Meanwhile, Eliza's growing celebrity brought notice of her comings and goings beyond the tight-knit expatriate community. One day not long after the publication of *Jinrikisha Days in Japan*, a summons came late in the afternoon saying the empress of Japan wanted to meet Miss Scidmore. Eliza scrambled to make herself presentable, then hurried to the train station with George. Less than an hour later she was received by the empress in her private apartment.[4]

In the summer of 1896, Eliza and her mother took up a gardening craze that was sweeping the country: the cultivation of *asagao*, or Japanese morning glories. "The Japanese have been doing quite wonderful things with them for—centuries perhaps—and this summer I went in for a little growing myself," Eliza explained in a letter to Robert Underwood Johnson at *The Century*.[5] These were not the puny vining flowers that grew in American backyards. Instead, the people who spent centuries elevating cherry trees to an art form had also used their horticultural genius to develop morning glories that were quite extraordinary. There were hundreds of varieties, in a wide range of colors. Some had blooms as big as saucers; others bore petals that were mottled, striped, shaded, spiked, corkscrewed, or frothy as a bamboo whisk. Many didn't look like morning glories at all, but resembled other plants, such as carnations, pelargoniums, peonies, and orchids. To Eliza, it all seemed magical.

When she mentioned asagao to other foreigners in Japan, they drew a blank. The fact that the plants bloomed mostly in the ungodly hours of predawn contributed to their limited appeal. Despite the drawback, Eliza found asagao so intriguing she set out to learn all she could about them. She rose early to visit flower shows and private collections. She and her mother propagated a hedge from the gate to their front door and then invited friends in to see the flowers in bloom.[6] Bent on introducing asagao to gardeners in America, she mailed packets of seeds to her editors and friends, and wrote an article for *The Century*.[7] Though the piece attracted wide notice in the press, growing Japanese morning glories took such dedication that they failed to catch on.

With her usual obsessiveness about things that caught her interest, Eliza continued studying asagao for years. She became such an expert that she lectured on the topic to the Japan Society in London and published a thirty-two-page paper, illustrated with drawings by a Japanese artist. She delivered the talk on the evening of May 9, 1900, to a gathering of the society's members at a hall in Hanover Square. When the program chairman, Arthur Diósy, addressed the audience, he said he had planned to present the evening's guest speaker as "Miss Scidmore, of Yokohama." Given her reputation as a great traveler, however, he changed his mind. He introduced her instead as "Miss Scidmore, of 'everywhere.'"[8]

***

In that summer of 1896, while Eliza was growing morning glories on the verandah and working on her book about Java, disturbing news began trickling into local newspapers and the diplomatic community. A horrific tragedy had occurred in the north of Japan, on the eastern coast of the main island of Hondo (now Honshu). "It will be some time, doubtless, before full particulars reach Tokyo of the cruel disaster that has overtaken the people living along the coast of Rikuzen and Rikuchu," the English-language *Japan Weekly Mail* reported on June 20.

Further accounts indicated that on June 15, an undersea earthquake had occurred far out in the Pacific.[9] The seismic disturbance triggered giant waves that built powerful momentum as they rolled across the sea. By the time they reached the Sanriku coastline at dusk, the waves had swollen into an enormous wall of water that slammed onto the shore like an angry fist, leveling everything in its path for 175 miles. Entire villages were wiped out. A ruggedly beautiful area of saw-toothed coastline and jutting bays, the Sanriku Coast boasted one of the best fishing grounds in the world because two ocean currents collided there, creating conditions that attracted abundant marine life.[10] Most of those who perished in the disaster were fishermen and their families in the Iwate and Miyagi Prefectures.

"We did not feel the seismic wave here in the least," Eliza wrote on July 3 to friends in America, "but we have been fed full of horrors by the accounts that come from the ruined northeast coast."[11] Details of the tragedy were slow to reach Tokyo because telegraph lines and operators had been washed away. A spine of mountains separating the coastal area from the mainland further impeded communications and emergency response. The Japanese government sent ships carrying surgeons, nurses, and soldiers with disaster supplies. Japanese news photographers made their way to the area, along with a couple of foreigners who provided eyewitness accounts for the press.[12] Within days, the *Official Gazette* started reporting the number of deaths and recovered bodies.

As the details emerged, Eliza drew on local newspaper reports and diplomatic dispatches to write an article for *National Geographic.*[13] Her reconstructed account of the events, published in September, introduced the magazine's readers to a Japanese word that would pass into the English lexicon over the next century with connotations of terror: *tsunami*, from the Japanese words for waves (*name*) breaking upon a harbor (*tsu*).[14]

Because Japan lies in one of the earth's most earthquake-prone zones, Eliza explained, residents along the Sanriku Coast had not been alarmed by minor shocks earlier in the day. Many spent the evening on the beach celebrating a local festival. Fishing fleets were out to sea when the earthquake struck ninety miles away in a deep-sea region known as the Japan Trench. The fishermen later reported feeling only a mild rocking as waves passed beneath their boats. Upon returning to shore the next morning, they found their homes reduced to rubble and their families wiped out.

Drawing on survivors' accounts, Eliza painted a vivid picture of the terror:

> Rain had driven them indoors with the darkness, and nearly all were in their houses at eight o'clock, when, with a rumbling as of heavy cannonading out at sea, a roar, and the crash and crackling of timbers, they were suddenly engulfed in the swirling waters. Only a few survivors on all that length of coast saw the advancing wave, one of them telling that the water first receded some 600 yards from ghastly white sands and then the Wave stood like a black wall 80 feet in height, with phosphorescent lights gleaming along its crest. Others, hearing a distant roar, saw a dark shadow seaward and ran to high ground, crying "Tsunami! Tsunami!" Some who ran to the upper stories of their houses for safety were drowned, crushed, or imprisoned there, only a few breaking through the roofs or escaping after the water subsided.

The disaster, the article noted, killed 26,975 people; grievously wounded 5,390 survivors; flattened 9,313 homes; and crushed or carried away 10,000 fishing boats. The exact numbers, Eliza told readers, came from a census system that was so exacting down to the local level that the human casualties and other damage was soon known. *National Geographic* published Eliza's article with several black-and-white photographs taken by the Japanese newsmen. One image showed a bloated body drifting downstream.[15]

Before the late 1890s, English-language references to the phenomenon of a tsunami had commonly used the term *tidal wave*, a technically inaccurate description because the destructive waves are triggered not by gravitational effects of the sun and moon but by undersea earthquakes or landslides.[16] Eliza's article suggests she understood the difference, as she began her piece by noting that "the great earthquake wave (tsunami) . . . was more destructive of life and property than any earthquake convulsion of this century in that empire."[17]

Several weeks after the catastrophe, Eliza herself saw a few traces of the disaster. As she told George Goode at the Smithsonian in a letter, she caught a glimpse of the region's devastation while traveling up the coast of Japan in early August to view a solar eclipse at an island off Hokkaido.[18] During the trip she met a "Mr. King," a dealer in sealskins, who described details he had heard from schooner captains whose ships were in their sealing grounds above the Tuscarora Deep around the time of the tsunami. A number reported unusual disturbances in the sea. In one case, a ship's crew recorded water-temperature fluctuations from 48 to 218 degrees Fahrenheit within a few miles. She had not read such details anywhere else, Eliza told Goode, and if deep-sea geographers wished to know more, she suggested, they could interview the schooner captains who wintered at Victoria, B.C. Meanwhile, she wrote up what she had heard for a short report that appeared with her article in *National Geographic*.[19]

***

Eliza planned to round out the year with a long trip to China. To clear her desk beforehand, she promised The Century Company the manuscript of her book on Java, which she had labored over for months. At the last minute, she held the manuscript back—again—to work on further revisions while on the road.

Her reservations about the book persisted even as she finally submitted the manuscript in the spring of 1897. She was abandoning it in resignation, she told her editors, after she had "rewritten and rearranged and muddled over it so, that I no longer have any perspective or judgment left."[20] In particular, she struggled over two chapters on the Dutch "culture system." One later reviewer who picked up on the book's weaknesses called her treatment of the issue "not quite so clear as one might wish," and questioned her conclusions that for all of its faults, the agricultural system had brought, as she put it, "incalculable benefits" to the islands by improving the living conditions of the native Javanese and putting "great stretches of jungle . . . under cultivation."[21]

Her general unease about the book reflected in part her own nagging awareness that she had not done as much reporting as she should have. As Eliza explained to her editors, she had hoped to revisit the island on her way home from China, to "go over my ground again and perhaps gather more material."[22] A cholera outbreak in the region scotched those plans. To compensate, she did library research in Hong Kong and Canton.

Heeding her instincts that the book needed more work may have spared her the anguish that followed. Trouble erupted even before the release of *Java, the Garden of the East* late in the fall of 1897. *The Century* ran two chapters to coincide with publication. Soon after they appeared, the magazine got an irate letter from a reader in Amsterdam.[23] An editor named R. A. Van Sandick detailed a long list of complaints, ranging from errors of fact to misspellings. He was most aggrieved, however, by the attitude of the author herself, whose "British imperialism" and "English mind" obviously caused her to "bitterly hate" the Dutch. Many of her statements were "laughable" and "ridiculous," he charged, calling her "untrustworthy" as a reporter.[24]

Eliza was stunned. She readily admitted that she made mistakes in the course of her reporting. But she approached her work conscientiously and welcomed corrections as a matter of record. The personal attack seemed something else altogether. Never in her many years in journalism had her integrity been so called into question.

"The Dutchman's opinions were, as you may imagine, a thunder clap of surprise," Eliza wrote to Clarence Buel, her editor in The Century's book division.[25] She had found in Buel, a seasoned New York newspaperman before joining the company, a sympathetic editor, and a friend. They agreed she had to figure out—fast—how to respond. The book was already at the printer.

With Buel's support, Eliza negotiated a week to look into the Van Sandick matter. She hauled out her notebooks and set out to check her information against his. Unfortunately, she found the Congressional Library temporarily closed; all its holdings were being transferred from the U.S. Capitol to a new building across the street.

Eliza raced around town hunting down sources in other collections. To her relief, she concluded that many of Van Sandick's complaints were overblown.[26] Some of the issues he raised were a matter of interpretation or contradictory sources. And yet, it embarrassed her greatly to discover that she had indeed made sloppy mistakes, some of them significant. Most egregiously, she reported that Dr. Treub, the head of the botanical gardens in Buitenzorg, had died and been replaced. But the world-famous botanist was still very much alive. Eliza had gotten the information from one of her travel companions on Java, a "Miss Easter" from Baltimore, and had foolishly failed to check it out.

To right her mistakes, Eliza insisted on having some of the book plates destroyed and corrected pages inserted at her own expense. She wrote a new preface, in which she clarified some points and extended an olive branch to Van Sandick by thanking him publicly for helping her to make the book as accurate as possible. *The Century* ran a rare paragraph of corrections to her Java articles.

Alone in her hotel room in Washington, Eliza worked herself into a state of despair over the incident. She was so mortified by the situation she felt "almost physically ill," she confided to Buel. "You must believe that this has been a most bitter and humiliating experience."[27] The worst part was the shame she felt at having let the magazine down. Accuracy was a key part of *The Century*'s stock-in-trade; its editors and fact-checkers went out of their way to get things right. Now she had broken that trust through her own hubris and carelessness. It had been arrogant to assume she could write an entire book on Java based on only a few weeks of travel on the island.

Van Sandick's criticism had clearly touched a nerve of self-doubt. Eliza was forty years old, yet her outpouring of emotion to Buel sounded like the voice of a Catholic schoolgirl pleading for atonement of her sins from a father confessor. "I truly believe," she continued in her letter to Buel, "that I was purposely blinded to my own errors that this exposure should teach me humility and charity to others—to make me realize how easy it is to see the mote in another eye, and not the beam in my own." Finding herself "in the pillory," she said, she felt deserving of whatever "severe punishment" the magazine might wish to mete out.

It was a remarkable confession from a proud, at times almost haughty, woman who had learned to conceal her vulnerabilities behind a mask of supreme self-assurance.

***

Eliza regained her equilibrium by burying herself in her work. She was writing a pair of long-delayed articles based on her India travels a few years earlier. She also planned to visit Alaska again in the summer of 1898, to report on the Stikine River for *National Geographic.*

On May 8, she was just about to leave for the Alaska trip when current events prompted thoughts of other another reporting subject. "Has anyone preempted the Philippines yet for the Century?" she inquired of her editors in New York.[28] Only a week earlier, Commodore George Dewey had led the U.S. Asiatic Squadron to a stunning victory in Manila Bay in the first major battle of the Spanish–American War. The blazing seven-hour attack, which started at dawn after the fleet's stealth entry into the bay under the cover of darkness, destroyed the ill-prepared Spanish flotilla and killed several hundred Spanish troops, but left the American force little damaged. One of the most wildly successful battles in U.S. naval history, the victory marked America's advent in the Pacific and made the country an international power for the first time.

In the patriotic fervor that followed, Dewey became an instant war hero. The swell of pride reached even remote Alaska. During her travels in Alaska that summer, Eliza heard reports of a man who had arrived in the north with a newspaper article describing the battle in Manila Bay. He made a thousand dollars reading it aloud at dance halls in Dawson at fifty cents or a dollar per person. Each reading took two hours, one observer told Eliza, because the crowds whooped and cheered so loud, even though the man "wasn't no great hand at reading no how."[29]

As Eliza told Robert Underwood Johnson at *The Century* on the eve of her Alaska trip, she thought she might visit the Philippines once she returned to the Far East later that year. Maybe there was a book in it. At any rate, she would not be able to go before the fall, as the situation in the islands would take some time to settle down.

On her way home from Alaska, while passing through British Columbia, Eliza learned that someone else had indeed "pre-empted" the Philippines. The June issue of *National Geographic* carried a major article about the Philippines by Dean Worcester, an ethnographer and photographer who traveled in the islands for zoological expeditions. He also had an article coming out in *The Century* that fall, timed to coincide with a book on the Philippines.[30]

During her later stay in Asia, Eliza continued thinking about a trip to the Philippines. But, as she told her editors, the little news from Manila that she got from the papers did not much pique her interest. The situation there seemed quiet.

In the end, she went after all. Though a hardy traveler on the road, Eliza had little tolerance for the biting chill of winter in Japan. In January 1899, after a monthlong bout of severe influenza, she planned an escape to a warmer place. She settled on Saigon, to see the temple ruins on the upper Mekong. At the last minute, she decided to go to the Philippines instead—as she put it, "just to know how much to believe of what I read."[31] The islands had come under U.S. military occupation in December, after Spain ceded its longtime colony to the United States under the treaty ending the war.

Eliza arrived in Manila at noon on Saturday, February 4, and checked into the Oriente Hotel on the waterfront. Several hours later, at dusk, the sounds of gunfire erupted. Fighting had broken out between U.S. troops and nationalist insurgents under the rebel leader Emilio Aguinaldo. Filipinos had hailed America's war with Spain, welcoming the prospect of independence after a century of colonization. But America's possession of the islands came as a bitter betrayal. Now, the open rebellion launched a conflict that would last more than two years, bringing more casualties than the Spanish–American War itself, as well as shocking reports of atrocities.

As darkness fell that evening in Manila, the fighting grew more intense. At midnight, U.S. troops arrived at Eliza's hotel to hustle her and the other Americans, including two dozen women and children, off to safety. They were taken aboard the transport ship S.S. *St. Paul*, anchored in the harbor alongside George Dewey's *Olympia* and a line of warships.

All the next day, on a hot Sunday, Eliza stood on deck watching—"as from an opera-box"—the spectacle of warfare.[32] Signal flags ran up and down the *Olympia*'s masts. On shore, the explosion of shells from the ships' giant guns cratered the earth, and smoke rose from burning houses. Eliza counted two dozen fires across the city. She found the heat of battle horrifying, but also grotesquely fascinating in its drama. "If I had planned and cabled, and bought a yacht, I could not have arranged better for a front-row seat at the great performance," she told Robert Underwood Johnson at *The Century*, writing a few days later from her hotel room.[33] She and the others had remained aboard the ship for two days, until they were put ashore on Tuesday so the *St. Paul* could leave with troops for another part of the islands where fighting had broken out. Later that week, a Major Bell took her and several others to see the area where he had led a charge of Pennsylvanians in the battle on Sunday.

At her hotel, Eliza received a cable from *Scribner's* requesting 4,000 words from Manila. Out of loyalty, she telegrammed her editors at *The Century* to see if they wanted her to send something. "Try," came the reply.[34] But a short time later, she was once again forced to take sanctuary aboard a ship in Manila Bay. "Twice I packed ready to fly, to the transports, but survived with an uncut throat," she wrote in a follow-up letter to *The Century* on February 28. "After a night of burning city wards, shooting, shouting, bullet-whistling and what not, I was more than glad to be a refugee again."[35] Since that frightful time, the city had fallen quiet as a cemetery.

She had been thinking about the requested magazine article, Eliza continued in her letter to Johnson. "I cannot give you the regulation war article," reporting on troop movements and such, she explained. She wanted to try writing something different: a more general view of war at close range—"the prose of it, the sorrow and injustice of it." Above all, "the absurdity of it." She would describe the ambitious young officers who pulled wires in Washington to get posted to the Philippines because they craved adventure; their wives who arrived by the shipload to take up garrison life. She would picture scenes straight out of Jacques Offenbach's comic operetta *The Grand Duchess of Gerolstein*, "when we drive out to 'the front' before dinner, and go out to the trenches every morning after breakfast." The military situation in Manila would be a picnic and a farce, she wrote, but for the suffering and misery inflicted on the small Pacific nation of people who wanted only to be free to live their own lives.

President William McKinley had tried to avoid armed conflict with Spain. But the sensationalistic "yellow journalism" of American newspapers, led by press baron William Randolph Hearst, had fueled a hysteria that pushed the United States into the war. Once the Philippines came under U.S. control, Americans differed sharply in their opinions about the islands' fate. Many endorsed annexation to prevent another power from seizing the islands and to give the United States a gateway for new markets in Asia, especially China and its more than 300 million people. Others saw America as having a Christian duty to "uplift" the Filipinos. As President McKinley himself said, explaining the annexation: "We could not leave them to themselves—they were unfit for self-government."[36]

Eliza agreed fiercely with the many anti-imperialists in America—among them, prominent figures like Mark Twain, Andrew Carnegie, and former presidents Benjamin Harrison and Grover Cleveland—who argued that the United States had no business being in the Philippines. As Eliza told Johnson at *The Century*, she hoped to show readers "the anomaly of the American in the tropics," the folly of "the wild westerner in the Far East."[37] Inflamed by her own passion, she closed on a dramatic note: "Oh! deliver us from the volunteer, the amateur, the dilettante soldier . . . that we might let those Philippines go."[38]

After returning to Yokohama in April, Eliza wrote a lengthy article on the Philippines, which she sent off to the magazine along with a handful of "Kodaks." She had agonized over the piece, she told the editors, wondering what was wise or safe to say. "I lost it utterly in Manila over American looting, brutality and want of civilization generally," she wrote.[39] During her stay, she heard horrifying stories about the cruelty of U.S. troops. They were said to be attacking and killing not just Filipino insurgents but also innocent civilians, including women and children. She wasn't sure what had already been reported in the press.

The Century may have found the article too polemical, as it never ran. The verdict was disappointing, but by the time Eliza got the word from New York, she had fixed her sights on other big developments in Asia. Reports were coming out of China about a coup d'état in the imperial palace. She needed to hustle back to Peking to see what she could find out.

13

Trouble in China

Though Japan remained Eliza's base in the Far East, China kept drawing her back. By the end of the century, she had visited the country seven times. In her eyes, China's shabby old capital of Peking was the most picturesque and interesting city in the East.[1] "The world certainly moves in some spots. Not here, however. Nothing is changed, except for a little more dirt and much disrepair, since Marco Polo's visit," Eliza told her long-time mentor at National Geographic, Gardiner Hubbard, in a chatty letter from Peking in the fall of 1896.[2] He had contacted her before she left Yokohama to request some prized water irises from Japan. Now she was writing to let him know the bulbs were on their way.

She had just spent all of October in Peking, Eliza informed Hubbard, "doing it at the rate of eighteen hours a day." During her stay she turned forty, a milestone birthday that passed without fanfare. Fired by her lifelong passion for discovery, she still had the doggedness of a traveler bent on not wasting a minute. Every night she retired to her hotel room exhausted but with the satisfaction of days well spent.

Eliza's fascination with China came from the same curiosity about the ancient civilization that had seduced adventurous travelers for centuries. On a more practical level, she recognized the travel-writing opportunities the country offered as a place still off the beaten track. Unlike neighboring Japan and India, China got relatively few tourists. The distances were long; the roads almost nonexistent. Most rural Chinese inns provided nothing more than a grimy bare room. As Eliza informed readers, neither Murray nor Baedeker, the leading guidebook publishers, had yet scouted out China, and the mass-tourism pioneer Thomas Cook & Son had "only touched at the edge of it in Canton."[3] The rare globetrotters who wished to venture beyond coastal cities like Shanghai and Hong Kong were on their own.

It was also her journalistic instincts that drove Eliza to China time and again. She sensed in the air a volatility stemming from a restiveness among many of the Chinese people. Japan's victory over the once-mighty empire—"the huge Humpty-Dumpty of the Far East," in Eliza's words—had left China shaken, demoralized, and badly in debt.[4] Long a target of imperial ambitions, China had faced the dominating presence of Britain and the loss of Hong Kong after the Opium War. It suffered further erosion of its autonomy and territorial integrity when Japan gained the upper hand in the region in 1895 and took control of the island of Taiwan. Now, several Western powers were seeking to exploit China's weakened state by moving in to advance their own interests. The humiliating subservience their country faced under the growing presence of high-handed foreigners and moralizing missionaries angered the Chinese at all levels of society. But the imperial government in Peking—insular and heavily bound by tradition—seemed ill-equipped to do anything about it. Uncertainty about Japan's intentions in the region added to the tension.

Something seemed bound to erupt, sooner or later. "The Pacific is not going to wait for the Twentieth Century to become the theatre of man's greatest activity," Eliza advised her editors at *The Century*.[5] It would be wise, she added, to save space in the magazine. Before setting off for China in the fall of 1896, she had requested all the money the magazine owed her, to help pay for what would be an expensive trip. She expected to be gone several months so she could see more than just Peking. "The eye of the world will have to be kept on this tottering old empire for the next decade or two," she told her editors, "and I want to know it personally."[6]

***

Eliza liked visiting Peking in the autumn, when the sparkling air and vast blue sky above the treeless plain reminded her of traveling in Minnesota or the Dakotas back home.[7] The 1,300-mile journey from Yokohama required endless transfers between steamer, train, riverboat, and even at times wooden cart. At Tientsin, an international port city along the coast, she hired a Chinese guide to accompany her on the seventy-mile trip to Peking. Reaching the city's massive fortress walls always left her awestruck. Peking was monumentally magnificent.

Nomadic Mongol horsemen of northern China had built Peking as a garrison town. The powerful Mongol ruler Kublai Khan was living there in splendor in the thirteenth century when he greeted Marco Polo during the Venetian trader's travels on what became known as the Silk Road. A great cultural flowering occurred under the Chinese Ming dynasty (1368–1643) that followed, as its emperors fused ancient knowledge with Western learning. But China's glory had diminished under the Qing dynasty that now held power in Peking. At the brink of a new century, the present Manchu rulers, mistrustful of foreigners and determined to maintain the traditions of China's past glory, were hunkered down in their opulent court life while much of the world rushed headlong toward modernization.

Peking, home to a million people, was really four cities, arranged like nested boxes, each with its own high defensive walls. Most visitors entered from the south, through a deep archway leading into the "outer" city of the Chinese. Two miles ahead, another set of gates opened into the Tartar City of the Manchus. That city, in turn, housed the Imperial City, an enclave of palaces, temples, government buildings, and gardens. At the very heart of Peking lay the mysterious Forbidden City, where the emperor resided in luxurious seclusion with his family and thousands of fawning eunuchs. In a tip she shared in her guidebooks, Eliza learned that some gatekeepers could be bribed into allowing foreigners to ascend the walls and walk for miles between the parapets, for a bird's-eye view of the city across the treetops. From all she had read, Eliza expected the famous yellow roof tiles inside the Forbidden City to gleam like gold. Instead, they had the dingy tint of mustard paste.[8]

Like other Westerners, Eliza stayed in the Legation Quarter, the three-quarter-square-mile district in the southeast corner of the Tartar City (east of present-day Tiananmen Square) that catered to foreign delegations from eleven countries. Eliza found welcoming hosts in the consular community, but she liked staying in a hotel because of the privacy and freedom from onerous social obligations. The Hotel de Peking, run by a Swiss proprietor, offered all the comforts of home. Outside its doors, the almost slum-like conditions were another matter. A fetid canal ran through the middle of the quarter, puddling the rutted main road with mud and sewage. Most diplomats and their staffs, including about 1,500 Americans, lived in formerly grand villas, or *fus*, once occupied by Manchu noblemen.

Despite her own enthusiasm for Peking, Eliza warned would-be visitors of the drawbacks. Officials had put many places of historical and architectural interest off limits to tourists. And the dirt and smells took some getting used to. Yet Peking offered "so much of novelty," she argued, that travelers with a week or two to spare should not miss it (Figure 13.1).[9]

Every morning, Eliza sallied forth to make the rounds of the city, usually in a "Peking cart," a small and boxy enclosed vehicle pulled by a soft-coated donkey. Folded inside the famously uncomfortable contraption, she strained to take in the street views from the tiny windows. Camels laden with trade goods plodded by, kicking up dust. At the city's central crossroads, marked by four grand arches (*pailows*), a long line of carts, wheelbarrows, and other conveyances sat backed up in a blockade outside the shops selling tea, medicine, silk, and curios. Mandarins of the imperial Chinese government strode by in long coats; squawking peddlers hawked their goods; and a babble of voices rose from the mix of people: Manchus and Chinese, Mongols from the plains, priests and pilgrims, Westerners, and travelers from other parts of Asia.[10]

Figure 13.1. Peking street scene, c. 1890s.
(Billie Love Historical Collection and Special Collections, University of Bristol Library)

One sight always prompted Eliza to stop and stare. She was bedazzled by the groups of handsome Manchu women going in and out of the shops in their richly colored gowns and towering T-shaped headdresses. As she told readers:[11]

> In my first breathless delight in each of these striking figures, . . . I berated all my traveled acquaintances, who, in harping on the dirt and the dilapidation, the offensive smells and sights, of Peking, had never told me of these Manchu women, with their broad gold pins, wings of blue-black hair, and great bouquets and coronels of flowers, the bewitching pictures in every thoroughfare.[12]

Though the Manchus had absorbed many customs of the Chinese, they rejected the brutal custom of foot binding, leaving the women to stride about in their full splendor.

One day Eliza got word that an elderly Manchu noblewoman wished to meet the foreign lady. When Eliza arrived at the dilapidated *fu*, whose owner had fallen on hard times, all the women and children turned out to greet her in their finery. Eliza took so many photos she used up all her film. On another visit, she herself became the center of attention when the matron invited friends in to join them. Eliza returned the hospitality by inviting the women to tea in her hotel, at which she served an American staple: chocolate layer cake. When they parted, the *tai-tai* ("madame") declared them "friends forever."[13] On a later trip to Peking Eliza went back to see the family, but they had moved, and the *fu* had been razed.

During her 1896 visit to Peking, Eliza and some companions from the legations made a four-day excursion to see the Great Wall at Nankow Pass and the valley of Ming tombs. Later, she and a "Mrs. Winslow" traveled by houseboat from Shanghai to Hanchow, cruising along China's Grand Canal past old provincial towns and crumbling battlements. They timed the journey to see the world's largest tidal "bore" at Hanchow Bay, in which a giant inrush of water raced along the Tsientang River in a thunderous roar. Eliza found the spectacle so thrilling she wrote an entire article about it.[14] On the final leg of their trip, the women sailed the Yangtze in bright December sunshine on a thousand-mile trip into the heart of China. They started the journey by steamer, then switched midway to a fifty-foot *kwatsze*—a cabin-topped flatboat—that carried them through the gorges and rapids of the Upper Yangtze. The large crew included "trackers" who scrambled across rocks and along the shores guiding the craft with ropes of braided bamboo.[15]

***

Eliza's reporting on China took on greater urgency in the fall of 1898, when reports coming out of the country indicated that the Imperial Palace was in turmoil after the empress dowager, Cixi, wrested power from the emperor. Everyone in the diplomatic community in Japan was waiting to see if heads rolled.

Little about China intrigued outsiders more than rumors surrounding Cixi, now the most powerful woman in the world as the de facto head of an empire of 300 million people. The sixty-two-year-old dowager had started life as an ordinary Manchu girl who became a favored concubine of the former emperor. She bore him a son who assumed the throne in 1861. After the young man's premature death, Cixi conspired to become regent to her nephew and adopted son, Guangxu, the designated heir. Though Guangxu currently held the title of emperor, everyone understood it was really Cixi—ensconced in the Summer Palace with an army of obedient eunuchs—who wielded power behind the scenes.

The coup d'état stemmed from Emperor Guangxu's desire to modernize China. In the wake of the Sino–Japanese War, public unrest simmered around the country amid calls for improved living conditions and policies to curtail the growing presence of missionaries and other foreigners. A quiet and introverted young man, the twenty-seven-year-old emperor had little direct knowledge of the outside world. But he developed an interest in Western ideas and institutions under a group of reform-minded advisers.

In the summer of 1898, Guangxu started implementing a sweeping program of changes he believed would make China stronger.[16] After a hundred days, Cixi—backed by a cadre of conservative allies—stepped in and put a stop to it. Some of the reformers were beheaded; others fled for their lives. Court eunuchs seized Emperor Guangxu and carried him to an island complex inside the Forbidden City, putting him effectively under house arrest. Though he would retain ceremonial powers, he became a puppet of the empress dowager for the rest of his reign.

On October 1, two weeks after the emperor's arrest, Eliza dashed off a letter alerting *The Century* to the dramatic developments. "China's the thing—the storm centre, the world's vortex now. Keep your eye on that," she wrote. Eager to get the inside scoop, she planned to visit China again in the spring of 1899. She had seen a good bit of the "outside" of China on her last visit, in 1896. "This time I hope to see the 'inside' of Peking," she explained, "unless my Manchu friends, who are returning to China to manage the ceremonies of the Empress Dowager's court, all lose their heads on arrival. It looks now as if the old civilization would have to go, and I want to see the reversal."[17]

Her "Manchu friends" were the Yu Kengs, a family the Scidmores had come to know years earlier when Yu Keng served as the Chinese minister to Japan. A man of progressive views, he and his wife, a woman of mixed Chinese and American ancestry, insisted on giving their children a Western-style education. That included their two daughters, Der Ling and Rong Ling. Eliza and her mother had found the adolescent Yu sisters charming—pretty, multilingual, and well versed in literature and the arts. Eliza saw great potential especially in Der Ling, a clever girl whom the Scidmores knew as "Lizzie" (short for her Christian name, Elizabeth).[18]

Though Han Chinese by heritage, Yu Keng identified as Manchu, and he admired the empress dowager. As news of the reforms in China spread, Yu Keng agreed to move back to Peking, where he and his family could be of service to Cixi as she sought to improve relations with Western powers.[19] Eliza knew she could count on the Yus' hospitality once they were resettled in Peking. Best of all, it would give her a direct pipeline to affairs inside the imperial court.

With China much in the headlines, Eliza hustled to finish several pieces for *The Century* based on the notes from her 1896 trip. Lacking a typewriter in Japan, she wrote out the material by hand and mailed off the sprawling mass of pages by diplomatic pouch, along with fifty of her own photographs to use in preparing illustrations. "Cut away as you please," she told the editors, "if you pay me for three articles."[20] She expected the usual rate of $125 apiece.[21]

Back in New York, the reviewing editor despaired at the formlessness—and excessive length—of Miss Scidmore's prose. One piece on her Yangtze River trip ran 15,000 words, with long digressions on the Taiping Rebellion and the ceramic industry; a section on tea ran twenty-three pages. "It has very much bored me reading it," the editor noted in an in-house memo.[22] Despite the harsh judgment, the staff got to work shaping the articles for publication.

When Eliza received word of *The Century*'s acceptance of the material, she got the added good news that the company agreed to her proposal for a book on China. She envisioned it as a companion volume to *Jinrikisha Days in Japan*.[23] Now, the trip she had hoped to make to Peking in the wake of the coup would have to wait. In the summer of 1899, Eliza sailed home to America to see the China book to publication.

***

The following spring, on May 10, 1900, the U.S. Army transport ship *Hancock* stopped in Yokohama on its way to Manila. Among the passengers going ashore was Helen Taft, whose introduction to Japan that summer would lead to her partnership with Eliza a decade later in bringing Japanese cherry trees to Washington.

Nellie Taft was traveling out to the Philippines with her husband and children. President William McKinley had named William Howard Taft to a commission charged with establishing a civil government in the U.S.-occupied islands. Taft had been cool to the offer, reluctant to give up the work he loved as a circuit court judge in Cincinnati. He expected Nellie to balk as well, given the long distance from home and health concerns for their three young children in the malarial tropics.[24]

To the contrary, she jumped at the chance to go. Much like Eliza, Nellie had dreamed of seeing the world while growing up. Well-bred but rebellious, she felt ambivalent about marriage, telling her diary: "I have thought that a woman should be independent and not regard matrimony as the only thing to be desired in life."[25] She eventually succumbed to the marriage proposal of the burly Yale Law School graduate once she realized that Will Taft, who admired his reform-minded and strong-willed mother, did not expect her to change her ways. When the Philippines offer came, Nellie said in her memoirs, "I knew instantly that I didn't want to miss a big and novel experience." She laughed off the comments of critics shocked by the couple's decision to take their children to "that pest-ridden hole."[26]

The *Hancock* docked for a week in Yokohama, where talk in the diplomatic community centered on the worrisome situation next door in China. For months, a secret society of martial-arts fanatics that Westerners referred to as Boxers had roamed the famine- and drought-stricken northern provinces spurring attacks against the foreign "white devils" who were undermining Chinese society and traditions. Mobs burned down Christian missions, destroyed railroads and other foreign property, and massacred foreigners and Chinese Christian converts. Thousands of people poured into Peking seeking protection from Western nations as the Boxers advanced toward the city.

On the day of the *Hancock*'s arrival in Yokohama, the *North China Daily News* carried a disturbing report that the empress dowager herself, eager to rout the country of foreigners, sanctioned the Boxers' actions.[27] Nellie Taft later recalled the sight of the U.S. Navy cruiser *Newark* anchored in the Yokohama harbor, ready to depart for China at a moment's notice to help quell the violence and chaos.[28]

During the layover in Japan, the American commissioners met the emperor in the Imperial Palace while Nellie and the other women had an audience with the empress, a tiny woman in Western dress who spoke through an interpreter. Days later, when it came time to leave, the Tafts' son Robert was diagnosed with diphtheria. He had to be quarantined, so Nellie arranged to stay in Japan for the summer with her sister Maria and the children. She set up house in a Victorian-trim bungalow on the Bluff, the leafy hilltop neighborhood of Yokohama favored by diplomats and other foreigners. One end offered a panoramic view of the Pacific; the other side overlooked a valley of bright-green rice paddies and groves of twisted pines. Nellie would retain fond memories of the sounds that emanated that summer from a Buddhist temple tucked in a hillside—priests chanting to the beat of little wooden drums, the low-pitched tong of a bell reverberating through the valley.

Leaving the children with their nurses, she and Maria shopped by jinrikisha and made excursions to Nikko, Kamakura, and Kyoto. When the heat grew stifling in late July, they packed up the household and decamped to Miyanoshita, the popular mountain resort near Lake Hakone, where Mount Fuji rose in the background.

One evening in Yokohama, Nellie dined with Mrs. Scidmore. Eliza was away at the time, finalizing her China book back in America, but her mother made a lively hostess. Eliza Catherine Scidmore's sociability was legendary among American travelers in the Far East. She no doubt shared with her visitor the story of how she had met William Taft's parents—and young Will himself as a boy in short pants—when the elder Taft served as secretary of war and then attorney general in President Grant's cabinet.[29] Though in her late seventies, Mrs. Scidmore seemed to Nellie "as bright and young as a woman of fifty." They would meet again on other occasions, leaving Nellie with an impression of the elderly white-haired woman as "a sort of uncrowned queen of foreign society" in the Far East.[30]

***

Eliza spent the spring in New York reading page proofs, before going to London to deliver her paper on Japanese morning glories. The China book was printed and ready for the binder when the editor contacted her to request a rewrite of the introduction. The book had been due out in the fall. In light of current events, The Century Company wanted to bring it out as soon as possible.

China: The Long-Lived Empire hit the market on June 27, just as events in China were dominating the headlines. A week earlier, the Boxers had put the Legation Quarter and Peitang Cathedral in Peking under siege. For fifty-five days, several thousand men, women, and children—diplomatic families, missionaries, Chinese Christians, and others—huddled in terror. Under the dire conditions, some eventually died of starvation. Meanwhile, Boxer-led mobs rampaged through the countryside and attacked foreigners in the port city of Tientsin. Eight nations scrambled to organize an allied force of 20,000 troops to quell the unrest and rescue their foreign nationals. But the relief came too late to save tens of thousands of victims, most of them Chinese. Many were gruesomely murdered.[31] Press reports of the atrocities inflamed growing hysteria in the West over the perceived threat of a "yellow peril."

The Century Company's decision to publish quickly made Eliza's book the first in a handful of new works on China. Amid the riveting headlines, it became a best-seller. Many publications picked up excerpts on the empress dowager, in which Eliza presented a more humanized portrait of the woman derided around the world as a cruel and tyrannical "Dragon Lady." Reviewers commended the book, at nearly 500 pages, for being revelatory but not pedantic. Miss Scidmore's latest volume, they concurred, benefited greatly from her spirited narrative style and graphic descriptions—"kodak-like pictures," one critic called them.[32] Eliza followed the book's success from Yokohama. "The local gossip here . . . has settled it that I have made $30,000 gold this summer out of the 'Long-lived Empire,'" she said in a letter that fall. "I hope a fraction of that guess will come true."[33]

While Eliza's book contained mostly straightforward travel reporting, her revised introduction included damning commentary on China that touched on common opinion of the day. In her early career as a newspaper reporter, Eliza had openly denounced attacks against honest and hard-working Chinese residents on America's West Coast. When it came to the country as a whole, however, she was fiercely critical of China for what she and many others regarded as its arrested state of development and progress. For centuries, nations in Europe had held up Chinese civilization as a model for advancement in many areas. Yet by the end of the nineteenth century, China conjured images of backwardness in the minds of Westerners.[34] China, Eliza wrote, had ossified and become "degenerate"—everything "dead, dying, ruined, or going to decay."[35] Cultural achievement and industry had been hollowed out; the people lived in revolting conditions and seemed to have little joy or lightheartedness. "Of all Orientals," she wrote, "no race is so alien."[36]

Still, Eliza disagreed with observers who predicted that China was headed for a "break-up." The empire had endured for 4,000 years, through many intervals of turbulence, and it would continue to survive. The big question was: to what end? Old, sick, and tired, China did not want to be dragged into the "fierce white light" of the new century, Eliza wrote.[37] But the "long-lived empire" would not have much of a future, she insisted, unless it radically changed its ways.

***

Scholars have written that Eliza Scidmore's reporting made her one of the influential writer-travelers who "opened" China to the West in the final years of the nineteenth century.[38] One historian considers her railway-company guide, *Westward to the Far East*, a forerunner of mass tourism in providing the first known itinerary of "must-see places" that led to what might be thought of as a Grand Tour of China.[39] Summarizing scholarly and popular works on China in her own book, Eliza singled out the narratives of two other women as "the best books of pure travel."[40] One of the authors, Constance Gordon Cumming, was a Scottish-born aristocrat and prolific writer who sailed the southern coastline of China and went inland to Peking in the 1870s.[41] Eliza also commended the work of the well-known British travel writer Isabella Bird, who made an 8,000-mile journey up the Yangtze to the edge of Tibet and on foot through Sichuan. Eliza reviewed Bird's 1899 account of the trip, *The Yangtze Valley and Beyond*, in *National Geographic*.[42]

Eliza's China legacy includes a campaign to fight the American use of an alternative spelling of "Peking" that became common at the turn of the century. With no uniform spelling of place names, countries followed their own style; even English-language usage was not consistent. As headlines from China increased, the name of the capital started appearing often as "Pekin." The U.S. government adopted the spelling in its own publications, a practice that irritated Eliza to no end.

Her activism on the issue flowed in part from her participation in a series of Oriental Congresses held in Rome. Organized in response to growing interest in "Orientalist" cultures of Asia and the Arab region, the meetings drew hundreds of attendees from around the world. In 1899, Eliza served as the secretary for two long sessions on China, Korea, and Japan. Among the topics, the delegates discussed the need for standardization of place names. Was it Chifu or Chefoo? Kau-lung or Kowloon? After much debate, the meetings ended with no firm resolutions.[43]

When Eliza tried to find out why the U.S. Board of Geographic Names had sanctioned the use of "Pekin," she was denied a hearing. In her typical bulldog persistence, she refused to let the matter rest. She marshaled opinions on the matter from English-speaking Sinologues. To the man, the experts agreed that "Peking" was the correct transliteration of the Chinese characters meaning "northern capital." Presented with the endorsements, the Board of Geographic Names eventually succumbed. In 1903, its members voted to change the style. Henceforth, "Peking" it would be.[44]

***

The aftermath of the Boxer Rebellion benefited collectors of Oriental art—Eliza among them. Old Chinese ceramics and other antiquities held special cachet for many Western collectors who valued such objects for their ancient lineage and imperial provenance.[45] And so it was that the Boxer-related chaos in China unleashed a frenzy of competition for what one newspaper called the "carnival of loot" that came available.[46] During the yearlong foreign occupation of Peking, soldiers moved through the city in a wave of plunder and destruction. Troops swept through the palaces from which Cixi and the court had escaped, snatching up imperial objects and smashing what they could not carry, and stole valuables from the homes of Chinese residents who had fled. The scramble for booty grew so intense that auctions were held outside the British Legation.[47]

Many of the foreigners regarded the pilfered goods as legitimate spoils of war. Some diplomatic officials condoned the removal of national treasures from the country as a way to ensure their preservation. Tons of items were shipped off to museums and private homes in Europe and America. Fine and sometimes rare pieces also started turning up at curio shops in Peking.[48]

Treasure-hunting put added bounce in Eliza's regular trips to Peking. By then, she already had the makings of a well-regarded Asian art collection that would grow to more than 200 objects. The outsized buying power and status she enjoyed as an American living and traveling in poor Asian countries enabled her to acquire many choice pieces at modest cost. She had started collecting early in her travels out of a personal love of beautiful things, such as Japanese embroideries. As her knowledge of Asian cultures increased, she branched out into ceramics and other items. One of her most precious possessions was a 300-year-old imperial-yellow Kangxi bowl. She had found it in a remote corner of Peking during her first trip to the city and she carried it in her arms on the long ride back to her hotel by sedan chair.[49]

Many avid collectors of Oriental art approached their acquisitions as a form of value, or a badge of social status. An expert at an auction house in New York that eventually sold Eliza's collection saw Miss Scidmore's choices as guided largely by an appreciation of their larger cultural meaning. "Every piece has its story," the expert said.[50] In an indication of the owner's discerning eye, it was noted, most of the items had been loaned out at one time or another to various museums.

Eliza developed her knowledge of Oriental art in part through long-time friendships with connoisseurs in America and Asia. In China, her sources included Alfred Hippisley, whose position as commissioner of maritime and customs aided his acquisition of "palace" porcelains and other rare objects, and Colonel Clement de Grandprey, a French military attaché in Peking who had a home at Versailles and steered Eliza to top Asian art shops in Paris.[51] She further expanded her expertise through a personal library of reference works that contained many volumes on Asian art and design.[52]

Prowling the local markets for Asian treasures became a lot more fun after Eliza met an American woman who shared her thrill of the hunt. It was apparently in India, in the winter of 1900, that Eliza first met Emily MacVeagh, a popular socialite in Chicago who possessed what her friends called a "passion for artistic decoration."[53] Fortunately, given her taste for exquisite objects, Emily had a deep purse to support her collecting habits as the wife of a wealthy banker.

Her late-in-life relationship with Emily MacVeagh offers a rare look at Eliza's capacity for female friendship, as well as the sentimental side of a woman who was heavily guarded in her public image.[54] A decade of long, gossipy, and often playful letters to Emily suggests almost a crush on Eliza's part. She was constantly solicitous of the gentle, auburn-haired Emily, whose health was delicate. Emily sent her friend flowers and clever gifts and recommended a podiatrist to treat Eliza's aching feet. Eliza acted as a buyer's agent for Emily in scouting out treasures in Japan and China. As Eliza advised Emily in one letter: "Hold fast to your court dress with the sleeveless overdress to match, as I learned that it is—or was—the ceremonial robe of the Emperor's wife."[55]

Over the years, Eliza grew so bold in her affection as to end her letters to Emily by extending "my dear love to you."[56]

***

The two women had first gotten to know one another when Eliza went to India to do more reporting for a book. As she explained to her editors before setting off, she had already covered the "preliminary tourists drudgery" during her round-the-world trip in 1894–5.[57] This time she wanted to be there during the Anglo-India social season, which had not been much written about. Every year, the British viceroy and other colonial officials gathered with India's native princes for a week of gala festivities in Calcutta during Christmas week. The crowds grew so large that clubs and hotels set up tents on their lawns to accommodate the overflow of guests. The visitors included many American and British travelers who came for the chance to mingle with what seemed like British royalty.

Emily MacVeagh was there in the winter of 1900 with her husband and friends from Chicago. They felt a special connection to the events through their association with Mary Leiter Curzon, a native daughter of Chicago who was now making a splash around the world in her role as the vicereine of India.[58]

Eliza went to India that winter hoping to exploit her acquaintance with the Curzons, after knowing Mary casually in Washington and socializing with the couple at the 1895 International Geographical Conference in London. Eliza enjoyed her time in Calcutta, where the social events mimicked "the life of London," though burnished with sunshine and local color. At the popular Viceroy's Cup, she found the crowds far more fascinating than the horses. The thousands of native Indians in white garments sported turbans so colorful in their mass effect that the scene made Eliza think of a vivid tulip bed at the height of spring.[59]

As for her encounters with the Curzons, Eliza was not disappointed. After registering her name in the visitors book and sending a note to Lady Curzon, Eliza received invitations to several official functions. The events included the grand Christmas reception held at Government House. As she entered the carpeted room where the Curzons sat greeting their guests, Eliza had visions of Jacques-Louis David's coronation painting of Emperor Napoleon and his wife, Josephine. The tall, swanlike Mary Curzon, in her elegant gown, long ropes of pearls, and a diamond necklace and tiara, seemed made for her official role. People who came to admire Mary in her grand position remarked on her graciousness and tact. To Eliza, it was Mary's sweet gentleness that made the biggest impression. In her private notes of the evening, Eliza wrote that when Lady Curzon greeted her at the Christmas party, "a radiant smile of welcome made me almost lose my head."[60] She savored her other encounters with the couple that week, as when she found herself seated next to the viceroy himself at a luncheon party in the countryside. They discussed Oriental art and cultural sites of ancient ruins across Asia.

Eliza stayed on in India to the end of January. She fraternized with other foreigners, joining them for excursions to museums and gardens and to visit the poet and philosopher Rabindranath Tagore. But her notes suggest a period of despondency during the trip that dampened her usual enthusiasm for such things. After planning an excursion "upcountry," she canceled at the last minute. She spent several evenings alone in her dreary hotel room, repairing a yellow dress.

From the very first sentence of her later book on India, Eliza warned readers that travel in the country was not for the fainthearted. "It can hardly be said with literalness that one enjoys India," she wrote. Its hotels were the worst in the world, and "never have I suffered with cold as in India."[61] While the country's native people were picturesque in their colorful dress and interesting customs, she admitted to finding them almost repulsive at times in the crush of multitudes.

Still, as a professional travel writer, Eliza understood her obligation to provide a fair-minded view of the country, a place of tremendous diversity with 240 languages, eight forms of religious belief, as much land as all of Europe outside Russia, and a culture 3,000 years old, not to mention a population of 295 million.[62] Indeed, Eliza assured readers, the delights were many for travelers willing to endure the hardships and discomforts. Having done considerable reporting on her previous trip to India, she managed to produce an informative and pretty thorough guide to the country and its major places, including Benares, Lucknow, Agra, and Delhi; Lahore, Simla, Jaipur, and Bombay, as well as the Himalayas. But her long-robust enthusiasm for life on the road was waning. The 400-page *Winter India*, published in 1903, would be her last full-length book of travel.[63]

14

Eyes on Japan

For a decade Eliza had refused all offers to do any more newspaper work. Then, a request in November 1903 came at the right time. She was between projects, and the money was tempting, especially since she had recently returned from a trip to China that left her, as she noted in a letter, frustrated by "all the heartbreaking things in Peking that I could not buy."[1] The *Chicago Daily Tribune* wanted a series of travel articles that winter from across the Far East. After the paper persisted, Eliza finally cabled her consent. Later, mortified by the outcome, she would blame her "mercenary spirit" for a brush with "yellow journalism" that almost gave her professional reputation a black eye.[2]

Things started out fine. Eliza sent the *Tribune* a batch of pieces based on her recent trip to China that fall. The journey through the Yellow Sea had included a brief visit to Korea. Sixteen years earlier, on her first trip to Korea, she and her mother had been forced to ride the sixteen miles from the seaport of Chemulpo (Incheon) to Seoul by sedan chair. This time, thanks to a wave of railroad building, she arrived in Seoul by train—and "not a little square box Japanese coach, but a big American car."[3]

Farther west, the steamer also stopped in the port cities of Dalny (Dalian) and Port Arthur, at the tip of the Liaotung Peninsula. After leasing the region from China, Russia was now constructing a commercial port and naval base for its Pacific fleet. In Dalny, Eliza watched giant cranes hoist chests of Hangkow tea from freighters. Thousands of blue-clothed Chinese workers labored away at huge public-works projects, and the streets teemed with every kind of conveyance, from jinrikishas to Russian *droschkys*. Eight olive-green Russian ships—"evil, sinister looking things," Eliza reported—were anchored in the harbor.[4] Construction also thundered in neighboring Port Arthur, where Russia had spent several years building its defenses and enlarging the shipyards. Large fleets arrived regularly from Odessa loaded with men and munitions.[5]

A decade after the Sino–Japanese War, the region was once again a storm center. This time the faceoff was between Russia and Japan, competing for control of the area as a gateway to their expansionist aims in Korea and Manchuria in northeast China. After the earlier conflict, China had ceded the Liaotung Peninsula to Japan as part of the spoils of war. But several European nations, concerned about their own access to China, fought the move, forcing Japan to give up its prize. Then, in an act of double-dealing, China leased the area to Russia. Outraged, the Japanese vowed to fight back against any more aggression by the West. Now Japan was on the brink of war with mighty Russia.[6]

Like most of the world, Russia assumed that if war broke out, it would easily defeat Japan. While Eliza was steaming toward China that fall, talks between the two countries had gone nowhere, and the situation was tense. She felt the journalistic rush that came from being caught up in momentous events, but also the palpable anxiety aboard the ship as it passed through the region. "We had a passenger list of official and distinguished Japanese, which would have made the ship a rich prize of war . . . had hostilities begun while we were in Russian waters," she reported in the *Tribune*. Fortunately, telegrams from Tokyo brought reassurances: "Many rumors, no prospect, of war."[7]

Reaching China, Eliza had found the circus playing in Peking for the first time. The empress dowager, at nearly seventy, reportedly went every day. During the foreign occupation after the Boxer Rebellion, scholars and scientists had rushed in to study the city's architecture and treasures, and a number of cultural institutions had been opened to sightseeing.[8]

Eliza toured the restored Summer Palace, nine miles northwest of the city. The interior, she told readers, seemed to reflect the empress dowager's taste for foreign comforts. The furnishings included a big feather bed, a bentwood rocking chair, and upholstered armchairs with cushioned springs. Photographs lined a wall. Cixi had commissioned images of herself as part of a public-relations campaign to improve her reputation after being savaged in the Western press for her role in the Boxer violence.[9] When Eliza leaned in for a closer look, the "quiet little woman" in the portraits hardly suggested a monster. As Eliza told readers, "She sits there as simply, as calm, and with as much gentle repose of manner as . . . a Whistler's mother in Manchu dress."[10] The emperor, for his part, was said to have no interest in modern pursuits that amused the rest of the court—studying English, learning to dance, and practicing the use of a fork and knife. He had retreated into near isolation, maintaining every detail of dress and decorum but engaged only half-heartedly in the duties of state.

Back in Japan that winter, Eliza took the pulse of public opinion. All the Mikado's subjects seemed to hanker for war with Russia. Japanese soldiers lounged in public parks, eager for the summons to arms. The nation had organized a first-class hospital and medical service, and several thousand Red Cross nurses stood by for duty. People saw the latest harvest—a double crop of rice—as a propitious omen of impending victory for Japan. Meanwhile, war with Russia was not expected to interfere with Japan's plans to exhibit at the St. Louis world's fair, as most of the displays had already been sent.[11]

***

After New Year's in 1905, Eliza continued her reporting for the *Tribune* during a winter escape to Siam (Thailand) and Ceylon (Sri Lanka). Monsoons raging in the China Sea forced her below deck with a book. During a two-week stay in Bangkok, the dreaded news came: the war had started. After a breakdown of negotiations, Japan had sparked the hostilities on February 8 in a surprise attack that overwhelmed Russia's eastern fleet at Port Arthur. Two days later, imperial Japan formally declared war on tsarist Russia.

As news of the war spread to America and Europe, the *Tribune*'s readers followed Miss Scidmore's fast-paced travels across Southeast Asia—to Shanghai, Hong Kong, Canton, and Macau, and to Penang and Singapore on the Malay Peninsula. She crisscrossed the region for several months, sending a steady stream of articles to the newspaper. Most were "evergreen" pieces on cultural topics, a mainstay of her journalism and a concession to the long mail delays across the Pacific.[12] Sprinkled throughout the pieces were newsy reports of war developments in the region. In total, the *Tribune* published 240 articles by Miss Scidmore in eight months—about one a day.[13] It was astounding productivity, even for a woman who boasted at times of her "super-capable self."[14]

The record was a lie. The intrepid Miss Scidmore had not blazed her way through the extraordinary feat of reporting as readers had been led to believe. In March 1904, midway through the assignment, Eliza discovered back copies of the *Tribune* at a colonial library, and her heart sank. The paper had badly misrepresented her. She had agreed to provide three travel letters a week, but the *Tribune*'s editors had taken great liberties with her copy. Bent on running a piece every day, to compete with a series in a rival paper, they cut the letters into pieces, rearranged them, and altered the dates, mangling the text "to the ruin of all sense and connection," Eliza wrote in dismay to her friend Emily MacVeagh. "They spelled my Peking without a 'g' and used that unspeakably illiterate word 'Jap' in their headlines."[15] To the unknowing reader the pieces read fine. But Eliza viewed the deception as a travesty.

There was more. The *Tribune* had also lied about her credentials. Crowing about its coup in landing such an esteemed correspondent, the paper described Eliza as "perhaps, the only American woman who has achieved a great reputation as a war correspondent." Her reports on the Boxer Rebellion had made her "the equal of any of her men competitors," the paper told its readers.

What's more, "if war breaks out between Japan and Russia, Miss Scidmore will go to the front."[16]

It was preposterous, Eliza fumed in a letter to MacVeagh. She had not covered the Boxer Rebellion, and "I am not going to the 'firing line' as you may well know, only back to Japan for cherry blossoms," then to China again. The whole episode was appalling. When she wrote to the editors in Chicago to complain, her letters went unanswered. In the end, she resigned herself to completing the assignment—feeling duty-bound to honor the contract, and perhaps, in personal remorse, to do penance for her sin of getting involved with the newspaper in the first place. As she told MacVeagh: "Having sold my soul to the devil or the *Tribune*, which is the same thing, I have to see them through."[17]

***

Returning to Peking for a month in the summer of 1904, Eliza wasted little time going round to the curio shops on Hatamen Street. A highly unusual—and monumental—object there had caught her eye on a previous trip. Now, with money in her pocket, she was back to claim it. The remarkable piece was said to be an authentic throne chair of the empress dowager, looted from the Summer Palace after the Boxer Rebellion. Nearly six feet high and made of Shitan wood, the seat bore open-work carvings of dragons and bats, and characters that translated as "The Hill of a Thousand Ages," the Chinese name for the Summer Palace. Eliza's determination to have it offers clues to her mindset as a collector. The object was too unique and culturally important to pass up, and she knew it would create a stir back in Washington. Though its bulky heft was likely a deterrent for most travelers and collectors, she could buy the massive item with little thought because she enjoyed shipping privileges through her brother's consular position. Once the chair reached Washington, it went on display for some years at the National Museum, until Eliza sold it at auction with the rest of her Asian artifacts late in life.[18]

While in Peking, Eliza stayed with her old friends the Yu Kengs in their villa near the Summer Palace. The family had recently returned from Yu Keng's posting as Chinese minister in Paris, where his daughters learned French and studied dance with Isadora Duncan. Now the cosmopolitan young women were serving as ladies-in-waiting and translators to the empress dowager (Figure 14.1).

When she arrived, Eliza found the Yu household recovering from an uproar involving their earlier houseguest. An American painter named Katharine Carl had been staying with the family after receiving a commission to do a portrait of Cixi that would go on display at the 1904 world's fair in St. Louis. But in June, Cixi had a huge falling out with the painter over payment for the work. In a fit of rage, "the Dowager threw Miss Carl out and the paint box after her," Eliza told Robert Underwood Johnson at *The Century* in

Figure 14.1. Lady-in-waiting Der Ling, center, with Cixi, the empress dowager of China.

(Photo by Xunling, Freer Gallery of Art and Arthur M. Sackler Gallery Archives, Smithsonian Institution, FSA.A.13)

a long gossipy letter.[19] The Yu women had suffered Cixi's wrath as well, but were recalled back to the palace a few days later.

The Century Company planned to publish Miss Carl's account of her days at the imperial court.[20] But the *real* literary coup, Eliza explained to Johnson, would be the memoirs of her friend Lizzie—Der Ling—given her family's long connections to the imperial court. Eliza had stressed to Lizzie over the years the importance of keeping detailed notes on everything she saw and heard at the palace. "There will be a book!," Eliza insisted to Johnson, "if Miss Carl doesn't fritter away and spoil the subject first."[21]

Apart from her own reporting in the Far East, Eliza regularly pitched her editors story ideas from other authors in the region with interesting tales to tell. In the case of two Japanese men who made historic journeys of discovery to Tibet and the Buddhist Holy Land, Eliza spent many hours massaging their manuscripts for submission to *The Century*.[22] When Der Ling eventually wrote a memoir about her experiences inside the Chinese court, Eliza helped shepherd it to publication. The 1911 book, *Two Years in the Forbidden City*, got good reviews but left Eliza disappointed. As she told Emily MacVeagh after receiving another chapter of the manuscript, "she still does not say anything, nothing vital or startling or novel." The account was "along the same complimentary level" as Miss Carl's, with little real insight into the empress dowager.[23]

Passing along another story tip, Eliza urged Johnson to solicit an article from Kang Yu Wei, a reform-minded Chinese intellectual and former associate of the emperor who had fled China in the wake of the deadly coup that elevated the empress dowager to the throne. Eliza had seen Kang Yu Wei in Hong Kong the previous winter. "It would not be a bad idea," she told Johnson, to have an article from him about the emperor "ready for use when something happens in Peking."[24]

Her premonition of troubling developments would prove chillingly accurate. "If death gives the Dowager two minutes warning of her own end she will see that the Emperor goes first," Eliza wrote. "You can see," she added, "how Peking is easily the most interesting place out here."

***

The year 1905 opened with convulsive joy in Japan, where Eliza spent the winter. Flags blanketed Tokyo as residents turned out to cheer Japan's rousing victory over the Russians at Port Arthur. "I took seven kinds of death-a-cold seeing Port Arthur celebrations—so have seen nothing but the ceiling since," Eliza wrote to *The Century* from her sickbed. General Aleksey Kuropatkin's surrender amid well-stocked fortifications had made Russia a laughingstock to much of the world. Eliza agreed with those who insisted the Japanese would never have allowed such a humiliating retreat. On a colored postcard to the magazine's staff, Eliza scrawled the words "Banzai! Banzai!" above a cartoon of Japanese soldiers carting away baskets of troops like coal. "Just wait a while," she wrote, "and Kuropatkin's army also will be shovelled into coal baskets and carted off—<u>and</u> the Baltic fleet." Really, she asked, "was there ever such a collapse of a big windbag as the Russian war so far? China did not show up worse in '94" (Figure 14.2).[25]

Figure 14.2. Annotated postcard from Eliza Scidmore to her editors at *The Century*, January 2, 1905.

(Century Collection, New York Public Library)

Despite the full-blown war, the Scidmores' Oakley relatives—Mrs. Scidmore's sister Jane and her daughter, Mary—arrived from Wisconsin for a visit. While Eliza and Mary went off to Tokyo for a week of dinners and other social events at the American Legation, their mothers took jinrikisha rides around town. Along the Bund in Yokohama, they passed hundreds of Japanese soldiers who stood lined up waiting to be shipped off to Manchuria.

At home, the ladies sat around knitting tummy warmers for the Japanese men fighting in Manchuria's bitter cold. Eliza and her mother had joined the local chapter of the Foreign Ladies Volunteer Nurses Association. The group worked closely with Japan's Red Cross, which boasted nearly a million members and a corps of 3,000 female nurses. Some had been trained by American volunteers like Dr. Anita McGee of Washington, an acquaintance of Eliza whose husband served for a time as president of the National Geographic Society (Figure 14.3).[26]

The war also attracted scores of Western correspondents. Among them was the American journalist George Kennan—the only respectable man among the whole lot, in Eliza's view.[27]

Figure 14.3. Foreign Ladies Volunteer Nurses Association in Yokohama, including Mrs. Scidmore, seated second from right, and Eliza Scidmore probably second from left.

(Courtesy of Royal Ontario Museum, Toronto)

Now in Asia reporting for *The Outlook* magazine, Kennan turned up one Sunday afternoon at the Scidmores' home with his wife. Just back from the warfront, he regaled his hosts with tales of the scenes he had witnessed at the fall of Port Arthur.[28]

An expert on Russia, Kennan had published a series in *The Century* years earlier exposing the cruel prison conditions of exiled dissidents in Siberia. After his visit in Yokohama, Eliza pondered the idea of doing something similar. On February 9 she queried her editors: "The Century once had much to say about Russian prisons. Would it care for something about Japanese prisons with Russians in them?"[29] Japan was establishing camps around the country that would eventually house more than 70,000 Russian captives—in temples, hotels, schools, and town halls. While the press in Europe carried Russian complaints about the treatment of its soldiers, her own sources, Eliza told the magazine, indicated that the conditions were quite humane, even generous in many respects.

Eliza saw the prison sites in Japan as a chance to shed light on an issue of world importance that had not been much reported on in the press. In 1899, at the first International Peace Conference in the Hague, twenty-six nations had adopted protocols for humane treatment of war prisoners. Among the provisions, the accord stipulated that POWS should be housed, clothed, and fed as if they were troops of the custodial country.[30] Eliza wanted to investigate Japan's response to the agreement.

After *The Century* cabled its go-ahead, Eliza and another woman—an American teacher—applied for the paperwork needed to visit several POW sites. Baron Jutaro Komura, the Harvard-educated head of the foreign ministry and a friend of the Scidmores, signed off on the permits.[31] Letters of introduction to the camp supervisors stated that Miss Scidmore was the sister of the U.S. Embassy's counselor and was writing for a famous American magazine, "so she should be given benefits and special treatment."[32]

The two women went first to Matsuyama, a castle town on Shikoku, a green and mountainous island in the south of Japan. The prison site held 3,500 men at the time, more than half of them sick and wounded. Many had lost limbs or eyes. The camp head assigned a convalescent Russian officer to accompany Eliza and act as her interpreter. The prisoners were permitted to talk freely, Eliza reported later, and to move about town with their guards. When they did, Japanese residents offered cigarettes, sweets, drinks, and other gifts. The men also enjoyed the local hot springs, swam at the beach, and played ball in the camp's courtyard. Their daily rations included meat and bread—brown bread, after Japanese officials learned it was their preference.[33] The wives of some of the Russian officers had been permitted to join their husbands, and Eliza with met with four of the women. At the well-equipped hospital, she trailed Frances Parmelee, an American missionary volunteering with the Red Cross.[34]

Eliza's favorable impressions during her weeklong visit accorded with those of Dr. Anita McGee, who spent some time at the camp, and of another American, Sidney Gulick, an American missionary and teacher in Japan. "How the [Japanese] government could have been more considerate in its treatment of the prisoners, it is difficult to imagine," Gulick wrote in a 1905 book describing his six-month stay in Matsuyama.[35] Various accounts of the POW situation there attributed the sympathetic spirit of the local people to their understanding that Japan was fighting not the Russian people but the government of Russia.[36]

Eliza also visited three other POW sites in Japan. But the week in Matsuyama was the most stirring. "It was history hot from the griddle, still smoking," she said in a letter to *The Century*. One day, teatime was interrupted by the arrival of three Russian officers who had come straight from Mukden, in Manchuria, where the most decisive land battle in the war had just been fought. The words tumbled from their mouths as they described developments at the front. As they did so, their comrades sat listening in sad silence. “For all that they are getting what they deserve,” Eliza wrote, “one has to feel sorry for some of them.”[37]

***

Eliza sailed home to America in the summer of 1905. It took several months to settle back in Washington while a real estate agent worked to find her a new house and her freight remained in transit. When she sat down in her hotel room to write the account of her POW visits, Eliza struggled with the piece. It was a complex issue to explain, yet *The Century* did not want a long article. Finally, with the magazine’s consent, she abandoned the assignment. She may have been thinking already of another way she could use the material to better effect.

Like everyone in Washington, Eliza followed press reports that summer on peace talks to end the war between Russia and Japan. Seeking a major role for himself in world affairs, President Theodore Roosevelt had put himself forth as mediator. A former assistant secretary of the Navy, he offered the naval shipbuilding port of Portsmouth, New Hampshire, as a site for the treaty negotiations.

With his war secretary, William Howard Taft, on a mission to the Philippines that summer, Roosevelt dispatched Taft to meet with officials in Japan to garner support for the peace talks. Taft was traveling with eighty Americans, including the president’s own irrepressible daughter, twenty-one-year-old Alice. When the ship sailed into Yokohama in late July, the Japanese greeted the party with flag-waving demonstrations and fireworks. Days later, amid news that the peace conference in America was about to get underway, crowds cheered Taft when he stepped onto the balcony of his hotel to wave goodbye. As the Americans sailed for home, Taft harbored knowledge of a confidential agreement he had made with Japan’s prime minister, Taro Katsura, in which they agreed that for the sake of peace in East Asia, their countries would respect each other’s territorial interests in the Philippines and Korea. The understanding would complicate relations as Japan's expansionist aims in the region grew ever bolder in the years ahead.[38]

Back in the States, President Roosevelt monitored the August peace talks in New Hampshire from his summer home in Oyster Bay, New York. Though Russia and Japan were both exhausted from the war and ready to end it, the negotiations snagged over the terms of settlement. Among its demands, Japan wanted a large indemnity from Russia, as it had received after its victory over China. Russia adamantly refused. When the Russians seemed ready to walk out, Japan's delegates bent to pressure from Roosevelt and agreed to drop their demand for an indemnity in exchange for some territorial acquisitions.

The war officially ended on September 5, 1905, with the signing of the Portsmouth Peace Treaty. But the terms left many Japanese embittered. It seemed they had paid a terrible price for meager rewards. Public outrage over the lack of an indemnity grew so strong that riots broke out in Tokyo and martial law was declared.[39]

Roosevelt and many other Americans had been pro-Japanese during the war and could not help but admire the small and scrappy Asian nation's victory over Russia. But Japan's military strength also provoked anxiety in the West. Before another year had passed, newspapers on both sides of the Pacific would be warning of war between the United States and Japan.

PART 3

A DREAM REALIZED

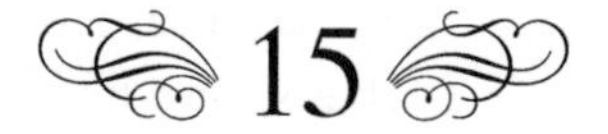

15

"The World and All That Is in It"

Upon her return to Washington in 1905, Eliza found the National Geographic Society enjoying a new lease on life. The January issue of *National Geographic* had rolled off the presses sporting a novel feature—one that would shape the magazine's signature format for the next century. Eleven pages of the magazine were devoted exclusively to black-and-white photographs, showing views of the remote Himalayan city of Lhasa, Tibet, taken by a pair of Russian explorers. Readers responded so enthusiastically the magazine repeated the innovative layout a few months later. Now, after years of sluggish growth, membership in the society was soaring as people sought to get their hands on the magazine. By the end of the year, enrollment in the National Geographic Society would triple to 11,500 members.[1]

Eliza followed the developments with keen interest. The magazine's increased focus on photography promised exciting possibilities for her own work. On a more personal note, it pleased her to see the society rebounding after her dear, late friend Gardiner Hubbard had struggled for years to attract interest beyond a loyal core of members who consisted mostly of government scientists and other Washington insiders. The society's public lectures drew as many as a thousand people, but enrollment in the organization had stagnated, and Hubbard saw the magazine as part of the problem. For too long, its contents had catered to the specialized interests of the existing members—serving up, as one editor later described it, "cold geographic facts, expressed in hieroglyphic terms which the layman could not understand."[2]

In his efforts to turn things around, Hubbard had consulted Miss Scidmore for advice. Perhaps, with her considerable journalistic experience, she had ideas on how to make *National Geographic* more appealing. Eliza addressed his concerns in letter during her trip to Peking in 1896. "I quite agree with you that it needs some shaking up or change," she wrote.[3] Though she had no immediate suggestions, she told him, she had been giving the matter some thought. To that end, she had tracked down, during her travels, the journals of other geographical societies—in Vienna, St. Petersburg, Paris, and London. By the time she returned to Washington, however, major changes to the magazine had already been made.

In her absence, John Hyde, an English-born statistician at the U.S. Department of Agriculture, had assumed the role of "honorary editor," a voluntary position. He and a small cadre of associates made revisions that included putting *National Geographic* on a regular schedule, after years of erratic publication. They also pledged greater attention to geographical subjects of national interest that reflected America's expanding reach into the world. The mix of editorial content would include articles, maps, diagrams, and occasional photos.[4]

The revamped *National Geographic* generated press attention when it debuted in January 1896 as "An Illustrated Monthly," selling for twenty-five cents a copy or two dollars a year. The tan-and-red cover listed "Eliza Ruhamah Scidmore" as one of the three honorary associate editors, along with society stalwarts A. W. Greely and W J McGee, though Eliza's extensive travel limited her direct involvement.[5] Despite Hyde's lofty intentions, the magazine remained a bare-bones operation. Its editorial "offices" consisted of a rented half-room "littered with old magazines, newspapers, and a few books of records."[6] With no money to pay contributors, Hyde continued to rely on members for content.

In her letter to Hubbard from Peking, Eliza commended Hyde's efforts and expressed the hope that the board would compensate him for his hard work. The changes he introduced were "a splendid beginning and a marked advance on what we had before," she wrote. Still, she thought more could be done. As she told Hubbard: "We need to have it make another leap and become a full-fledged, serious standard magazine of geographic literature."

Clearly, *National Geographic* was coming along. But any discussion Eliza might have had with Hubbard about further improvements soon became moot. Not long after their correspondence, he died at the age of seventy-five.

***

Eliza greatly felt the loss, as she had quite enjoyed the old gentleman's company, both professionally and personally. From the time she first served as a secretary of the society, she had worked closely with Hubbard on many issues important to the organization. He had included her regularly in delegations he organized to represent the National Geographic Society at international conferences.

His death left the society adrift: leaderless, confused about its direction, and nearly $2,000 in debt.[7] Eager to keep her husband's legacy alive, Hubbard's widow, Gertrude, pleaded with her son-in-law Alexander Graham Bell to take over as president. But Bell did not want the job. His lively inventor's mind swirled with ideas and permutations, centered most recently on an obsession he shared with his good friend Samuel Pierpont Langley, secretary of the Smithsonian Institution, to develop a flying machine that could carry people aloft. The Bells' summer estate near Baddeck, Nova Scotia, on Cape Breton Island, offered an ideal setting to test their ideas using tetrahedral-shaped box kites.[8]

After much cajoling, Bell finally succumbed to his mother-in-law's request to take over the National Geographic Society. "Only in order to save it," he wrote in his diary.[9] In January 1898 he became the organization's second president.

Eliza had known the Hubbards and Bells since her days as a young reporter covering the social scene in Washington.[10] The relationship blossomed into a warm friendship during her involvement in the society. Her personal fondness for the family was evident in another letter to Hubbard. Writing to him during her trip along the Upper Yangtze in the fall of 1896, she closed on a wistful note that suggests an outsider's yearning for the intimate joys of family life that the Hubbard-Bell clan epitomized. "I trust you are all well," she wrote, "and as enviably happy as you and your family colony by Dupont Circle always seem to be."[11]

After Hubbard's death, Eliza stayed in touch with the family. Her knowledge was helpful when, in the fall of 1898, Bell packed up his wife, Mabel, and their daughters, Elsie and Marian (known as "Daisy"), for their first trip to Japan. The emperor had requested a meeting with the great inventor, and everywhere the tall, white-bearded figure went, people treated him reverently. Impatient with social formalities, he chafed at the incessant banquets and found it agonizing to bend his large, bulky body into a sitting position on the floor for hours at a time. It also made him uncomfortable to be pulled up and down hills in a jinrikisha powered only by a small Japanese man. The Bell women, on the other hand, enjoyed themselves thoroughly. Elsie and Daisy began calling their father "Daddy-san": honorable father.[12] One afternoon while his wife and daughters went off on their own, Bell accepted an invitation from Eliza's mother to join her for "tiffin" (lunch) at the Scidmores' home in Yokohama.[13]

Childless herself, Eliza had watched the Bell girls grow into lively and attractive young women. Late in the summer of 1899, she wrote to Mabel Bell before heading off to Berlin for the Oriental Congress. She was going as a representative of the National Geographic, and after learning that Bell himself would not attend, Eliza hoped either of his daughters might accompany her. "Pray let one of them go," she urged their mother.[14] In her eyes, the Bell family were like American royalty; they ought to be represented. Besides, she pointed out, the experience would be educational. International leaders including the king of Italy and Emperor Wilhelm of Germany were expected to attend. But at nineteen and twenty-one, both of the Bells' daughters were preoccupied with far more engaging matters. Elsie was heading into a flirtatious courtship that would lead to her marriage, while the younger and more bohemian Daisy was developing a serious interest in art.

Now and then during her husband's frequent absences, Mabel Bell invited Eliza to dine at the family's home at 1331 Connecticut Avenue NW, just below Dupont Circle. Because her deafness made it difficult to follow conversations in large gatherings, Mabel shrank from the whirl of Washington society and preferred casual entertaining at home. She appreciated Eliza's dedication to the organization her father had held so dear. In one letter to her own perpetually disorganized husband "Alec," Mabel chided him about his procrastination as a lecture deadline loomed. "I think you ought to write somebody about your slides," she told him. "Shall I ask Miss Scidmore?"[15]

In later years, as their social circles and charitable interests overlapped, Eliza would see a great deal of Mrs. Bell and her by-then-married daughter Elsie. Daisy, meanwhile, came to share Eliza's interest in Oriental art. In 1905, they both advocated separately for the Smithsonian's acquisition of the wealthy businessman Charles Freer's vast collection, which included thousands of Asian artifacts and works by American artists such as James McNeill Whistler. That winter, the Smithsonian dispatched several regents, including Alexander Graham Bell, to Detroit to meet with Freer, who was seeking a permanent home for his collection. Bell and his colleagues were men of science. They knew little about art and failed to appreciate the collection's importance. Fortunately, Daisy had accompanied her father, and was able to convince the men that the Smithsonian ought to accept Freer's offer.[16]

By the end of the year, however, the Smithsonian had yet to follow up. Its sluggish response alarmed the art community in Washington, while President Teddy Roosevelt himself expressed his outrage in a letter to the Smithsonian. Eliza delivered her own punch in an open letter to the editor, published December 27 in the *Evening Star.* "It has seemed incredible," she wrote, "that the Smithsonian regents had not promptly pursued Mr. Freer and secured a definitely signed deed or compact at the first suggestion or promise of such a gift months ago." No other collection in Europe or America rivaled it, she pointed out. The next day, the *New York Times* picked up on her arguments about the Freer collection, noting in a headline that "Miss Scidmore Thinks the Smithsonian Should Have It." Two decades later when the Smithsonian unveiled the treasures in a specially built gallery on the National Mall, the collection would be valued at $23 million.[17]

***

Still writing for *The Century* and other publications, Eliza took a more active interest in *National Geographic* as it became more appealing under changes introduced by Alexander Graham Bell. In his eyes, the magazine needed to be more entertaining and journalistic, to better compete with leading periodicals of the day. A man who read encyclopedia articles for relaxation, he wanted *National Geographic* to make readers curious about the world and its wonders—to teach people things they didn't know, in language they could understand. As he explained to the young man he hired in 1899 to help run things: "'The world and all that is in it' is our theme, and if we can't find anything to interest ordinary people in that subject we better shut up shop and become a strict, technical, scientific journal for high-class geographers and geographical experts."[18]

His new assistant, a twenty-three-year-old prep-school teacher and Amherst College graduate named Gilbert Hovey Grosvenor, had grown up in Constantinople as the son of an American professor. Bell paid the young man a hundred dollars a month, made him an associate editor, and told him to think up ways to improve the magazine and grow the society's membership. Among his innovations, Grosvenor introduced a process by which existing members would "nominate" friends and relatives for membership in the society, to make the affiliation seem selective. The ploy worked. In a single year enrollment doubled to almost 2,500 members.[19]

Transforming the magazine proved more challenging. John Hyde, now heading an advisory board of twelve editors, resented the brashly confident young man who had Bell's ear. Because Bell greatly admired *The Century* and the fast-growing *McClure's* magazine, the publication committee sought the advice of the publisher S. S. McClure himself. His solution was radical: move the magazine to New York, switch to newsstand sales, and change the name, "because people hate geography," he said.[20]

Grosvenor was in England on his honeymoon—he married Bell's daughter Elsie in 1900—when he got wind of the plans. With his bride, he rushed home in time to block the move and keep *National Geographic* in Washington, its fate tied to that of the society. Bell backed his new son-in-law by naming him managing editor. Upon his retirement as president of the society in 1903, Bell made Grosvenor, at twenty-seven, his successor (Figure 15.1).

Figure 15.1. Editor Gilbert H. Grosvenor in his office at *National Geographic*.

(Courtesy of National Geographic Society)

At his father-in-law's urging, Grosvenor began publishing more photographs in *National Geographic*. Bell believed that the right kind of photos—"dynamical pictures" showing life and action; "pictures that tell a story"—would make people want to read the articles.[21] Grosvenor scrambled to acquire images from every possible source: photographic plates from government agencies, stock pictures from agencies, photos taken by scientists, travelers, and explorers. *National Geographic* had increased its use of photography during John Hyde's tenure as editor, going so far as to run its first photograph of a bare-breasted woman, showing a Zulu bride in South Africa.[22] But it was Gilbert Grosvenor who made the breakthrough that would define the magazine for the next century.[23]

The dramatic shift that began with the January 1905 issue occurred when Grosvenor got a call from the printer saying eleven pages had come up blank. With no manuscript at hand to fill the hole, Grosvenor reached for a packet of black-and-white photos that had recently come across his desk: fifty images of Lhasa, Tibet. On impulse, he decided to run eleven of them, each one a full page. The move was risky, as photographic plates were costly to make. Besides that, the magazine had never before printed standalone photos. Grosvenor would say later that he all but expected to be fired. Instead, the issue turned out to be wildly popular.[24] Encouraged by the response, he ran thirty-two pages of photos three months later in the April issue, scenes from the Philippines provided by his cousin William Howard Taft, then the secretary of war. That issue too was a big hit.

Suddenly, the magazine had a winning formula—though the new format had its detractors. Two board members grew so irate over a photo series on wild animals, shot using a remote-control flashlight camera, that they resigned in protest. "Wandering off into nature is not geography," one huffed. Grosvenor, they complained, was "turning the magazine into a picture book."[25] Despite the criticism, there was no going back. By 1908, photographs would fill half the pages of *National Geographic*.[26]

Like others at National Geographic, Eliza had to adjust to Grosvenor's assertive style. He was an editorial greenhorn whose only publishing experience had been helping his father compile textbooks; she was a seasoned journalist, and twenty years his senior. But Eliza had to admit that the young man seemed to have smart instincts. She herself welcomed the new emphasis on photography. Her job as a professional travel writer was to tell vivid, compelling stories that brought places alive for readers. Photos enhanced that experience by giving people a more direct way of seeing.

Eliza herself had been traveling with a camera for years, only to see her images converted into drawings and engravings by clients such as *Harper's* and *The Century*. But the industry was changing. Thanks to technological advances, *National Geographic* and other periodicals were finding it easier and cheaper to print "halftone" photographs.[27] While *National Geographic* embraced the coming of photography, Eliza's long-time client *The Century* remained a holdout; its chief editor, Richard Watson Gilder, considered such illustrations "vulgar" and stuck with engravings.[28] Later, when *National Geographic* started printing in color, Eliza lamented *The Century*'s failure to innovate. As she told Grosvenor: "The colored proofs you send are wonders—whatever can The Century and those other people be thinking of to let you go on doing such things under their noses? They surely have not learned yet what 'takes.'"[29]

Pressed for good material, Grosvenor looked to Eliza for sources of photographs. "Major Powell-Cotton is your man for African pictures—of pygmies, elephants, lions, hippos, etc.," she wrote from London after attending a lecture of the Royal Geographical Society. "He shot big game, was clawed in the back by lion, etc., and gave . . . a most interesting address last night."[30]

It had been more than five years since Eliza herself published a major article in *National Geographic*. Now Grosvenor commissioned her to write a series of six articles on "wonders of the East"; each would be "profusely illustrated," the magazine told readers.[31] In a ten-month period from December 1906 to October 1907, the magazine ran five stories under Eliza's byline—the first of eight articles in total that she would publish in the magazine during Gilbert H. Grosvenor's tenure. "The Greatest Hunt in the World" (December 1906) described the King of Siam's weeklong drive to round up wild elephants and drive them to the royal family's kraal. Other subjects included the massive gathering of Hindus on the Ganges River in Benares (February 1907), and Eliza's account of her visit to the sacred city of Koyasan—"the Japanese Valhalla"—and its cryptomeria-draped graveyard atop Mount Koya (December 1906).

The articles appearing under Eliza's name included some of her own photographs, and her legacy as the *National Geographic*'s first female photographer is not in doubt.[32] The loose attribution practices of a century ago, however, make it difficult in many cases to determine with certainty which images Eliza herself took and which came from other sources. The elephant hunt story, for example, carried seventeen full-page black-and-white photographs of a quality that suggests a fairly high level of technical skill—yet there are no photo credits.[33] An April 1907 photo essay on "Women and Children of the East" featured two dozen photographs "made and collected by Eliza R. Scidmore," the text noted, while an article on Ceylon that appeared under Eliza's byline in February 1912 included photographs identified as being "from Eliza R. Scidmore." Eliza's correspondence with Grosvenor makes it clear that she regularly submitted some of her own photos, but also acquired negatives and prints from other sources and even hired photographers on occasion to accompany her.[34]

An example of Eliza's own photographic work in the field comes from a March 1907 article on her trip to Ceylon to report on archeological explorations at the ancient fortress of Sigiriya, or "Lion Rock." In it, she related an experience she had photographing female tea pickers who were among the thousands of Tamils brought from India to work the plantations. As Eliza and her party were climbing a steep hill lined with terraces of tea bushes, a group of Tamil women in red and white head cloths caught her eye. Eliza stopped to take photographs, only to create a local stir when one of the women reacted with alarm to the stranger's intrusion. In the kind of encounter that was typical of Western travelers in the tropics, Eliza used terms widely regarded today as condescending and culturally insensitive in describing how the young woman—a "black beauty, with rings in her nose and a mite of a black baby astride her hip"—pulled her veil over her face and started wailing. Eliza tried to appease the young mother by offering her a coin. The money failed to do the trick, Eliza told readers, though other women in the fields who had watched the incident rushed over begging to be "kodaked."[35]

***

Given their strong personalities, Eliza and the young Mr. Grosvenor were bound to clash at times. One day in the fall of 1907, Eliza was at home in Washington when she opened an issue of *National Geographic* and came upon one of her photographs. Annoyed, she shot off a letter to Grosvenor. "I need hardly tell you that I am greatly surprised at the use that has been made of my material of late," she told him. Months earlier, she reminded him, she had temporarily withdrawn the related article. Now he had published one of the photos she submitted with the manuscript. "I made my position quite clear I am sure," she wrote. "I must ask that you will not make any further use of my material until this matter can be frankly discussed and disposed of in a way satisfactory to us both."[36]

Grosvenor was taken aback. "My Dear Miss Scidmore," he replied a few days later. "I have been hoping to see you, as I do not understand your note of November 25. I assure you I want to use your articles and photos in the manner most agreeable to you." He had used the photo as an "appetizer" for readers, he explained, "as a hint of the good things to come in the magazine.[37] Eager to make amends, he invited her to attend the upcoming annual dinner as a guest of the society. Eliza thanked him for the invitation, but declined, in a note of high-handed self-righteousness.

"Being a member, I have no right to such a benefit, and shall pay for my ticket as I always have done and as other members do," she wrote.[38] She enclosed a check for five dollars.

In time, as they put their heads together planning stories, Eliza and Grosvenor developed a warm friendship. He sponsored her for membership in the exclusive new Chevy Chase Club, just north of Washington, and put up the money for the $200 initiation fee until she returned to town.[39] By then, Eliza was spending about half her time abroad. "I will keep my eye out for the N.G.S.," she told Grosvenor in a letter from Nagasaki, on the eve of a move to Seoul with her mother and brother.[40] He assured her how much he valued her judgment and initiative. "We miss you very much this winter," he wrote. "You were always so full of ideas and suggestions and you really ought to live here permanently."[41]

But it was her extensive travels, in fact, that made Eliza such an asset to Grosvenor in his mission to reshape *National Geographic*. Under its successful new format, the magazine had a voracious appetite for photographs, and Grosvenor counted heavily on Eliza for picturesque material from the Far East. Planning her next trip to China in the fall of 1909, Eliza wrote to Grosvenor with a couple of ideas. "I can get you some <u>very</u> stunning pictures of Chinese architecture," she promised.[42] She had in mind a series on Chinese temples, another on stone bridges. Given the nature of the material, she noted, why print in black and white? Color printing "is coming to stay," she argued, "and you might as well take a first flyer on some of those red and yellow temples among green trees, <u>with</u> snow on the ground."[43]

Though color photography was still too new for wide commercial use, color-tinted black-and-white photos were easy to obtain.[44] For years, Eliza had often had lantern slides colored for her lectures. And in fact, Japan had some of the best colorists in the world, a vestige of the "souvenir photography" industry that arose with the growth of globetrotting, when travelers compiled elaborate albums of photographs from various ports of call to document their journeys.[45] An entire street in Yokohama specialized in the trade of photographic tinting. Eliza herself had a favorite Japanese technician to which she sent images to be colored.[46]

So it was, in her letter to Grosvenor, that Eliza proposed hand-tinting of the photos she hoped to acquire in China. Grosvenor replied by expressing his own enthusiasm about color printing, especially as the methods were getting better all the time. Still, the technology was tricky and expensive, and the results unreliable. "I have been for a long time anxious to have some colored work, but the proper subjects have not yet materialized," he explained. Her idea for the China photos sounded good, though he was not yet ready to commit to something definite. At any rate, he added, he would "not dare to attempt to reproduce them in colors unless you were here to direct the engraver."[47]

A year later, in 1910, a development occurred that made Grosvenor change his mind. He received a batch of photographs so exquisitely hand-colored that printing them in black and white would not have done them justice. The images, from China and Korea, had been taken by William Wisner Chapin, a wealthy amateur photographer from Rochester, New York, during a round-the-world tour. Grosvenor ran twenty-four color pages of Chapin's photographs in the November issue of *National Geographic*, despite the considerable cost.[48]

And so the leap to color was made to spectacular effect. Now Grosvenor was hungry for more. Discussing story ideas with Eliza in the spring of 1912 as she headed off once again to the Far East, Grosvenor suggested they repeat the theme of her earlier piece on "Women and Children of the East," using different photos. This time, he hoped to publish the images in color.[49]

Arriving in Asia that summer, Eliza hustled to acquire photographs for Grosvenor. In September, she wrote him from Kyoto: "Herewith 31 pictures of 'Women and Children,' mostly children, as you see. I have had them made uniform in size and strongly colored, so that you can cover yourself all over with glory with another number in color and thereby catch a few thousand more subscribers."[50] The same day he received them, Grosvenor sent Eliza a letter glowing with excitement. "I am delighted with the pictures, which are exquisite in coloring and interesting in character." It was a pleasure, he added, to send her a check for $200, on top of the $100 advance he had already paid. "I hope before you return to America you will secure many Chinese and Korean colored photographs for us and also some more Japanese."[51]

A few weeks later, Eliza mailed a much thicker packet from Seoul, where her brother George was now serving as the U.S. consul general. The envelope contained ninety-six more photographs, all of Japanese subjects. "It took search and some expensive and heartbreaking experiences before I found this photographer," she told Grosvenor.[52] She organized the images by themes, such as rice, tea, and straw braid, wrote titles on the back of each, and numbered them in order. She couldn't help crowing about her diligence. She had been "rather aghast" at agreeing to send a large selection of photos by January 1, Eliza wrote, "but fortunately I have been able to more than do it."[53] She promised to scour Seoul for photographs on Korea, but cautioned Grosvenor not to expect anything comparable in quantity and quality to the ones from Japan (Figure 15.2).

Meanwhile, there was the matter of money. The effort she had spent acquiring the two batches of images from Japan—a total of 127—certainly exceeded the $300 she had been paid, Eliza told Grosvenor pointedly. He agreed. As he well realized, her contributions were invaluable in his efforts to make *National Geographic* a trendsetter in pictorial journalism.

In a letter of reply, he sent an additional check for $150. Altogether, the payment came to $450. "This is very much more than we have ever paid for photographs before, but they are worth it to us, and I hope you will find the amount satisfactory," Grosvenor told Eliza. "They are wonderful," he assured her, "and exactly what I wanted."[54]

Figure 15.2. Japanese children drawing on cast-off sliding paper screens, from National Geographic Society's Eliza Scidmore Collection.

(Hand-colorized image; courtesy of National Geographical Society)

16

Field of Cherries

Around noon on March 27, 1908, Eliza donned her hat, stepped out the door of her house at 1837 M Street NW, and set off down the street in her pleasant Dupont Circle neighborhood, just south of the mansion-lined Massachusetts Avenue. Up and down the sidewalks, dark tree limbs throbbed with bright green shoots. In a few weeks, Washington's reputation as a "city of trees" would be on full display as leafy canopies cast their shade.[1] Today it felt as though summer had already arrived. An unseasonable heat wave had descended on the city, pushing the temperature to 82 degrees Fahrenheit by eleven o'clock.[2] Men and women passing by tugged at their high collars and patted their brows with handkerchiefs. Eliza resigned herself to what was clearly going to be an uncomfortably warm afternoon.

At fifty-one, she no longer moved with the lightness of the young reporter who had walked these streets decades earlier. After thirty years of tramping about the world, her feet hurt. On some mornings, her eyes betrayed strains of the insomnia that would grow more chronic with time.[3] Still, advancing age had not chiseled away the doughty sense of purpose that seemed baked into her character. So it was as she headed across town on this Friday with a look that suggested a woman who was not just out for a springtime stroll.

Her destination was Franklin Park, at Thirteenth and K Streets NW. It was Arbor Day, an event that had started a generation earlier in Nebraska when settlers planted a million trees across the territory in a single day.[4] Now, the tree-planting holiday was common in American towns and cities. The conservation-minded President Theodore Roosevelt heartily endorsed Arbor Day as an opportunity to teach schoolchildren the benefits of trees and forests. As he declared in a proclamation: "A people without children would face a hopeless future; a country without trees is almost as hopeless."[5]

David Fairchild, a botanist at the U.S. Department of Agriculture, had helped arrange tree-planting ceremonies that day at all the city's schools (Figure 16.1). Because Franklin School lacked a proper schoolyard, Col. Charles Bromwell, whose office managed the city's public parks under the U.S. Army Corps of Engineers, had granted special permission for the students there to plant a tree in the park across the street.[6] Fairchild would be giving a speech that afternoon at the school to cap the day's events, and Eliza planned to attend.

Franklin Park, which filled an entire block in the heart of Washington, was an urban oasis of walking paths, mature trees, shrubs, and flower beds organized around a central fountain.

Figure 16.1. Botanist David Fairchild in his office at the U.S. Department of Agriculture.

(USDA National Agricultural Library)

Many prominent residents lived in the surrounding townhouses and apartment buildings. Franklin School itself was a source of civic pride. Built as a model public school after the Civil War, the red-brick, German Renaissance building had spacious light-filled classrooms and other features so innovative for the time that the design won an architectural prize at the Centennial Exposition.[7] The school's Great Hall, which could seat a thousand people, drew large crowds for concerts and lectures. During Eliza's early reporting years in Washington, Alexander Graham Bell made history at Franklin School by testing one of his early inventions there. In 1880 Bell used his "photophone" to transmit, on a beam of light, a voice message from the roof of the school to the window of his laboratory across the park, in a precursor of future wireless communications.[8]

On this day in 1908, David Fairchild, Bell's son-in-law, had planned a very different kind of demonstration at the school. Eliza came out to hear what he had to say because he would be talking on a subject dear to both of them. The saplings the children were planting that day were flowering cherry trees, donated by Fairchild and his wife. He had ordered them all the way from Japan.

***

Eliza knew Fairchild through the National Geographic Society and her long friendship with the Hubbards and Bells. An earnest young man in his late thirties, with a thick brush mustache and owlish spectacles, Fairchild had joined the family three years earlier by marrying Bell's younger daughter, Marian. The match made Fairchild not just Bell's son-in-law but also the brother-in-law of Eliza's editor at *National Geographic*, Gilbert H. Grosvenor, who was married to Bell's older daughter, Elsie May.

A native of Michigan who hailed from a line of distinguished educators, Fairchild had developed his love of trees and plants as a boy exploring local woods and streams.[9] At age ten he moved with his family to Manhattan, Kansas, where his father became president of an agricultural college. It was there, in adolescence, that one of the most formative experiences of young Fairchild's life occurred when the family hosted a world-renowned guest. During a lecture tour in America, the British naturalist Alfred Russel Wallace stayed at the Fairchilds' home. David sat enrapt listening to their charming, white-haired visitor's tales of his years of travel in the tropics—first in the Amazon, later in the Malay Archipelago—gathering evidence to support the theory of evolution through natural selection that he and Charles Darwin were exploring independently. Wallace's tales sounded magical. For years to come, Fairchild would harbor fantasies of travel and exploration that shaped his own future.[10]

Mentored by an uncle who taught botany, Fairchild gravitated to the new science of plant diseases. After college he moved to Washington to work in the U.S. Department of Agriculture's new division of plant pathology. Still keen to travel and see the world, he arranged research fellowships in Italy and Germany. But it was the tropics, with a plant kingdom vaster and more diverse than anywhere else in the world, that most excited his imagination.

Luck arrived in the form of a benefactor. During an Atlantic crossing he met Barbour Lathrop, a wealthy bachelor and gadfly traveler who spent his days cruising the globe from one port to the next. Lathrop saw promise in the young botanist and agreed to bankroll a period of research for him at the botanical garden on Java. Fairchild went in 1895, not long after Eliza's own travels on the island. In a career that later took him many times around the world, Fairchild would say that his first experience of Java "held more interest than any voyage I have ever taken."[11] His enthusiasm for the region led later to a deep appreciation of Eliza's *Java, the Garden of the East* when it came out in 1897.

Fairchild capped his time on Java by joining Lathrop on a long steamship cruise in the region. They visited Bangkok and Singapore, Australia and New Zealand, Fiji and Hawaii. At every place they went ashore, Fairchild rushed off to the local markets to investigate the local produce, which included many foods and other plants completely unknown back home. In a series of follow-up voyages with Lathrop, Fairchild came to imagine a novel career for himself as a "plant explorer" who could roam the world in search of plants that might be adopted in America.[12]

Fortunately, his former boss in Washington saw merit in the idea. Agriculture Secretary James Wilson believed the country needed to expand its crop diversity; that meant finding new species and varieties beyond the nation's borders.[13] Wilson established an office of Seed and Plant Introduction at the USDA and tapped Fairchild to head it. The seeds and cuttings Fairchild and his fellow "plant explorers" went on to collect over several decades gave Americans many thousands of new foods and crops, such as nectarines, dates, avocados, soybeans, mangos, and pistachios.[14]

As Fairchild and Eliza got to know one another in Washington, he grew to appreciate her own considerable knowledge of flowers and horticulture. They talked about interesting plants she came across in her world travels. He found her judgment "so keen and valuable" that he eventually arranged her appointment to the list of official "collaborators" who aided the bureau's plant-collection efforts. She was sure to be an asset in providing leads, he told his supervisor, as he could think of no one who was "so widely known or knows so many people in China and Japan as Miss Scidmore."[15]

***

Eliza had already been familiar with Japanese cherry trees for quite some time when Fairchild started studying them soon after the turn of the century. Though he visited Japan a number of times, he apparently never actually saw the trees blooming in their native habitat.[16] As a botanical connoisseur, however, he recognized the exceptional beauty of their blossoms just from pictures.

On one of his odysseys with Lathrop, they landed in Japan in the fall of 1902. As Fairchild recalled years later, he noticed that all the photographs he took had a "picturesque touch." Especially striking to him was the character of the trees he saw lining the streets, shading houses, and filling the parks. They were, he learned, the country's famed cherry-blossom trees. And although they were not in bloom at the time, they conveyed nonetheless a spirit that Fairchild decided he wanted to better understand.[17]

He was sent off to see a man, Magoemon Takagi, who was hailed locally as one of the country's leading authorities on flowering cherries. An expert gardener, Takagi had devoted his life to saving cherry tree varieties threatened by the upheaval of Japan's modernization; three generations of his family had preserved at least seventy-eight different varieties. It was from trees the Takagi family planted along the banks of the Arakawa River, near Tokyo, that Japan would later acquire grafts for the cherry saplings sent to Washington in 1912.[18]

Fairchild visited Mr. Takagi at his home. As they sat beside one another on tatami mats, the tiny Japanese man slowly unveiled a series of finely detailed watercolor drawings, which left Fairchild stunned. Here were exquisite examples of what he had only read about in botanical volumes and general books about Japan. "I have rarely been so thrilled," Fairchild wrote, "for I had no idea of the wealth of beauty, form and color of the flowering cherries."[19] With Takagi's help, he chose thirty of the best varieties. He and Lathrop had samples of the trees shipped to a USDA plant-introduction garden in Chico, California, but it was not an ideal place, and most of the saplings died in the hot sun.

As he aimed to learn more about Japanese cherry trees, Fairchild was aware that they already existed in some places around the United States. But there were no mass plantings, and only a few varieties were known. Most cherry trees in the country were individual specimens that had been brought in by naval officers, scientists, and others who developed a love for them while living in the Far East.

In the spring of 1904, Eliza attended a cherry tree-viewing party with Fairchild and others to celebrate the blooming of a particularly beautiful cherry tree in Washington. It grew on the grounds of a home on Massachusetts Avenue occupied by Charles Marlatt, a friend and former university classmate of Fairchild in Kansas. Now colleagues at the USDA, the men organized the gathering as a "strictly Japanese affair," with a traditional tea ceremony. Fairchild's later account of the event shows that he considered it a great honor to have Miss Scidmore there that day, as an expert on Japan and the author of what he called "the greatest guidebook to Japan ever written."[20] Among the other guests was Marian Bell, who would soon become Fairchild's wife.

Fairchild took up his study of Japanese cherry trees in earnest soon after his marriage to Marian in the spring of 1905. One day the newlyweds were out driving—in a motorcar Marian's grandmother had given them as a wedding gift—when they came upon a tract of land for sale. The forty-acre property, ten miles north of Washington, in the new suburb of Chevy Chase, Maryland, was heavily wooded, with Rock Creek flowing through it. Fairchild saw the spot as ideal for a country home and garden that would allow him to study foreign plants for possible cultivation in the United States.[21] Flowering cherry trees topped his list.

Soon after the couple acquired the property, which they named "In the Woods," Fairchild wrote to Uhei Suzuki, president of the Yokohama Nursery Company, placing an order for 125 flowering cherries of twenty-five varieties. When the trees arrived in the spring of 1906, Fairchild planted them with the help of a young Japanese gardener named Mori, who came to work for him. They cleared space among a patch of pines and cedars to create a *sakura-no*—a "field of cherries."[22]

The following spring, when it became clear the cherry trees would do well in the local climate, the Fairchilds started urging their friends and neighbors to plant some (Figure 16.2). Under Fairchild's guidance, the land-development company for Chevy Chase acquired 300 of the trees for local streets. Arbor Day offered a perfect opportunity to spread the word more widely. So in the winter of 1907–8, Fairchild worked with a local nature-studies teacher, Susan Sipe, to arrange the tree-planting ceremonies at

Figure 16.2. Marian Fairchild among cherry blossoms at "In the Woods," c. 1910.
(Archives of Fairchild Tropical Botanical Garden)

D.C. schools the following spring. He ordered 150 cherry saplings from Japan, requesting rush delivery so they would arrive in time.

On the morning of March 26, 1908, the day before Arbor Day, Miss Sipe and several others escorted eighty-three boys out to Chevy Chase by streetcar. They all tramped single file through the woods, where hepaticas, blood-roots, and dog-tooth violets were coming up under the dead leaves on the ground. At a clearing, the boys collected the young trees from Fairchild and Mori and got instructions on how to plant and care for them.

That afternoon, Eliza also traveled out to "In the Woods" to visit the Fairchilds and see their cherry trees.[23] They discussed their mutual interest in seeing Japanese cherry trees planted in Washington. At some point during the visit, the *sakura-no* that David Fairchild and Mori had planted led to the idea that something similar—a "field of cherries"—might be created along the new paved roadway that flowed past the Tidal Basin in Potomac

Park, near the Washington Monument.[24] The "Speedway," people called it because it was reserved from 4 to 6 p.m. on Saturdays for "speeding purposes" to accommodate the growing number of automobiles in the capital.[25] Before Eliza left to return to the city, Fairchild told her about the illustrated lecture he planned to give at Franklin School and invited her to attend.

When the program got underway the next day at 1:45, Fairchild mounted the stage before an audience of school board members, city officials, and members of the public. Eliza had arrived and taken her place—a plain-looking, middle-aged woman who hardly stood out among those who did not recognize her as a celebrated author and well-respected figure around town. In his lecture, Fairchild showed slides of Japan and sang the praises of the country's picturesque cherry trees that graced the country's waterways and roadsides. Since he had found the trees to be well suited to conditions in the Washington area, he explained, he urged the city to take them up in local parks and gardens.[26]

As the *Evening Star* noted in its coverage, the young cherries Mr. Fairchild and his wife had given to D.C.'s schools were a start. But in fact, the distinguished botanist had come that day with a bigger idea to propose. In his wind-up, Mr. Fairchild projected some images of the new Speedway (Figure 16.3). The area looked rather desolate, as it was still little used. He proposed that the city transform the roadway into a "field of cherries." In doing so, he told his audience, the avenue would one day be so spectacular when the trees bloomed that people would come to Washington just to see it.[27] He nodded toward Miss Scidmore to acknowledge the approval of his distinguished guest, a noted authority on Japan.[28]

The program that afternoon was historic. Fairchild's presentation at Franklin School brought the first public airing of the idea Eliza had been pushing behind the scenes for two decades, though in slightly different form. Nothing in the record indicates how she felt at being consigned largely to a bystander's role that day. Over the years, both she and David Fairchild would refer only obliquely

Figure 16.3. Cars and carriages on the paved "Speedway" in Potomac Park.
(Library of Congress)

to the other's role in the events that transpired—an indication, perhaps, of a relationship that grew strained in some respects.

Eliza and the Fairchilds had come to their love of Japanese cherries through different avenues. They also harbored different visions of the trees' place in Washington. David Fairchild, in his practical, rational scientist's perspective, saw the trees as an exciting new ornamental plant with the potential to beautify streets and parks, a view his artistic wife shared. Eliza understood those benefits, though she never lost sight of her own hopes of creating, in the nation's capital, "such a riverside avenue as the Mukojima" and other similarly scenic spots in Japan, where crowds could gather in the springtime to mingle beneath cherry blossoms.[29]

Despite their differences, Eliza and the Fairchilds were now partners in a shared goal. Yet a fundamental problem remained. As Fairchild put it, how could they break through the "barrier of conservatism" in the Army Corps of Engineers office in which the men in charge of the city's parks slammed the door on any suggestion of planting Japanese cherry trees in Potomac Park?[30]

***

On April 25, a month after the Arbor Day event, Washington readers opened the *Evening Star* to discover the answer to a long-held secret. "Perhaps the most-talked-about book in Washington drawing rooms this winter has been 'As the Hague Ordains,'" the paper noted.[31] The title referred to a book published anonymously a year earlier. "Every writer who has been in Japan and many who have not been there" had been charged with its authorship. Now, with a new edition coming out, the publisher Henry Holt and Company had decided to reveal the author. "Miss Scidmore Wrote 'As the Hague Ordains,'" the headline announced.

Written in diary form, *As the Hague Ordains* tells the story of a Russian officer's wife who travels to Japan during the Russo–Japanese War to nurse her wounded husband at a prisoner-of-war camp. It was Eliza's first published novel, though some people read it as a true first-person account. She had written the book, Eliza told the *Star*, over two months late in the summer of 1905, soon after returning from Japan. After abandoning her magazine article for *The Century* based on her reporting from the POW camps, she had channeled her voluminous notes into a work of fiction, thinking the approach would make the subject more comprehensible to readers. The title *As the Hague Ordains* came from language in the prisoner-of-war accords adopted at the 1899 International Peace Conference in the Hague.

Eliza apparently modeled the book on a work she had read a few years earlier. In *Ground Arms*, an Austrian novelist and peace activist, the Baroness Bertha Félicie Sofie von Suttner, employed a similar journal format to tell the story of a German noblewoman's suffering as war in Europe tore her family apart. As Eliza told her friend Emily MacVeagh in the summer of 1902, after reading the book: "I was much wrought up over 'Ground Arms' as I am emphatically with Baroness von Suttner on the subject of war."[32]

As the Hague Ordains opens soon after the first major land battle of the Russo–Japanese War. The narrator, Sophia von Thiell, a middle-aged, childless woman of English–Russian heritage, is in St. Petersburg when she receives a telegram. It informs her that her husband, Vladimir, a Russian colonel and diplomatic adviser in the Manchuria campaign, was badly wounded. Now he is at a prison hospital at Matsuyama, on the southern Japanese island of Shikoku.

Sophia's account of traveling to her husband's bedside to care for him becomes a meditation on her feelings about the war and its combatants. While living in Tokyo during a previous marriage, Sophia developed a deep affection for Japan and its people. In Russia, she has had to suppress those feelings at Red Cross meetings amid the other ladies' scorn for the "barbaric" Japanese. Secretly, Sophia has grown disenchanted with life in Russia: "It seemed to me that the whole thing was a sham, a thin veneer of western civilization, a clever imitation up to a certain point. The government denied too much to the people, and the want of education in the masses appalled me."[33] On the way to Japan with her German maid, it rattles Sophia to see people in major cities of Europe and America cheer headlines about mighty Russia's licking. At the hands of an upstart Asian nation of "little brown monkeys," no less. "Brobdingnagians on horses fleeing from Lilliputians on foot," she despairs, in Eliza's reference from *Gulliver's Travels*. "Oh, the shame of it!"[34]

Many of the novel's descriptions draw heavily on Eliza's journalistic work, and the account of Sophia's stay in Matsuyama offers clues to Eliza's own daily life in Japan over the years. Sophia savors her tiny house and its pleasant verandah that overflows with pots of plants, including Japanese morning glories (*asagao*). She welcomes visitors who come by to admire her Japanese screens and other treasures; donates money to local families impoverished by the war; joins townspeople who gather at dark on the hillside to celebrate the moon's birthday.[35] Told over the course of a year, the story is melodramatic, the text heavy with exclamation marks. Eliza's research is evident in long didactic passages of dialogue that refer to developments in the war. Despite the novel's flaws, it became a bestseller. Soon after the announcement of Eliza's authorship, *As the Hague Ordains* underwent a sixth printing.[36]

In her interview with the *Star*, Eliza admitted that before her identity was revealed, a few people had confronted her about whether she wrote the book. "But a little bluster usually put inquiries to rout."[37] As for the conditions described, she had heard foreign residents in Japan express outrage at the generous treatment of Russian POWs. When she repeated the criticism to a Japanese cabinet minister, Eliza said, he questioned what good could come from ill treatment of brave men who had suffered enough. She herself denounced the very idea of POW camps as "wholly illogical." The loss of many thousands of prisoners had not seriously crippled the Russian military. Yet the heavy burden of caring for the soldiers humanely made Japan's war minister essentially "the head of a great chain of foreign hotels." During her tour of prison camps, the scene in Hamadera, near Osaka, annoyed her. There, she watched thousands of soldiers from Port Arthur idle away the days—"feeding and feasting and playing games, even dancing to pass the time"—while the Japanese toiled in the fields and earned a pittance pushing bread trucks long distances to feed the men.

***

As the Hague Ordains appeared at a time of great transition in U.S.–Japan relations. Since Japan's military defeat of Russia in 1905, countries in the West had grown increasingly uneasy about Japan, especially with regard to its intentions in Korea and Manchuria. For its part, the United States was anxious to protect its own emerging interests in the Pacific. After a half-century in which America and Japan had been close friends and allies, the relationship was growing strained.

Racist attitudes in America contributed to the tension. Decades earlier, the Chinese workers who arrived to help build the country's railroads were subject to prejudice and attacked as a perceived threat to the local labor force. Now the hostility was spreading toward other Asians, inflamed by mounting hysteria about a "yellow peril." In 1906, the San Francisco school board triggered an open crisis by extending its segregation of "Oriental" students in public schools to include Japanese and Korean children.

The move outraged the Japanese. While President Teddy Roosevelt denounced the "idiots" in California responsible for the policy, foreign ministers in Japan and the United States worked to salvage their countries' relationship by adopting a so-called Gentlemen's Agreement.[38] Under the 1907 pact, the Japanese agreed to voluntarily limit immigration to America in exchange for repeal of the school segregation order in California.

Just when it seemed the tension might ease, a rabble-rousing congressman from Alabama named Richmond Hobson threw fuel onto the fire by declaring publicly that war with Japan was imminent. America was vulnerable, he charged, because of its weak military preparedness, especially in its naval strength. With a "yellow wave" moving eastward over the Pacific to the shores of America, he claimed, control of the sea was "the white man's only chance for maintaining his supremacy and his civilization."[39] Though many Americans dismissed Hobson and his kind as cranks and alarmists, the views got a sympathetic hearing in the press.

To calm the talk of war and demonstrate America's naval readiness, Teddy Roosevelt announced he would send a fleet of sixteen U.S. battleships—all painted white—on a fourteen-month goodwill tour around the world.[40] Meanwhile, the Japanese ambassador to America, Baron Kogoro Takahira, sought to ease anxiety in his own country by insisting that war between Japan and the United States was "quite unthinkable."[41] In the spring of 1908 Takahira extended an invitation for the U.S. fleet to visit Japan.

The arrangements were being made when the news of Eliza's authorship of *As the Hague Ordains* broke in the press.

The book reinforced the positive feelings that people in Japan felt toward Miss Scidmore and her work. She had written favorably of their country in *Jinrikisha Days in Japan* and her other reporting, and she never hesitated to express her outrage publicly in U.S. attacks against Japan. In gratitude for her loyalty, the emperor of Japan agreed that winter to award Miss Scidmore the country's Order of the Precious Crown.[42] As the nominating documents submitted by the minister of foreign affairs said of her: "She prides herself as a good friend to Japan."[43] The medal Eliza received consisted of a five-starred, brightly enameled cross with insets of large pearls. A friend would recall, after her death, that she wore it "when the appropriateness of the occasion overcame her modest inclination to parade her honor."[44]

17

An Ally in the White House

A hundred names. That was the number Eliza had in mind when, early in 1909, she sat down in her temporary room at the Shoreham Hotel, on K Street NW, and began making a list. She had recently returned to Washington after her latest trip to China and a visit with her mother and brother in Nagasaki. Among the business she took up after arriving back in town was the unfinished matter of Japanese cherry trees for Potomac Park.

Eliza had been hopeful of developments after David Fairchild's Arbor Day program a year earlier. But there had been no progress, and by now, the idea she had first thought of two decades earlier was something of an obsession. The intransigence of the men in charge of the city's parks made her all the more determined to prevail.

Fortunately, she had thought of a way to get around them. She would seek the support of everyone she could think of who had lived or traveled in Japan; those who had seen cherry trees in bloom and "sipped the Emperor's champagne" at palace garden parties. She would ask them all to pledge a dollar a year to buy flowering cherries for the city. "If we could give 100 trees every year," she said in a later account of her plan, "in 10 years there would be a great showing in Potomac Park—a rosy tunnel of interlaced branches, a veritable Mukojima along the river's bank."[1] Backed by such support, how could obstinate bureaucrats refuse such a gift?

She started her appeals by approaching Admiral George Dewey, an old friend from their days in Asia. He and his wife had settled in Washington after his service in the Far East. Eliza was a frequent guest at their home on K Street. Dewey "thought well of the scheme," she said later, noting that his thick white mustache bobbed in laughter as she described her encounters with the "hardened old Westpointers" who dismissed her and her cherry tree idea out of hand. Despite Dewey's amusement, she wrote, "he dropped me into a gulf of gloom when he said, 'It would be just like those straightbacks not to accept or plant those trees when given to the park—say they haven't room, perhaps!'"[2] He pledged his full support for her cherry tree campaign.

The names on Eliza's list included another prominent couple in Washington: William Howard Taft and his wife, Helen. Eliza knew they had warm feelings for Japan and its people. The Japanese, in turn, admired Taft, now the president-elect. He "understands us thoroughly," a visiting Japanese official told the American press. So long as U.S. affairs were entrusted to a man like Taft, "harmonious relations would be maintained."[3]

Eliza and the Tafts moved in the same circles of Washington society. Eliza encountered them now and then at public and private functions. Earlier that year, she and the Tafts had been among the several thousand guests invited to the White House for the last diplomatic reception hosted by President Teddy Roosevelt and his wife, Edith.[4]

Though Eliza shared the buzz of optimism in Washington over the incoming Taft administration, she planned to skip the inauguration in early March. As she explained in a letter to Emily MacVeagh, she had decided to vacate her hotel room temporarily and spend the week in Atlantic City "to get away from the confusion and tobacco smoke." But she was leaving with a light heart. It delighted her to know that by the time she got back, she would have her dear friend nearby. The press had just announced that Taft chose Emily's husband, Chicago banker and lawyer Franklin MacVeagh, as his treasury secretary. The winter was shaping up to be so interesting, Eliza told Emily, that she would have been sorry to miss it had she "stayed out in the East."[5]

The winter would become memorable for other reasons. With William Taft's inauguration, Eliza wrote years later, "good luck came at a stroke."[6] Though she had hoped to seek the couple's support at some point for her cherry tree campaign, she hardly expected the chance to come so soon. Or to find such a powerful ally in the White House.

***

Washingtonians looked forward to an unusually bright social life in the capital under the Tafts. "It is no secret," said the *Baltimore Sun*, "that Mrs. Taft will dispense almost constant hospitality during her regime as first lady of the land, for she has such a host of personal friends both in society and out of it."[7] The stylish Nellie, a girlish woman with an upswept "Gibson Girl" hairdo, had entertained often as a Cabinet secretary's wife, and the affable and portly Will Taft was well liked. One rumor had it that the new first lady planned to bring back afternoon receptions for ladies, a popular tradition the clannish and elitist Edith Roosevelt had abandoned.

Nellie Taft wasted no time putting her own stamp on things in the White House. In the first few weeks, she changed the menus, the décor, the domestic staff, even some official protocols. She expanded the social calendar and guest lists to make the mansion accessible to a wider range of visitors, instead of catering solely to Washington's elite. And she vowed to make presidential parties more fun. She would serve her famous "champagne punch," she insisted, despite the protestations of temperance activists. She ruffled feathers but won people over by her intelligence and grace.[8] As the *New York Times* said of her: "Mrs. Taft has brains and uses them."[9]

As she assumed the routine duties of first lady, Nellie quickly grew bored. The White House had a well-trained staff who kept things running smoothly. And so, soon after breakfast on the morning of April 1, only a month after her husband's inauguration, Nellie summoned the president's chief aide, Major Archibald

Butt, to her office. Mrs. Taft was "quite excited," Butt recalled, as she briefed him on a project she had in mind.[10] She wanted to build a bandstand in Potomac Park and hold concerts there during warm weather. More broadly, she spoke of wanting to create an area in the nation's capital that might rival famous boulevards around the world: London's Hyde Park, the Malecón in Havana, the Bois de Boulogne in Paris.

Her inspiration came chiefly from a park she had enjoyed in Manila when her husband served as governor-general of the Philippines. The Luneta, as the area was called, was a treeless, open-air plaza facing the bay. Although the park itself was not impressive, Nellie loved the gaiety that occurred there every evening at dusk. As breezes cooled the tropical air and band music floated across the water, thousands of people of different nationalities and social classes circled the park's elliptical drive in carriages or on foot, greeting one another and gossiping while children played on the patches of lawn.[11]

The experience led Nellie to imagine a similar place in Potomac Park. Now she had a chance to follow through. "I determined, if possible," she said in her 1914 memoir, "to convert Potomac Park into a glorified Luneta where all Washington could meet, either on foot or in vehicles, at five o'clock on certain evenings, listen to band concerts and enjoy such recreation as no other spot in Washington could possibly afford."[12]

After her meeting with Major Butt that morning, Nellie called for a car, and they drove down to Potomac Park. Butt, a tall, bulky man with a tidy mustache, had been a favorite of the first family in Teddy Roosevelt's administration and would also grow close to the Tafts.[13] On his outing that morning with Mrs. Taft, he marveled at her decisiveness. In barely an hour, she decided to hold the concerts twice a week, on Wednesdays and Saturdays; picked a date to launch the season; and chose a spot for the bandstand alongside the Speedway, where Nellie liked going for drives with family and friends. She and her husband both adored automobiles (Figure 17.1). Already, Nellie had arranged to replace the

Figure 17.1. President William Howard Taft and Helen Taft in 1912, driven by military aide Archie Butt.
(Harris & Ewing Collection, Library of Congress)

presidential horse-drawn carriages with a fleet of new motor cars, including a small "runabout" for herself.[14]

Back at the White House, Major Butt briefed the president on his wife's plans for Potomac Park. To William Taft, the idea sounded just like Nellie. Everyone who knew her admired Nellie's executive skills. Early in their marriage, when Will Taft traveled for long periods as a circuit judge, Nellie raised their three children while channeling her love of music into the creation of a symphony orchestra in Cincinnati. She became a critical helpmate to her husband in the Philippines. Though William Taft would be excoriated over the years for his insensitive reference to "my little brown brothers," he and Nellie won the trust and affection of many Filipinos for their egalitarian attitude and sympathetic spirit. Nellie raised the wrath of U.S. military officials by insisting that Filipinos be included in social functions at Malacañang Palace. In one effort to learn more about the country and its people, she traveled for weeks by horseback through the northern Luzon jungle with a military escort, sleeping in pup tents and abandoned huts.[15] Nellie, as the president well knew, was a force to reckon with.

Taft understood his wife's thinking on Potomac Park because the couple shared a keen interest in its development. They saw the new 700-acre park as a key element in a major planning document for the capital that they endorsed.[16] The 1902 report, known as the McMillan Plan, laid out a coherent vision for the growth of federal buildings and monuments around the National Mall, inspired by L'Enfant's original city plan as well as "city beautiful" ideas that grew out of the 1893 Chicago World's Fair and its neoclassical White City. Among its recommendations, the report also called for a vast network of public parks that would give citizens open places in which to enjoy the benefits of nature and recreation.[17]

As war secretary, William Taft had followed progress on the new Potomac riverfront park his Army engineers were creating.[18] In doing so, he conferred frequently with Captain Spencer Cosby, the senior man in charge of the Potomac flats reclamation.[19] Two weeks after his inauguration, the president named the dapper, dark-eyed officer—still a bachelor in his early forties, and a popular "society man" in Washington—to head the Office of Public Buildings and Grounds.[20] Because the position included serving as administrator of the Executive Mansion and a de facto aide to the president, Cosby was already, a month into the job, a steady presence at the White House.

Now it fell to Colonel Cosby—the new job came with a promotion in rank—to follow up on the first lady's plans for Potomac Park. Once Major Butt passed along the go-ahead from the president, Cosby ordered his staff to begin designing and building a bandstand on the spot Mrs. Taft had selected. They would need to work around the clock, he stressed. She wanted to hold the first concert on April 17. The date was only two weeks away.

***

Barely forty-eight hours after Nellie Taft's meeting with Major Butt, news of her plans hit the press. Eliza and other Washingtonians opened their newspapers that weekend to find several articles on the White House's plans for Potomac Park. "Esplanade in Park," the *Washington Post* announced on Saturday April 3. "Taft Orders a Promenade to Rival Famous Boulevards."[21]

The articles quoted Colonel Cosby at length. Many residents with images of the formerly nasty Potomac flats had no idea just how pleasant that part of town had become, he explained. Though by no means finished, the new park had trees, flower beds, and fountains; foot paths, bridle trails, and macadamized roads. Plans called for adding sports fields, including baseball diamonds, tennis courts, and polo grounds. "The president desires this to be the great playground of the public," Cosby said.[22]

Now, he continued, Mrs. Taft had requested that a riverside area along the Speedway be turned into a place where people from all classes of society could gather to enjoy band music. Concertgoers could arrive on foot from the nearest streetcar; take in the music from their carriages or cars; dock their boats by the seawall along the river. While he himself would be in charge of the project, Cosby explained, all the credit for the idea belonged to the first lady. She was to be consulted regularly, "and her suggestions will be carried out to the letter."[23]

To Eliza, the news must have seemed too good to be true. Mrs. Taft's plans sounded much along the lines of her own long-time vision of creating a public gathering place along the river where people could ramble beneath cherry blossoms. It was a perfect marriage of interests. She knew Mrs. Taft had lived in Japan for a few months, in a pretty bungalow on the Bluff in Yokohama, and had summered at Lake Chiuzenji. As Eliza would say years later, Mrs. Taft "knew Japan."[24] Here was a chance to solicit her support for the planting of Japanese cherry trees in Potomac Park.

Across town, David Fairchild was apparently thinking along the same lines. He wrote many years later, in what may have been a draft chapter of his memoir, that Eliza visited the couple again that spring to see their cherry trees in their "first real blooming." Miss Scidmore "was surprised and delighted at the growth the trees had made," he said.[25] Once more they discussed the idea of cherry trees along the Speedway. The matter must have been fresh in both their minds that weekend.

Eliza owed much of her success in life to a talent for seizing opportunities whenever they came her way. This time was no different. On Monday morning April 5, she wrote Mrs. Taft a letter urging her to incorporate cherry trees in her park project. Though a copy of the letter has not survived, later developments suggest that Eliza proposed a spot along the Potomac, in keeping with the Japanese custom of planting cherry trees along waterways.[26]

The response was swift. "My Dear —," Mrs. Taft wrote two days later, on Wednesday, April 7. "Thank you very much for your suggestion about the cherry trees. I have taken the matter up and am promised the trees."[27]

Could it really be happening? *I have taken the matter up and am promised the trees.*

Instead of the area Eliza had proposed, Nellie Taft explained that she "thought perhaps it would be best to make an avenue of them, extending down to the turn in the road, as the other part is still too rough to do any planting. Of course, they could not reflect in the water, but the effect would be very lovely of the long avenue. Let me know what you think about this."

Any thoughts Eliza might have had were immaterial by then. Even before replying to Eliza's note, the first lady had galloped ahead. She had summoned Colonel Cosby and expressed her desire, as he told his staff, to have "magnolia and cherry trees, and other similar early blooming trees, planted in Potomac Park."[28] For years the Army officials in charge of the new park had fiercely resisted any interference in their own landscaping plans. Now they were backed into a corner. There was no question the first lady's wishes would prevail.

Cosby arranged a meeting at 10 o'clock on Wednesday April 7, for Mrs. Taft to discuss her ideas with the head gardener, George Brown. Eager to move quickly, she prodded the men to acquire as many flowering cherries as they could find at U.S. nurseries and have them set out as soon as possible.[29]

I have taken the matter up and am promised the trees.

Eliza's joy was so great she could not help sharing the news with two Japanese men she encountered in Washington the day after receiving Mrs. Taft's note. Kokichi Midzuno, the Japanese consul in New York, and Dr. Jokichi Takamine, a renowned Japanese scientist living in America, had come to town together for social events that week organized in honor of two Japanese commissioners traveling in the United States to drum up support for their country's first world's fair.[30] A frequent guest at Japanese diplomatic functions, Eliza may have talked with Midzuno and Takamine at a dinner the Japanese ambassador hosted on April 8.

Dr. Takamine, a distinguished-looking man in his fifties, commended attention everywhere he went (Figure 17.2). The son of a samurai physician, tutored in English and trained in "foreign science," he had achieved fame—and great wealth—from a string of revolutionary breakthroughs in biochemistry. Most notably, he had isolated the hormone adrenaline, giving medicine a powerful new drug, epinephrine, for use in surgery and in treating

Figure 17.2. Japanese chemist Jokichi Takamine.
(Williams Haynes Portrait Collection, Science History Institute, Philadelphia)

maladies such as allergic reactions.[31] Takamine and his American-born wife had a home on New York's fashionable Upper West Side. Highly active in civic affairs, he was a cofounder of the city's Japan Society, which had been founded in 1907 to promote friendly relations between the United States and Japan amid the escalating tensions between the two countries. Eliza herself would soon be elected a member.[32]

When she briefed the two men on Mrs. Taft's plans for cherry trees in Washington, Eliza would say years later, Dr. Takamine grew excited by the idea. Immediately, he offered his support. "Will you find out if Mrs. Taft will accept 1,000 cherry trees for her Mukojima?" he inquired. "In fact," he added, "I had better give 2,000 trees. She will need them to make any show."[33]

It was Mr. Midzuno, Eliza said, who suggested to Takamine that he donate the cherry trees on behalf of the city of Tokyo. For one thing, the approach was more in keeping with Japanese values that frowned on extravagant gestures by an individual that could be perceived as showing off.[34] As a savvy diplomat, Midzuno likely understood as well the goodwill implications in making the trees a gift from Japan to the people of the United States.

Takamine agreed to Midzuno's idea. If they could not get Tokyo to go along for some reason, he said, he would seek the help of the Japanese community in New York.[35] As Takamine told Eliza, he himself had wanted for some time to have flowering cherries planted in Riverside Park, along the Hudson River, near his home. But city officials had rebuffed his offer. Maybe once cherry trees were planted in Washington, he could get New York to accept some as well.

On Monday, April 12 four days after her talk with the two men, Eliza went to the White House to discuss Dr. Takamine's offer with Mrs. Taft. She arrived after lunch, when the grounds staff were cleaning up after the popular Easter Egg Roll that had been held that morning on the White House lawn.[36] Later that day, Eliza wrote to Midzuno in New York, reporting on her meeting with the first lady:

> Dear Mr. Midzuno:—
>
> I had a long talk with Mrs. Taft this afternoon and she will be more than pleased to have the cherry trees offered her as a gift from Japan, for her Mukojima.
>
> She has taken it up with great enthusiasm, and has already had her head gardner [sic] communicate with Japanese florists in New York, and has had Col. Cosby, the Superintendent of Public Buildings and Grounds, follow up the gardner.
>
> She was greatly surprised to know that her interest in the sakura no hana is so appreciated by yourself and friends.
>
> I said [as though speaking] "From the Mayor of Tokyo," the cherry blossom avenue is so entirely Mrs. Taft's own affair, that it is sure to be a success.[37]

Meanwhile, Cosby and his staff had followed through on Mrs. Taft's orders to try and obtain Japanese cherries from domestic sources. Though the trees were still relatively rare in America, the men found a sizable quantity at a nursery in West Chester, Pennsylvania. The same day Eliza met with Mrs. Taft, Cosby placed an order for ninety "Japanese double-flowering cherry trees," each six to twelve feet high, at a cost of $106.[38] When they arrived a few days later, he had them planted along a section of the roadway in Potomac Park, in time for Mrs. Taft's first band concert, though over time the trees would not survive.

Curiously, it was not until April 13 that Cosby finally answered a letter from David Fairchild that had landed on his desk at the War Department two weeks earlier. Fairchild had written around the same time Eliza wrote to Mrs. Taft.[39] In the letter, Fairchild told Cosby that his wife, Marian, had often mentioned, while riding on the Speedway, "a favorite plan of hers" to donate Japanese flowering cherries for the roadway. Now he wanted to offer fifty "choice trees" from Japan for the project Mrs. Taft was said to be planning in Potomac Park. He himself would see to the trees' delivery in time for the planting season the following spring.

In his response, Cosby apologized to Fairchild for the delay. He had "wished to look into the matter as thoroughly as possible."

And in fact, he explained, he had already arranged to purchase flowering cherries from domestic sources, in accordance with the first lady's wishes. Of course, they would "not be as choice as the ones you propose to send us."[40] But he would make sure Mr. Fairchild's cherry trees were handled with care when they arrived.

The simultaneous letters suggest that Eliza and David Fairchild may have been working in concert—or possibly in competition with one another. Fairchild knew of the earlier resistance Miss Scidmore had faced from the city's park officials. Perhaps he thought he could succeed where she had failed, backed by the authority of his position at the USDA and his generous personal offer to donate cherry trees. In the end, however, it was Eliza who had the greater advantage. Though her friendship with Mrs. Taft was only casual, the two women felt a warm connection and understood one another at a certain level; they were alike in their bold thinking and "can-do" attitudes. Together, they possessed the will and the force it would take to give Washington its now-famous Japanese cherry trees—despite the stumbling blocks they encountered along the way.

After her meeting with Mrs. Taft at the White House, Eliza remained in touch with Mr. Midzuno in New York about the offer of the cherry trees. He stressed, in a letter in late April, that their conversation in Washington had been "just a preliminary step."[41] An official offer of cherry trees from Japan would have to come from his embassy. He had cabled his superiors in Tokyo about the situation. In the meantime, they would have to wait.

***

The Saturday of April 17 brought perfect spring weather. Just before 5 o'clock, Nellie Taft plopped a large purple hat on her head, pulled on her white gloves, and climbed into a small motorcar, a landaulette, with her husband for the short drive down to Potomac Park.[42] She had been nervous about the public band concert in Potomac Park. But when they got there, she was delighted to find the area packed. Thousands of people crowded the grounds. Vehicles of every kind sat banked two deep on the Speedway, which she and Colonel Cosby agreed should be renamed Potomac Drive.[43]

In a sentimental nod to her years in Manila, Nellie had chosen the Constabulary Band of the Philippines to give the first concert. She had helped formed the group while her husband was serving as governor-general of the Philippines. The band, whose director, Walter H. Loving, was a black musician classically trained at the Boston Conservatory of Music, had performed at William Taft's presidential inauguration.[44] As the band played a selection of American and international music, carriages and automobiles chugged along the Speedway, the well-dressed ladies calling out to those they knew.

In its coverage of the event, the press noted that a cablegram had reached Washington earlier that day. It was from the mayor of Tokyo, offering a gift of a thousand flowering cherry trees for the new drive.[45]

That spring, Nellie Taft introduced another innovation inspired by her time in the Far East: she decided to throw a series of garden parties. There was no lovelier setting than the south garden of the White House, with its wide lawns, great fountain, shady trees, and two long terraces. Invitations went out to 750 people for the first reception, on Saturday afternoon May 7. An hour before the event was set to start at 5 o'clock, heavy rain soaked the tent and forced everything to move inside. Happily, the next garden party, held in perfect May weather a week later, came off without a hitch. But it was to be the last merry White House function for some time.

On May 17, the Tafts set off on an afternoon cruise down the Potomac to Mount Vernon with family and friends. The day was frantic for Nellie. Her father was in town. The Tafts' younger son had just had his tonsils removed that morning. And, as usual, her husband Will was running late.

The boat had not yet reached Alexandria, Virginia, twenty miles downriver from Washington, when Nellie fainted. Everyone assumed the collapse came from exhaustion. But as she regained consciousness, it became clear the problem was much graver. After they all hurried back to the White House and Nellie was carried inside, the doctors who examined the first lady concluded she had suffered a stroke.[46]

It would be a year before she regained her health and resumed her normal activities in the White House.

18

Up in Smoke

Dr. Takamine's offer of cherry trees to Mrs. Taft had been the spontaneous act of an individual. Once he and Mr. Midzuno, the Japanese consul in New York, agreed to arrange the gift from the city of Tokyo, the dynamic changed. Delicate cherry blossoms—the very emblem of Japan—became an instrument of international diplomacy, endowed with symbolic importance amid tense relations between Japan and the United States.

Though rumors about the trees started appearing in the American press, no official word had come by the time Eliza sailed to Japan in June of 1909. The summer abroad turned out to be chaotic, which made it impossible to get any writing done, she complained in a letter to Emily MacVeagh.[1] Only three weeks after her brother George's transfer from Nagasaki to Kobe, he got new orders from Washington that made them all glum: he was being sent instead to Korea, now a protectorate of Japan, and soon to be formally annexed. While George appealed the transfer and requested a leave of absence, they all scrambled to adjust. Eliza and her mother arranged to stay on in Japan before returning together to Washington later that year.

They were still in Kobe in late summer when Eliza saw in the local press the first official news of the cherry tree gift to Washington. The Tokyo Municipal Council had decided on August 18 to present 2,000 cherry trees to Mrs. Taft "and other ladies in America," after learning of their intention to import flowering cherries for an area near the Potomac riverbank. Tokyo authorities agreed to send ten-foot trees of a dozen prized varieties, some with poetic names like "Large Lantern," "Milky Way," "Weeping

Chrysanthemum," and "Tiger's Tail." One press account noted that "the mayor of Tokyo appears to have interested himself especially in this matter."[2] Speaking for himself, the mayor, Yukio Ozaki, would say years later that he urged the gift of the trees "in appreciation to the government of the United States for their kindness shown to Japan during the Russo–Japanese War" (Figure 18.1).[3]

After reading about the Tokyo Council's vote, Eliza wrote immediately to Robert Underwood Johnson at *The Century* proposing to write about cherry blossoms. She filled him in on her frustrating attempts over the years to have flowering cherries planted in Washington. As she explained: "From Col. Ernst in the Harrison administration to Col. Brownwell in the T[eddy] R[oosevelt] time, all those court chamberlains & grounds keepers would have none of me."[4] Now that it was finally about to happen, she had in mind an article, similar to her piece on Japanese morning glories, that would educate Americans about flowering cherries. It could run in March or April, she suggested, "when the first big planting party will be coming off in Washington."

In September, soon after Johnson cabled his consent, Eliza hurried off to Kyoto to do research and arrange watercolor

Figure 18.1. Tokyo mayor Yukio Ozaki.
(Bain Collection, Library of Congress)

illustrations for the article. She found the subject engrossing. "You have no idea how much I have discovered about the Japanese cherry tree and its 100 variations," she wrote to Emily MacVeagh.[5] In an update to Johnson, she extolled the beauty of the trees and advised him to act fast if he had any interest in acquiring some. "They are jumping in price here, the nurseries are being drained, and when the Century lets loose in the spring and the planting has taken place in Potomac Park they will be quoted at orchid prices."[6] She herself, she noted, was already arranging small shipments to friends back in Washington at their request.

***

After the trip to Kyoto, Eliza made a "flying visit" to Peking in October to keep up with developments there.[7] She had hoped to go later, in November, to report on the official funeral of the empress dowager, who had died a year earlier, from a stroke, at the age of seventy-three.[8] Cixi had ruled for nearly fifty years, after usurping power from her nephew, Emperor Guangxu. Mysteriously, the thirty-seven-year-old emperor died a day before Cixi did. Eliza was traveling in China around that time, and she found Peking abuzz with questions about the emperor's demise. Having been in poor health for some time, had he succumbed to disease—possibly Bright's disease, as foreign physicians speculated? Or was he perhaps poisoned?[9]

Eliza found it necessary to alter her China plans in the fall of 1909 because she had promised her mother they would be back in Washington by Thanksgiving. Before leaving Asia, she wrote to Gilbert Grosvenor at the National Geographic Society. Though she could not cover the dowager's funeral, might he be interested in a general lecture on Manchuria? She would plan to present lantern slides. "If I can get the pictures colored they will be utterly, utterly gorgeous," she promised.[10] Grosvenor jumped at the offer. The society, he noted, would pay its usual honorarium of seventy-five dollars.

With Manchuria much in the news, they both knew a lecture on the subject would attract a large audience. Manchuria was the ancestral home of the late empress dowager and the Qing dynasty. Equally of interest, the area was a storm center of world politics since Japan's victories over China and Russia. In fact, trouble was brewing there again, as Japan charged ahead on a contested railway line and other developments that would give it greater advantages in the region. The situation made Western powers nervous, and it put America in a bind, as its longtime ally seemed to flout the "open-door" policy in China that the United States had brokered. Japan's "game of grab" in Manchuria, as one editorialist called it, threatened to renew strained relations between Japan and the United States.[11]

When Eliza delivered her lecture on Manchuria a few months later, she showed many of her own photos, along with images of the imperial family and other subjects—some hand-colored—acquired from photo agencies and other sources.[12] She expanded her talk into an article for *National Geographic* that focused heavily on Mukden (Shenyang) and its impressive art museum (Figure 18.2).[13] Housed in the former palace compound,

Figure 18.2. Street scene in Mukden, from Smithsonian Institution's Eliza Scidmore Collection.

(National Anthropological Archives, Smithsonian Institution)

the museum held acres of imperial treasures. The porcelains alone, most covered in dust and stacked high on the floor or stored in bins, totaled thousands of pieces—blue-and-white vases; plates and rice bowls in imperial yellow, powder blue, and "chicken-liver red."

Among the personal relics of the Manchu emperors that were on display, one object in particular had caught Eliza's eye. "Here it is!" she wrote in excitement to Emily MacVeagh at the time, enclosing a photo. "The rosary or official necklace of the Emperor Kienlung in the yellow satin box, in which it has lain for more than one hundred and twenty-five years."[14] During her many years in Asia, Eliza had become knowledgeable about such "rosaries." In Japan, the stringed beads were imbued with Buddhist symbolism and used in many rituals; Chinese dignitaries wore elaborate necklace variations as part of their court costume. Thanks in part to her friendship with temple priests, Eliza herself acquired a couple dozen of the rosaries, which she loaned to the National Museum in Washington.[15] The one she came upon in Mukden, crafted from lapis lazuli, coral, and sapphires, with a giant ruby medallion inset and 108 half-inch pearls, topped them all.[16]

***

Meanwhile, back in Washington, the summer of 1909 had passed with no official word on the cherry trees promised to Mrs. Taft. Rumors led to hysterical claims in the press. "20,000 Trees Offered Taft," the *New York Times* announced, inflating the number by ten times.[17] The matter had risen to high-level diplomatic talks between the U.S. secretary of state and the Japanese ambassador. Finally, on August 30, the Japanese Embassy issued a statement saying that "the City of Tokyo, prompted by a desire to show its friendly sentiments towards its sister Capital City of the United States," decided to offer 2,000 cherry trees to Washington.[18] The *Washington Post* predicted that driving through Potomac Park one day with cherry trees in bloom would "put Hyde Park to shame" and outshine Berlin's famous avenue of lindens.[19]

Cherry tree fever also hit New York. That fall the city was set to celebrate the three hundredth anniversary of Henry Hudson's discovery of the Hudson River and the centennial of Robert Fulton's paddle steamer. In honor of the occasion, Dr. Jokichi Takamine, the Japanese scientist behind the gift to Mrs. Taft, persuaded city officials to accept 300 cherry trees for Riverside Drive, along the Hudson River. The Mikado himself offered to send the trees, it was reported, along with Japanese gardeners to ensure that the saplings got a good start on American soil.[20]

The Outlook, an American magazine of news and opinion, hailed the promised cherry trees as a sign of renewed cordiality between Japan and the United States after the "fierce war cries" in both countries a year earlier.[21] In another positive development, a privately organized mission of fifty Japanese businessmen and civic leaders arrived in Seattle in September to attend the Alaska-Yukon Pacific Exposition and tour America for three months. Invited by U.S. Chambers of Commerce, the group came chiefly to promote trade and commerce. But the members of the delegation also took seriously their role as "citizen diplomats" aiming to strengthen friendly relations with Americans.[22]

The group received a warm welcome in the cities they visited, though the "little brown men"—as many newspapers around the country described them—were hounded relentlessly by reporters who pressed them for comments on the talk of war between Japan and the United States.[23] The delegation's leader, Baron Eiichi Shibusawa, a banker and Japan's leading industrialist, echoed his government's insistence that Japan and its people posed no threat to the United States. President Taft signaled his faith in Japan's friendly intentions toward America by going to Minneapolis to meet the group during their cross-country trip.[24]

While the developments in America were unfolding, authorities in Japan were preparing the 2,000 cherry trees for Washington. They selected trees from "ten of the most representative kinds."[25] The trees were reinforced with bamboo canes, grouped into bundles of ten, and wrapped in canvas for protection during the long trip across the Pacific. To coordinate the transit and other arrangements, a heavy stream of letters flowed between all the parties involved: the Tokyo mayor's office; the U.S. Agriculture, War, and State departments; the Japanese Embassy; and the White House. The shipment of trees left Japan on November 24 and was due to arrive in Seattle in mid-December. The president of Nippon Yusen Kaisha (Japanese Steamship Company) agreed to transport the trees at no charge. Mayor Ozaki himself signed off on the declaration of goods when the shipment left Yokohama. To ensure a personal hand in the gift, Ozaki delegated a member of Shibusawa's commercial mission, Torakiro Watase, an agronomist and Tokyo alderman, to act as his agent for the trees when the group reached Washington.[26]

During the preparations, the secretary of agriculture, James Wilson, stressed in a letter to Colonel Spencer Cosby the "great importance" of having the trees thoroughly inspected once they reached Washington.[27] The step was necessary to make sure the imported trees carried no foreign pests or diseases that might pose a threat to American farmers. As he noted, most U.S. states had quarantine or inspection laws for imported plants; the District of Columbia did not. As such, cooperation between the USDA and Cosby's Office of Public Buildings and Grounds was crucial to make sure the cherry trees got a thorough check.

The USDA also put in-house botanist David Fairchild on standby to help with the inspection, given his experience growing Japanese cherries at his home outside Washington. Fairchild would say years later that the physical appearance of the imported trees concerned him a great deal when they arrived. But he had "not dreamed" that the greatest nightmare would be caused by his own colleagues at the Department of Agriculture.[28]

***

Eliza and her mother arrived in Washington in November around the time the Japanese trade mission was turning around to head home. U.S. officials and local citizens had feted the group nonstop for ten days. The activities included an automobile ride along the roadway in Potomac Park to view the area where Mrs. Taft's 2,000 cherry trees would soon be planted.[29] The women in the delegation were taken to the top of the Washington Monument to enjoy the spectacular views.[30]

While still abroad that summer, Eliza had arranged to have her Washington house, on M Street NW, freshly painted after the last tenants moved out. She and her mother were newly settled in when they celebrated Mrs. Scidmore's late-November birthday with a cake and eighty-six candles.[31] The homecoming was an occasion for sentimental reflection. It had been nearly fifty years since Mrs. Scidmore arrived from Wisconsin with young Lillie and George, eventually to make Washington their adopted home. They had all spent half their lives since then living on and off in Japan. Mother and daughter both loved their life in the Far East, but every year the absence of family and friends in America felt more acute. As Eliza Catherine Scidmore told relatives, "Japan has many attractions, but after 24 years . . . you tire and want to go back."[32]

Now in her fifties, Eliza too was beginning to feel the toll of so much travel and dislocation. Being back in Washington reminded her of all she missed. Weeks earlier she had headed off to China with her heart only half in it. "I only want to go to America and get there as speedily as possible," she wrote to Emily MacVeagh. "This time surely," she added, she was ready "to get done with and shake off the Far East" for good.[33]

She had made similar declarations before—always in vain. In twenty-five years, she had reported on a broad swath of Asia, in four books of travel and hundreds of articles. Still, there was always more to explore, some new drama to follow. At the end of the day, it was chiefly her work that gave meaning to her life. And for some time that work had been deeply rooted in the Far East. The pull seemed irresistible, as much as she might wish to settle down. Washington and Japan offered competing attractions. Fortunately, a part-time home abroad with her brother offered the freedom to come and go at will, without having to choose between one place or the other.

In the second week of December, Eliza wrapped up her cherry tree article for *The Century*. She had uncovered so much fascinating material it was "an agony to amputate and prune," she complained to her editor.[34] But she managed to cut the piece to 4,161 words. In it, she described how the planting of 2,000 cherry trees in Washington would one day make Potomac Park "a worthy cousin" of Japan's most celebrated cherry blossom sites, like Mukojima in Tokyo and Arashiyama near Kyoto. Though Americans associated cherry trees with the juicy red fruit, she explained, a Japanese cherry was not a cherry tree in the normal sense: "It does not have to work for a living, and produce a crop for the market. When its burst of beauty is over, nothing more is expected of it. Its whole strength is well and wisely spent in flowering, and it rests in peace until the season rolls around again."[35]

That same week, on the afternoon of December 11, Eliza went to the White House to meet with Mrs. Taft for an update on her plans for cherry trees in Potomac Park. Nellie had spent the summer recuperating from her stroke at a large house the Tafts rented in Beverly, Massachusetts, along the North Shore. The family managed to keep it out of the papers just how severe her illness was, with the press laying her withdrawal to exhaustion.[36] While the president and other family members drifted in and out, the first lady went "absolutely nowhere," even forgoing the motorcar drives she loved, the *Washington Post* reported.[37] Secluded from the public eye, Nellie spent her days working to recover the use of her paralyzed right arm and restore her impaired speech. When she returned to the White House that fall, she resumed her official duties with the help of her sisters. The Tafts' daughter, Helen, took a leave of absence from her studies at Bryn Mawr to lend a hand.

Visions of "her" cherry trees blooming soon in Potomac Park boosted Nellie's morale at a tough time. Just before her meeting with Miss Scidmore, Mrs. Taft received word that the cherry trees from Japan had arrived in Seattle and would soon be forwarded to Washington. "She is rejoiced at her success," Eliza said of the first lady in a letter to *The Century*'s editors the next day when she sent off the final revisions of her cherry tree article.[38]

Mrs. Taft's instincts, Eliza was pleased to discover, seemed spot on. Nellie had met two weeks earlier with Colonel Spencer Cosby and his landscapers about planting arrangements for the 2,000 cherry trees. To her dismay, she learned that the men had interspersed the ninety domestic cherry trees acquired the previous spring among a couple hundred elms, newly planted to correspond with the elms already bordering other roadways in the park.[39] Nellie was irate, and insisted the elms be taken up. "My dear Mr. President," Cosby wrote contritely to her husband on November 29. "In accordance with your request, I have directed that the elms recently planted between the cherry trees along a part of Potomac Park be removed at once."[40] He promised to have a new plan ready for the first lady's review within a few days.

The final planting scheme, Eliza happily learned, included parallel rows of cherry trees around the Tidal Basin—a plan she sketched in her letter to *The Century* (Figure 18.3). The result would be a pleasing effect of "double reflections" in the water.[41] During cherry blossom season in Washington, when a halo of pink and white blossoms surrounded the Tidal Basin, Americans would enjoy a scene as lovely as anything Japan had to offer.

***

As the year wound down, Christmas week brought a blizzard that blanketed the East. Washington residents turned out for a winter carnival and sledding on the Speedway. Eliza and her mother were hunkered close to the fire on New Year's Eve when they opened the *Washington Star* to the latest news on the cherry trees from Japan. The article trumpeted Eliza herself. "Her Labor of Love," a headline declared in large black type, adding that "Miss Scidmore Procured the Cherry Trees for Us." The article was the

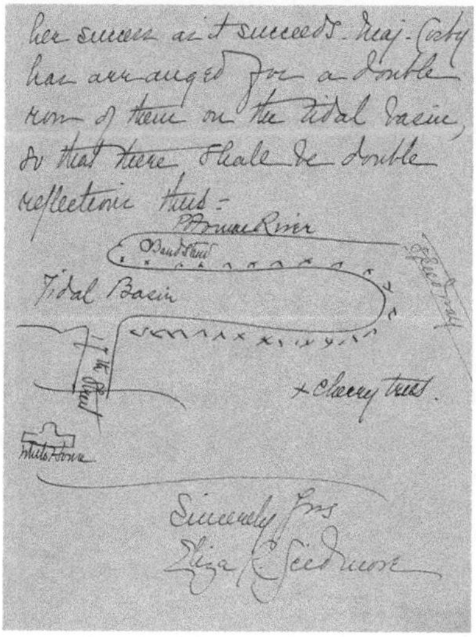

Figure 18.3. Eliza Scidmore's sketch for cherry tree plantings around the Tidal Basin, in a 1909 letter to *The Century* magazine. (Century Collection, New York Public Library)

first public acknowledgment of Eliza's critical role in securing the "splendid gift" from the city of Tokyo.[42]

The reporter described Eliza's "persistent efforts" to have Japanese cherry trees planted in Potomac Park and how, by winning the support of both Dr. Takamine and Mrs. Taft, she had outmaneuvered park officials who "manifested an indifference to her request." Miss Scidmore, the article noted, had recently returned from Japan with a list of the cherry blossom varieties being sent to Washington. The trees, due to arrive the first week in January, would bear flowers ranging from dark pink to greenish white, with petals as delicate as stars or as lush as peonies. One pale-pink, single-blossom variety had a name—*mikurima-gayeshi*—that translated colorfully as "looking backward from the carriage." While the trees would not bloom that spring, Eliza explained in the interview, all of them should reach their glory within three years. In sending its prized cherry trees by the thousands to Washington and New York, she said, the annual blooming would be, in the words of the Japanese donors, "a perpetual reminder of the friendship between the two peoples."

Eliza's own article on the trees was set to run in *The Century*'s March 1910 issue. When a check arrived in the mail, she was not pleased. "It is a little awkward," she scribbled in reply, "but I must say frankly that the cheque for $125 for the 'Cherry Blossom' article is not satisfactory to me." She had accepted that rate for years, she said, in the belief it was the magazine's standard payment, only to learn that some favored contributors were paid more. She expected the same treatment and would hold onto the check until she heard further. "I trust you will pardon my frankness," she wrote, "but it is surely the best way."[43] The magazine agreed to pay seventy-five dollars more.

When the article appeared in March, it carried a note to readers. The planting of cherry trees in Washington that Miss Scidmore referred to had not occurred after all. Just as *The Century* went to press in January, a tragic turn of events made Eliza's article woefully obsolete.

***

William and Nellie Taft ushered in 1910 by hosting their first New Year's Day reception at the White House. On the cold but brilliantly clear day, more than 5,000 people filed through the mansion. Though getting better every day, Nellie was still in fragile health when she made an official appearance. Standing at her husband's side, in a white satin gown and flat-soled shoes, she mumbled greetings to disguise her impaired speech and clutched a bouquet of lilies of the valley to avoid shaking hands. She stayed an hour for the presentation of diplomats, then retired to watch from an adjacent room.[44]

Eliza spent New Year's Day with her friend Emily MacVeagh, assisting at the treasury secretary's reception. It marked the start of a busy week, with social events that included a private dinner at the MacVeaghs' in honor of the vice president and his wife.

On Friday evening Eliza gave her lecture on Manchuria at the Masonic Temple Auditorium.[45]

That week also brought the arrival of the cherry trees. On January 6, a mix of snow and sleet was falling in Washington as the bundled trees were transferred from train cars and hauled across town by horse-drawn wagons to the USDA's storehouses, near the grounds of the Washington Monument.[46] The trees would be kept there until early spring, then transplanted to Potomac Park and other spots around town.

First, as agreed ahead of time, they had to be inspected. Two days after the trees arrived, David Fairchild went to check on them. He noticed problems at once. For one thing, the trees were large, which meant they were old. They had likely been selected to make a showier display of blooms, he concluded, though younger, smaller trees stood a better chance of surviving the transplant to foreign soil. Even more worrisome, the roots had been cut back severely, possibly for ease of shipping. Fairchild advised Colonel Coby to have the branches of the trees pruned, in the hope it would help their survival.

More devastating news was yet to come. After Fairchild's initial review, a team of USDA scientists spent several days inspecting the trees. What followed became controversial, with some people charging years later that the incident was largely a political move by the man leading the inspection: David Fairchild's childhood friend Charles Marlatt, now acting chief of the USDA's Bureau of Entomology. For years Marlatt had been pushing for Congress to enact a federal plant quarantine law, without success. The 2,000 cherry trees from Japan offered a highly visible test case for his argument that the importation of foreign plants posed a threat to American agriculture in harboring potentially harmful nonnative pests and diseases.[47]

On January 19, Marlatt submitted a report to the secretary of agriculture that summarized the team's findings on the trees. The results were alarming. The scientists reported heavy damage from a host of pests, including scale, root-gall worms, a deeply embedded wood borer, and a half-dozen other insects. The trees were so badly infested, Marlatt wrote, that fumigation would hardly kill all the pests. The situation left the USDA with only one option, he stressed: "the entire shipment should be destroyed by burning as soon as possible" to protect American growers.[48]

The advisory came as a huge blow. President Taft and the first lady had taken a personal interest in the cherry trees. The public anxiously awaited their arrival. And the trees' destruction would be a terrible affront to the Japanese, who had so generously donated them and for whom the cherry trees held strong nationalistic meaning.

Word of the unfortunate situation went out to U.S. and Japanese officials. The letters included copies of the USDA's report, to underscore the severity of the perceived threat to U.S. agriculture. After a week, when the news had sunk in, there was nothing left to do but act. On January 28, with a heavy heart, President Taft gave his consent to have all the cherry trees from Japan immediately destroyed (Figures 18.4 and 18.5).

The sky was gray, and the streets of Washington covered in slush on the bitterly cold Friday when the order was carried out. Workers hauled the 2,000 cherry trees to the grounds of the USDA's propagating gardens and stacked them in long rows. Batches of the trees were clumped into teepee-like piles, then set ablaze, along with their bamboo canes and wrappings. Supervisors in dark wool coats and felt hats roamed the scene for hours, as the bonfires smoldered, and the Washington Monument towered in the background. Twenty-four hours later, all that remained of the much-anticipated trees, and the hopes that went with them, was a pile of ashes. Nellie Taft's cherry tree project had just gone up in smoke.[49]

Reporting on the trees' destruction, the *Washington Star* cautioned readers that the move should not be seen "as an act of discourtesy to Japan."[50] To show there were no hard feelings over the incident, a number of Japanese officials stated publicly that their country would arrange to send another consignment of

Figures 18.4 and 18.5. U.S. Department of Agriculture inspection of cherry trees from Japan in January 1910, and subsequent burning of the shipment.
(U.S. National Arboretum Collection, USDA National Agricultural Library)

cherry trees—this time in much better condition. Yukio Ozaki himself apparently discussed the matter personally with Colonel Cosby during a visit to Washington in the spring of 1910.[51] In a later letter to Cosby, Ozaki would say graciously that he and his fellow Japanese were "more than satisfied" that the infected trees had been handled as they were, "for it would have pained us endlessly to have them remain a permanent source of trouble."[52]

Amid another flare-up in U.S.–Japanese tensions, however, sending more trees was no sure thing. In February 1911, Congressman Richmond Hobson intensified his "yellow peril" rant by declaring, in a House debate over a naval appropriations bill, that war with Japan was near.[53] Years later, in his memoir, David Fairchild would recall Miss Scidmore's telling him of a meeting that summer in Mayor Ozaki's office in which it was suggested that if Americans felt as much animosity toward the Japanese as some public statements seemed to indicate, then the matter of the cherry trees ought to be dropped altogether. "Fortunately," Fairchild wrote, "more generous counsels prevailed."[54]

19

"Mrs. Taft Plants a Tree"

Frigid temperatures made January 26, 1912, a dreadful night to go out. Still, the foul weather did not keep several hundred National Geographic Society members and their guests from trekking across town for the annual dinner.[1] Eliza was among the distinguished-looking men and women in formal evening dress who streamed into the Willard Hotel, on Pennsylvania Avenue, which had the swankiest banquet venue in Washington.

Inside the cavernous ballroom, massive crystal chandeliers cast a glittering light, and the long, white-clothed tables groaned under elaborate place settings and splashy floral arrangements. As people drifted in to take their seats, Eliza greeted friends and colleagues. The sea of faces included many well-known public figures: senators, ambassadors, and military heroes; publishers, scientists, and explorers like Admiral Robert E. Peary, whose recent North Pole expedition the National Geographic helped support. Clearly, the event was shaping up to be what the press called "the most interesting and significant dinner of the winter."[2] For three years in a row, Eliza's editor and the society's president, Gilbert H. Grosvenor, had scored a publicity coup by getting his second cousin William Howard Taft to attend. This time the president had to cancel at the last minute because of a lingering cold.[3]

For Eliza, the evening represented an anniversary of sorts. Twenty years earlier she had so impressed the fledgling society's members with her energy and talent that they elected her corresponding secretary. Over the years, as she held a succession of leadership positions in the organization, she never wavered in her devotion. For tonight's event she had used her influence to

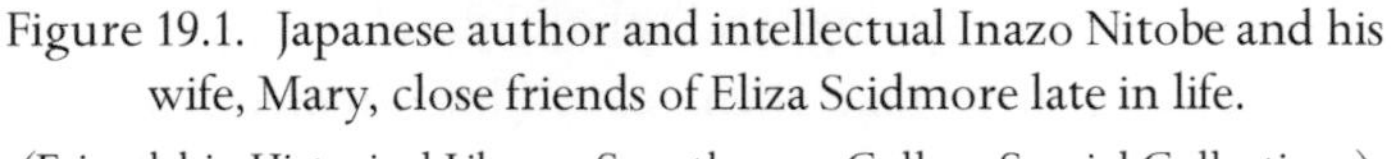

Figure 19.1. Japanese author and intellectual Inazo Nitobe and his wife, Mary, close friends of Eliza Scidmore late in life.
(Friendship Historical Library, Swarthmore College Special Collections)

arrange one of the key speakers, the renowned Japanese professor and intellectual Inazo Nitobe (Figure 19.1).[4]

Eliza had followed Nitobe's work for a long time, since first quoting from one of his early books in her own 1891 *Westward to the Far East*.[5] She had gotten to know him and his American-born wife, Mary, in Japan, and she entertained the couple regularly on their visits to Washington. This winter, she had seen a good deal of them while Inazo Nitobe was in the area delivering a month-long lecture series at Johns Hopkins University in Baltimore. The Carnegie Endowment for International Peace had brought him to America on an extended tour around the country to help neutralize the tide of ill feelings between Japan and the United States.[6] He would be speaking on the same topic for this evening's National Geographic Society program, which took as its theme the peace of nations.

As the business portion of the dinner got underway, the sight of a tall, familiar bearded figure bounding onto the stage brought the clink of crystal and the din of conversation to a hush. Sixty-four-year-old Alexander Graham Bell was the evening's toastmaster. Gazing out over the room, Bell puffed with pride. "There has never been in the history of the world a scientific society that has increased in influence and power as the National Geographic Society," he declared to thunderous applause.

Bell cataloged the society's success. Membership had exploded to 107,000, with thirty percent growth in the last year alone. The public lectures, which drew as many as 2,000 people, had to be given in afternoon and evening sessions because the number exceeded the capacity of the largest halls in town. The society even had its own headquarters now, a small but handsome building at Sixteenth and M Streets NW that Mabel Bell Hubbard and her family built in honor of her late husband. And of course, Bell continued, there was the crown jewel: the magazine. From a "valuable technical journal that everyone put upon his library shelf and very few people read," he boasted, *National Geographic* had become "the greatest educational journal of the world." Distributed to members and some schools, it reached more readers than *The Century*, *Harper's*, or *Scribner's*.[7]

A half-dozen speakers followed Bell's presentation. The society paid honors to the British ambassador, James Bryce, for his expeditions in Trans-Caucasia and other regions. O. H. Tittmann of the U.S. Coast and Geodetic Survey described trigonometric surveys being conducted around the world to fix geographic boundaries and provide accurate measurements for a uniform world map. Miss Mabel Boardman, who succeeded founder Clara Barton as head of the American Red Cross, briefed the audience on the organization's latest relief efforts in many countries.

None of the speakers generated more curiosity than Dr. Nitobe, who was famous to many in the audience as the author of *Bushido: The Soul of Japan*.[8] The small volume, written in English, had created a buzz at the turn of the century as one of the first books that explained the Japanese national identity to readers in the West. Nitobe described the core values of Japanese society as rooted in the ancient samurai code of *bushido*, or the "way of the warrior." Its values, Nitobe wrote, stressed not just physical prowess but also moral and ethical behavior—a system he likened to medieval chivalry. Widely translated, *Bushido* became a best-seller as people sought to better understand how the tiny nation of Japan could have forged itself—and so quickly—into a modern industrial and military power. President Teddy Roosevelt was so taken with the book that he bought dozens of copies and handed them out to friends and colleagues.[9]

Gifted in the English language from a young age, Nitobe had decided early in his career that he could use his talent to aid the flow of ideas between Japan and nations in the West. He would make himself, in effect, a "bridge across the Pacific." He developed a strong humanitarian outlook shaped in part by Quakerism, which he embraced while studying at Johns Hopkins University. His wife was from a prominent Quaker family in Philadelphia.[10]

Now, as he prepared to address the National Geographic Society, Nitobe enjoyed a reputation as an eloquent spokesman on the need for cross-cultural understanding to improve relations between the United States and Japan. Some people in the room had bought tickets to the banquet just to hear his talk. Sitting with her dinner companions, Eliza watched as her friend—a small mustachioed man with round eyeglasses and a formal bearing—rose from his seat and took his place at the podium.

Nitobe got quickly to the point. America and Japan, he declared, needed to stop the hysteria of anger and mistrust that was poisoning their long-time friendship. Japan, he explained, had a shrill form of everyday language known as *ki-iro-no-koye*; translated literally, it meant a "yellow-colored" voice, indicative of excitement or lunacy. It was the ugly voice of hateful and warmongering *jingoes*. "Self-respect," he said, "demands that we close our ears to it, whether it proceeds from that or this side of the Pacific Ocean."[11]

In a quiet but dramatic gesture, Nitobe nodded toward Alexander Graham Bell. Could the great scientist not invent a "new kind of telephone," one that translated messages for better understanding between those who spoke to one another in different tongues? Geography, with its cross-cultural focus, had an obligation to resolve linguistic confusion, he argued. And so, he had come that evening with a challenge. He wanted to know: How would geography and its practitioners respond? What did those in the audience intend to do to counter the negative impressions and increase accord between people and nations "when, as we say, jingoes speak in yellow-colored voices?"

***

On March 14, two months after the National Geographic dinner, Eliza took the train to New York, accompanied by the new Japanese ambassador, Viscount Sutemi Chinda, and his wife, the Viscountess Iwa Chinda.[12] They were headed to a series of events sponsored by the Japan Society. Eliza was to be feted that evening as a guest of honor on the eve of a two-week floral exhibition at which she would be lecturing on Japanese gardens. The Japan Society also planned a welcome dinner that week for Chinda, who had presented his credentials to President Taft at the White House only two weeks earlier.

The Chindas were the latest of many Japanese envoys Eliza befriended over the years. A frequent guest at Japanese and Chinese diplomatic functions in Washington, she returned the hospitality with occasional teas or luncheons for those posted to Washington.

Eliza had been especially fond of Chinda's predecessor, Baron Yasuya Uchida, and his wife. Madame Uchida, a graduate of Bryn Mawr, spoke English fluently and was so charming she became a popular Washington hostess in her own right.[13] A Chicago friend of Emily MacVeagh recalled fondly a merry luncheon at Miss Scidmore's at which the Uchidas were among the handful of guests. Everyone chatted knowledgeably—but not too seriously—about a host of things. Though Miss Scidmore's home on M Street was furnished quite simply, the many Oriental objects made the place seem like a small jewel box. "Her house speaks to one, in every nook and corner, of the mysterious East," the visitor recalled. She found the hostess herself "a woman of keen vision and broad interests—forceful, sincere, buoyant." Miss Scidmore had an air about her that made her seem "quite apart from the rushing crowd, though more or less in it."[14]

After the Uchidas ended their assignment in Washington that winter, Eliza transferred her loyalty to the Chindas. The viscount was already a familiar figure to her. He had served as vice minister of foreign affairs during the Russo–Japanese War and was now a rising star in his country's diplomatic service. The U.S. public showed strong interest in the handsome, mustachioed new ambassador, especially after it was reported that he had studied in America several decades earlier and played both golf and bridge.[15]

With U.S.–Japanese relations so delicate, newspapers tracked the Chindas' every move, from the time they left Japan and traveled through Hawaii, San Francisco, and Chicago to reach Washington's Union Station, amid a great fanfare, on the evening of February 22. Reporters pressed Chinda constantly for his comments on the political situation. Time and again, he stressed his country's peaceful intentions toward America. As he would say at the Japan Society dinner in New York, the long friendship between the United States and Japan was "not like a hothouse flower that has to be forced into flowering artificially." Geographically the two nations were neighbors, with the same great sea washing their shores. "The Pacific is happily named," he said, "in that it characterizes the friendly relations of our respective countries."[16]

***

For society-goers in New York, the two-week Japanese exhibition was billed as one of the season's cultural highlights. The Japan Society had arranged the installation of a Japanese garden in the glass-roofed belvedere of the Astor Hotel, complete with a traditional tea house. The daily activities included tea ceremonies, lectures on Japanese gardens and flower arranging, and performances of classical Noh theater. Members of the society could attend free, and the general public for a dollar a person.[17]

Eliza was there to educate visitors about the principles and practices of Japanese gardens. Unlike in America, she explained, the Japanese did not create gardens as places in which to grow flowers. Instead, Japanese gardens were meant to be idealized landscapes in miniature. Strict rules guided the placement of every element—rock, tree, bridge, and lantern—to achieve perfect scale and harmony. Eliza had researched the subject extensively for an article in the April issue of *The Century*, on "The Famous Gardens of Kioto." She was also scheduled to give an illustrated National Geographic lecture on the subject at the end of March, around the time the replacement cherry trees from Japan were due for planting in Washington.[18]

On March 21, Eliza was at the Japan Society exhibition when Nellie Taft arrived for a tour. The first lady had come to town with one of her sisters, on what turned out to be one of the coldest days New York had seen in years. They had a busy day planned: the Japanese garden show in the morning, a women's industrial exhibition at the Grand Central Palace in the afternoon, and a night at the theater, for which they had tickets to see "The Quaker Girl." It pleased onlookers to see the first lady looking so good. At one point during her rounds, she fell forward while stepping out of a push chair, and the crowd gasped. But she had simply tripped on her skirt. She picked herself up and continued on.[19]

When Mrs. Taft and her party arrived at the Astor Hotel, a news photographer was there to get a picture of the first lady at the Japanese exhibition. Lindsay Russell, president of the Japan Society, arranged himself in the group portrait with a half-dozen ladies, including his wife. Nellie Taft was given the place of honor, seated at front center, with Miss Scidmore next to her (Figure 19.2). Surrounded by the other women, who looked trim and stylish in their wool suits, fur stoles, and broad-brimmed winter hats, Eliza appears awkward and old-fashioned in her elbow-length white gloves and a fussy dark dress with a lace bodice insert. Her hat—a high-crowned affair, trimmed with a

Figure 19.2. Eliza Scidmore and Helen Taft at a Japan Society exhibition in New York City on March 21, 1912.
(From *New York Herald*)

thick band of artificial flowers—seems especially out of kilter, though she had likely chosen it as suitable for spring.

The photograph offers a rare portrait of Eliza late in life. The image is also significant because it documents a moment between her and Mrs. Taft only a week before they would meet up again in Washington—on the latter occasion for what would become a historically important event alongside the Tidal Basin in Potomac Park.

***

During the flurry of Japan Society events, Dr. Jokichi Takamine was also anticipating the arrival of the cherry trees from Japan. Inspired by the gift to Washington, a Japanese–American friendship society had arranged to donate 3,000 cherry trees to the city of New York around the same time. Takamine was a key figure behind the effort but had to push the matter to the back of his mind for the time being. As a cofounder and one of the top leaders of the Japan Society, he was caught up in the whirlwind of social events. He himself hosted a reception for Ambassador Chinda, which gave the Takamines a chance to unveil their new home on New York's Upper West Side.[20] The couple had spent a fortune remodeling the interior of the three-story townhouse at 334 Riverside Drive, near General Grant's tomb. Guests arriving for the evening entered rooms elaborately made over in a Japanese style, with carved teak furniture, an inlaid dining table, gilded grillwork on the walls, and a huge bronze temple lamp.[21] Before long, Takamine would be able to gaze out the windows of the upper floors to see flowering cherry trees across the street in Riverside Park.

Five weeks earlier, a shipment of more than 6,000 cherry trees had been packed on board the S.S. *Awa Maru* in Yokohama. Now the trees were on the final leg of their journey to the United States. Half of them—3,020, to be exact—were bound for Washington. The rest would be sent on to New York, for Riverside Park and other sites around the city.

After the earlier debacle of the infested trees, Japanese officials had responded with stoic grace. Though still resentful of racist attitudes that demonized the Japanese in America, they well understood the symbolic value of planting their country's national tree in the heart of the U.S. capital. What better demonstration could Japan make of its desire for continued friendship with the United States and its people? Besides, the matter was one of honor. It seemed only right to rectify the unfortunate mishap.

To produce a substitute batch of cherry trees free of blemishes, mayor Yukio Ozaki and the city of Tokyo had looked to horticultural experts to handle things. Late in 1910, the scientists collected scions, or budded branch cuttings, from twelve choice varieties of flowering cherries growing along the banks of the Arakawa River. The cuttings were fumigated with hydrocyanic acid gas and placed in cold storage for several weeks, then grafted onto understock trees selected for their robust growing qualities. A year later, after the young trees had shed their leaves, they were removed from the ground, fumigated again, and prepared for shipment to America.[22] Professor Yoshinao Kozai, director of the Imperial Agricultural Experiment Station in Tokyo, wrote to U.S. Department of Agriculture officials in Washington explaining all the precautions that had been taken to ensure the health of the young trees. He himself, he noted, had supervised their cultivation.[23]

On February 14, 1912, Yukio Ozaki personally oversaw the dispatch of the trees to America. Nellie Taft got word of their imminent arrival in a letter from the mayor's wife. Born of Japanese–English parentage, Yei Theodora Ozaki was multilingual and a translator of Japanese short stories and fairy tales. Her husband, she wrote in her impeccable English, had shipped the 3,000 cherry trees in the hope that they "will form an avenue in Washington as a memorial of national friendship between the U.S. and Japan." Stressing the special care the saplings had received, she added that "We hope to hear of them blooming in the salubrious Washington climate reminding you of all Japan's faithful devotion and admiration for all her friends and tutor, America."[24]

The mayor himself also followed up. His own letter, addressed to Colonel Spencer Cosby, the Office of Public Buildings and Grounds superintendent who was working closely with Mrs. Taft on the project, included a detailed account of the dozen cherry tree varieties that Japan had shipped.[25]

***

After returning from New York in late March, Eliza scrambled to juggle her obligations in Washington. She and a group of friends, including Mrs. Alexander Graham Bell, spent many hours organizing a booth in the spacious second-floor ballroom of Rauscher's restaurant, on Connecticut Avenue, for a charity bazaar to benefit the Working Boys' Home.[26] She also had to finish preparing her National Geographic lecture on Japanese gardens, scheduled for March 29. Many in the Japanese diplomatic community were expected to attend.[27]

Meanwhile, spring struggled to arrive. The grass was greening around town and wildflowers poked from the underbrush in Rock Creek Park. But temperatures were ten degrees below normal for that time of year, falling some nights to below freezing. Heavy rains in the region were pushing the Potomac to the flood stage. While ads for Easter fashions filled the newspapers, residents venturing out for the evening donned warm coats and winter hats.[28]

Somber events added to the gloom. On Saturday March 23, throngs of people stood for hours in a downpour to watch a long, slow funeral march of caissons along Pennsylvania Avenue, escorted by Army and Navy detachments. All business ceased for the day as the city turned out to pay homage to victims of the *Maine* explosion in Havana harbor, the event that had triggered the Spanish–American War fourteen years earlier. Congress had passed a bill to raise the wrecked vessel from the bottom of the sea. Now, the last bones retrieved from the battleship were coming home for burial in Arlington Cemetery. President Taft was among those who stood by in the drizzling rain until the last coffin was lowered into the ground.[29]

The cold weather continued into Monday when residents awoke to fresh snowfall. Then, happily, a break came by midweek, as the days grew steadily milder.[30] On Wednesday March 27, the temperature had reached the sixties at midday when Nellie Taft climbed into a motorcar and set off from the White House. The fair and slightly breezy day was perfect for an outing, one she welcomed as a diversion from the gloom in the White House, where everyone was in a slump over William Taft's reelection efforts. In a stunning development, Teddy Roosevelt had recently thrown his hat into the ring as a candidate for president, on a third-party ticket. Adding salt to the wound, he was attacking William Taft in speeches around the country. The president felt devastated by the betrayal of a man he had considered a good friend. For her part, Nellie was livid. She had never trusted the egomaniacal Teddy, and his treachery proved the soundness of her instincts.[31]

Today, on a balmy afternoon, she had something to take her mind off all the unpleasantness for a while. She had arranged with Colonel Cosby to be on hand when the first of the flowering cherry trees from Japan were planted in Potomac Park. The project dear to her heart, delayed for years by setbacks, was about to see the light of day.

The trees had reached the capital the previous afternoon, after arriving in Seattle and being shipped across the country in insulated rail cars. In Washington, they were delivered to the Agriculture Department's propagation gardens and storehouses, near the grounds of the Washington Monument. In a letter informing the agriculture secretary, James Wilson, of the trees' arrival, Colonel Cosby urged that the required inspection get underway immediately, as the trees needed to be "set out in Potomac Park without delay."[32] He had ordered his men to have the grounds ready.

The inspection continued into the morning of March 27. Midway through the process, Wilson sent Cosby a reassuring note. The trees, he reported, seemed "singularly free from injurious insects or plant diseases."[33] Nellie and the president had been waiting anxiously for the verdict. The official word came around noon, when the inspectors completed their review and cleared the trees for planting.

Given the uncertainty of the situation—and with memories of the previous debacle perhaps still fresh in everyone's minds—the White House had not organized a major public ceremony to mark the planting. Yet protocol dictated that the gift from Japan be properly acknowledged. And so, Mrs. Taft had arranged for a few people to join her for a small and private dedication. Arriving at the spot Colonel Cosby staked out, on the northern bank of the Tidal Basin, Nellie greeted the Japanese ambassador, Sutemi Chinda; his wife, the viscountess; and Miss Scidmore. Cosby stood by with men from his grounds staff, who had dug rows of holes around the perimeter of the Tidal Bain so hundreds of the trees could be planted expeditiously over the next few days.

Cosby nodded for everyone to gather, and they huddled closer. The Washington Monument rose above them in the background, the sun's midday arc casting shadows across the southeast grounds of the landmark. After the days of heavy rain, Eliza and the others felt the cool dampness of the earth beneath their feet. Some thirty feet away, in the gray-green waters of the Tidal Basin, light breezes stirred gentle whitecaps that glinted in the sunlight like silver petals.[34] At Cosby's go-ahead, Mrs. Taft took the brand-new spade handed her and plunged it into the ground at the designated spot. Workers rushed forward to plant the sapling. When Nellie finished, she stepped back and invited Madame Chinda to plant a second tree.

At the conclusion of the brief ceremony, Nellie presented the viscountess with a bouquet of long-stemmed American Beauty roses.[35] While hardly as significant a gift as the 3,000 cherry trees sent from Japan, the flowers were a source of local pride, and not just for their patriotic name. A new variety growing popular in international markets, the rose had been developed right there in Washington, in the private garden of the historian George Bancroft at his home overlooking Lafayette Square and the White House.[36]

On that now-historic day of March 27, 1912, no photographers were on hand in Potomac Park to record the event. Press reports were brief. "Mrs. Taft Plants a Tree," the *Washington Post* announced the next day, atop an article of three short paragraphs.[37] The *Evening Star*, in its similarly brief coverage, noted that the cherry trees were expected to bear slight blooms the following year, though it would take two or three years before they were radiant with color.[38]

Most of the 3,000 cherry trees from Japan were planted in Potomac Park, though some were planted elsewhere around town, including in Rock Creek Park and on the grounds of the White House and the Library of Congress. A portion of the shipment was set aside in a government nursery for later planting, to replace trees that did not thrive. About 1,800 trees of the Yoshino variety—two-thirds of the total consignment—were planted in double rows around the Tidal Basin (Figure 19.3).

Because the typical lifespan of a flowering cherry is about forty years, nearly all the 1912 trees have been replaced, with perhaps a hundred or so of the original ones thought to survive. A plaque now marks the site of the first plantings, amid a small stand of

Figure 19.3. Double rows of Japanese cherry trees around the Tidal Basin, in the winter of 1922.

(Harris & Ewing Collection, Library of Congress)

heavily gnarled trees.[39] Over the years, many of the trees surrounding the Tidal Basin have been lost to flooding, erosion, beaver attacks, and root damage from the heavy foot traffic of tourism. At least four fell victim to vandalism, chopped down one night not long after Japan's surprise attack on the United States at Pearl Harbor on December 7, 1941.[40]

With the recovery of relations between the United States and Japan, the two countries have engaged for several decades in regular cherry-tree exchanges, as well as reciprocal gifts of other plants, such as American dogwoods. The 1912 planting ceremony has been reenacted often during the National Cherry Blossom Festival, which today draws more than 1.5 million visitors to Washington each year. Under the nurturing of the U.S. Park Service, the springtime cherry tree extravaganza has withstood the test of time—a testimony to the long friendship between the United States and Japan, and to the vision of Eliza Scidmore and her allies who saw possibilities for transforming a corner of the U.S. capital into something out of the ordinary.

20

War and Peace

Eliza was in Yokohama in the summer of 1914, soon after her brother's transfer from Korea, when World War I broke out in Europe. Though the United States would remain on the sidelines for more than two years, Japan joined almost from the start, as a partner of Britain and the other Allies. Eliza reported on the developments late that year in *The Outlook*, an American magazine of current affairs.[1]

The Japanese people dreaded the entanglement, she wrote, as they saw no need for young men to lose their lives over a European cause. Nor did they harbor any real animosity toward Germany. But key decision-makers in the government prevailed by arguing that Japan had an obligation to support Britain, under an alliance the two countries had made a decade earlier to safeguard one another's interests in Asia. After declaring war on Germany in August, Japan acted quickly to help rout German ships from Pacific waters. Japanese-led forces also moved into China's Shandong province and forced the surrender of Tsingtao, a seaport leased by the Germans.

In her article, Eliza commended Japan's readiness to prove itself a loyal and honorable ally of nations in the West. So, she told readers, "let us hope that we have done with those senseless catch-words, 'the Yellow Peril.'" And besides, though few people realized it, Japan was also providing humanitarian assistance. Back in Washington that winter, Eliza briefed a Red Cross conference—attended by President Woodrow Wilson himself—on Japan's offer of aid to help treat war victims in Europe. Several countries had accepted, and medical units were now being dispatched, complete with surgeons, nurses, interpreters, and portable hospitals. Eliza went to New York in early January to speak at a Japan Society luncheon for one of the teams, on its way to England.[2]

With Americans jittery about Japan's involvement in the mostly European war, Eliza's article in the December *Outlook* got wide attention in the press.[3] Called on to explain things, she denied, as some people charged, that Japan's move was a calculated attempt to draw the United States into a conflict over supremacy in the Pacific. Nor, she insisted, should Japan's occupation of the Shandong province be viewed as a deliberate flouting of the U.S.-sanctioned "open-door" policy in China. As she explained, Japan planned to return the seized territory after the war.

But events made Eliza's trust seem badly misplaced when news broke a few weeks later that Japan had secretly issued an ultimatum to China. In what became known as the "twenty-one demands," Japan claimed territorial concessions and special privileges in the Shandong province, Manchuria, and other areas that would, in effect, give Japan a huge measure of control over China. The backlash in the West was fierce. In the face of wide condemnation Japan eased its demands, yet the damage was done. Though Japan would remain a partner of the Allies, the incident provoked ill-will and mistrust that would linger for years.

***

Eliza's reporting for *The Outlook* gave her journalistic lifeline at a time when her other magazine work was drying up. The war's disruptions coming on top of editorial changes at *The Century* and *National Geographic* brought a fall-off in assignments from what had been two of her steadiest clients.

At *The Century*, the death of its longtime chief editor, Richard Watson Gilder, in 1909 elevated Eliza's own editor at the magazine, Robert Underwood Johnson, to the top position. His tenure did not go smoothly. *The Century* had gradually lost its edge with the appearance of livelier and less expensive mass-market periodicals like *McClure's*, *The Cosmopolitan*, and *Munsey's*. Johnson hoped to transform *The Century* into a fresher and more modern magazine that could better compete, but he failed to get the board's backing. He resigned in 1913 and went on to serve as the American ambassador to Italy.[4]

Meanwhile, at *National Geographic*, editor Gilbert Grosvenor was expanding his roster of freelance contributors from Asia. When Eliza pressed him in 1913 to send her to Java on assignment, he informed her that "a special trip on our account" would not be necessary. While he would be "more than glad" to consider any photos she might want to submit, he explained, he was already working to acquire Java material from another source—William W. Chapin, whose high-quality photographs from China and Korea had influenced the magazine's adoption of color printing.[5]

Eliza soon found her hands tied anyway, after a devastating development early in 1914. During her brother's transfer from Korea to Yokohama, the ship carrying his possessions sank. As Eliza told Grosvenor, in sad resignation: "My own camera and negatives are at the bottom of the sea, with the 25 cases of my brother's household effects. So!"[6] She had lost not only her own work but also many prints and negatives acquired from other sources in Japan and Korea. "Young Japan," which appeared in the July 1914 *National Geographic*—with eleven images "By Eliza R. Scidmore"—would be her last article in the magazine.

A year later, the coronation of Japan's new emperor gave Eliza a chance to do once again the kind of in-depth cultural reporting that had made her books so popular. In the fall of 1915 she went to Kyoto and Tokyo to cover the event for *The Outlook*.[7] As she explained to readers, the term *coronation* was misleading, used mainly for the benefit of foreigners; there was no crown nor any passing of the title from a higher authority. The official rites consisted chiefly of the new emperor proclaiming to ancestral deities his accession to the throne.

This time, however, the enthronement was truly historic. Never before had the events been held in public. Forty-seven years earlier, the Meiji emperor had assumed power in a simple private ceremony that took place amid civil strife and threats from abroad. Now, with Japan becoming a world power, the ascension to the throne of Yoshihito (Emperor Taisho) as the one hundred twenty-third ruling descendant of the Japanese imperial family was treated as a grand affair of state. Diplomats and other foreign dignitaries attended as special guests. Millions of Japanese—from princes and priests to soldiers and commoners—took part in several weeks of ceremonies, many modeled on old court rituals.

Eliza described in detail the amount of planning that went into making sure everything came off without a slip. Ornamental nail heads in the palace floor marked where each person should stand; kimono-clad crowds sat silently in choreographed rows for two miles or more along the routes through which the emperor passed. Upon his triumphal return to Tokyo, the main boulevard was awash in banners and storefront displays that echoed the rich color of his coronation robes: the soft red orange of the sun rising in a mist.

***

With her mother feeling poorly, Eliza stayed on in Japan in 1916. She acquired a puppy, which cheered up everyone in the house. Mrs. Scidmore, now ninety-two, had complained for some time of not feeling like her old self (Figure 20.1).[8] A winter cold lingered through the summer. Then, in September, she suffered a bad fall. Two weeks later, just when she seemed to be on the mend, she experienced heart failure. "In reality she died then," Eliza wrote to family in America, "but by all the terrible drugs and resources of medical science her poor heart was kept beating and her lungs gasping." Her death came at 7:30 in the evening on October 5, after several nights of thrashing about with bad dreams and delirium. "Much of the sting was for us taken away," Eliza wrote, "to have her hopeless struggle ended and to lie in peace at last."[9]

Figure 20.1. Mrs. Scidmore not long before her death at the age of ninety-two.
(Wisconsin Historical Society, WHI-91610)

Mrs. Scidmore's death generated headlines in both America and Japan.[10] The *Japan Gazette* reverently reported her passing as "the oldest foreign resident in Japan"—having lived there thirty-five years—and a much-beloved figure in the Far East whose "fame for hospitality crossed the Pacific." At her "Wednesday afternoons," an obituary writer noted, she had enthralled guests with her lively storytelling and animated spirit.[11] The overflow crowd of mourners at her funeral included many high-level Japanese officials. As Eliza informed family, her mother had decided, in the last year of her life, to be buried in Japan where she had many friends. Twice she visited the Foreign General Cemetery with one of the Scidmores' servants to point out lots and monuments she liked.[12]

George and Eliza grieved in their own ways. George, his health already affected by "overwork and worry," grew so distraught he went for several weeks of rest at a hot-springs resort in Japan, under a doctor's order.[13] Eliza tackled the condolences and other postfuneral affairs with the help of friends who came to stay, including "Miss Denton," a distant cousin of the Scidmores who taught school in Japan.[14]

Finally, bearing the burden of her loss privately, Eliza sailed home to America at the end of the year. She spent a period of mourning in Washington with her good friend Ida Thompson, before moving into the Hotel Grafton for a socially quiet winter.[15]

***

The mood in the capital was sober. Strong public sentiment against America's involvement in the war in Europe had kept the United States neutral for two and a half years. Now, under German aggression that included the sinking of U.S. ships, attitudes were changing. The dreaded moment came on April 6, 1917, when President Wilson obtained authorization from Congress to declare war on Germany. Eliza went to Europe that summer to report for *The Outlook* on Red Cross relief efforts. In Switzerland and France, she visited makeshift hospitals and sanitariums where thousands of sick and invalid prisoners and other war victims were being treated.[16]

Back home, Eliza felt compelled to aid the war effort more directly, as her mother had done after arriving in the capital during the Civil War. She joined the women's section of the Navy League to train for national service.[17] With many "Washington girls" taking jobs at the State Department and federal agencies, a local society writer reported that Miss Scidmore—"a real Washingtonian of the old 'cave dwelling' set"—insisted on doing her part.[18] Eliza became a yeoman (F), the Navy's designation of its wartime female recruits, and went to work in the paymaster general's office.[19] Her service was cut short by the armistice in November 1918.

Having reached the peak of her success, Eliza took up residence in the Stoneleigh Court apartments, a luxury complex at Connecticut Avenue and L Street, a few blocks north of the White House.[20] Now in her sixties, and almost frumpish at times in her well-worn dresses and out-of-date hats, Eliza could seem a relic to those in the younger and more stylish circles of Washington society. Yet she had a legion of admirers. Her name appeared often on the guest lists of people at the highest ranks of power. The city's wealthiest socialite, Mrs. Thomas Walsh, was a good friend.[21]

George Washington University, an institution with deep ties to the city, expressed in tangible form the high esteem in which Miss Scidmore was held. In 1919, a year before passage of the Nineteenth Amendment that would give most American women the vote, the school awarded her an honorary doctorate of letters. She received the honor at commencement exercises on June 18, in recognition of her achievements as:

> authoress of numerous works on the countries of the Far East; gifted with a power of keen observation, graphic description, and sympathetic interpretation; ever seeking to promote that intimacy of acquaintance which dispels ignorance, removes prejudice, banishes fear and prevents hatred, thus aiming to make the pen mightier than the sword.[22]

The award recognized the fruits of a career Eliza had embarked on a half-century earlier when America was a much simpler place. As a young reporter she had ventured to Alaska on a rustic mail steamer; ridden trains cobwebbing their way across the West. Now, the country stood at the brink of a new era that would bring enormous changes across every level of society.

In an experience befitting the age of modernization that lay ahead in the postwar 1920s, Eliza capped the decade with another milestone: she rode in an airplane for the first time. On a late-autumn day in 1919, she joined a group of Washington friends and VIPs who turned out for flight demonstrations at Bolling Field. Infused with the curiosity and daring she had never outgrown, she climbed into the cabin of a Curtiss Eagle, an eight-passenger biplane being developed for civil aviation.[23]

As the plane puttered through the clear, blue sky above Washington, the views below revealed a city that sprawled in every direction—beyond the core monuments and buildings, the tidy green rectangle of the National Mall, the gleaming silver ribbon of the Potomac and the busy main boulevard of Pennsylvania Avenue. The rugged town Eliza had first seen as a child was no longer recognizable. Since then, it had grown into the handsome capital of a nation that was now, only 150 years after its founding, one of the world's great powers.

***

Of all that defined her, Eliza never lost her fundamental urge to be engaged in issues of the day. While other people took up pleasures like golf or bridge late in life, she made it her business in the 1920s to keep up with world affairs. And to travel as much as her money and health allowed.

For several years in a row, she carved out time in late summer to attend the Institute of Politics in Williamstown, Massachusetts, held on the leafy campus of Williams College. Started in 1921, the program drew several hundred men and women—foreign leaders, academics, journalists, and activist citizens—to a month-long forum of lectures and round-table discussions designed to increase understanding of major problems around the world. Eliza bunked with the other women attendees in a dormitory, like at summer school.[24]

The new League of Nations provided another object of study. From the start, Eliza had been a staunch advocate of the League, formed in 1920 as an outgrowth of the Treaty of Versailles that ended World War I.[25] In her lifetime she had witnessed the tragic effects of wars on three continents; an international body to promote lasting peace between nations seemed a hopeful development for humanity. As her letters made clear, Eliza would never overcome her disappointment that the United States failed to join. President Woodrow Wilson had championed the League, but a stroke prevented him from making the push he needed to secure the backing of the U.S. Senate. Japan, however, became a charter member, after participating in the Paris Peace Conference as one of the five major Allies in the war.[26] The appointment of her Japanese friend Inazo Nitobe as one of League's undersecretaries gave Eliza a direct pipeline to developments in Geneva, Switzerland, where the League established its headquarters. In her later years she visited the Nitobes in Geneva, and met up with them occasionally in Nice, France, where they liked to vacation.

With the war at an end, Eliza rediscovered Europe. In 1921 she made a sweeping tour of a half-dozen countries. She returned a year later on a pilgrimage to the Scidmore (Scudamore) ancestral home. The manor house, church, and other monuments of Holme Lacy, in the English county of Herefordshire, were "very, very impressive," she told her cousin Mary. She could not say the same of the local inn, which was so dreary she fled to Oxford for two restful nights at the Randolph Hotel before returning to the "roar of London."[27]

Family connections seemed dearer than ever after the heavy losses during the war. It made all the more heartbreaking the tragic news Eliza received that fall. She was at home in Washington when a telegram arrived from the State Department on November 27, 1922, informing her that her brother George had died in Yokohama. He was sixty-eight. George had become one of America's oldest and longest-serving consular officers, in a career that included nearly forty years in Asia. Though modest and retiring in nature, "he was popular wherever he went," one obituary noted. "There was probably no man in the Consular Service in the Far East who was better known and more appreciated" (Figure 20.2).[28]

George died during Thanksgiving preparations in the American community. Several months earlier he had fainted during the dedication of a war memorial outside the Foreign General Cemetery in Yokohama. With the Prince of Wales as the guest of honor, George—stubbornly self-reliant, like his sister—had insisted on handling many of the details himself. Doctors laid his collapse to sunstroke and fatigue.

Those who rushed to his aid had to carry him only a short distance to his home. After their mother's death, George had

Figure 20.2. George Scidmore, who spent most of his long U.S. consular career in the Far East.
(Wisconsin Historical Society, WHI-91612)

moved into a house atop the Bluff, near the U.S. Naval Hospital and across the street from the cemetery where Mrs. Scidmore lay in the section reserved for Americans.[29] George and his two oldest friends in Japan—W. W. Campbell, general agent of the Pacific Mail Steamship Company, and E. W. Frazar, a lawyer who became the executor of George's estate—spent many hours over drinks and cigars reminiscing about their sailing days and other good times. "There are very few places on the coasts in this part of Japan which he did not visit in his yachts," a local newspaper said of George after his death.[30]

George's health had weakened after the fainting incident. That fall, he suffered an abscess of the teeth and jaw that required surgery and confined him to bed for a month. Though he remained in good spirits and seemed to be steadily improving, he died suddenly of heart failure early one morning.[31] His body was cremated the next day. Following Masonic rites and a brief memorial service, he was laid to rest alongside his mother, amid a colorful display of 200 floral wreaths sent from across the region.

Back in America, Eliza struggled to absorb the reality of her brother's death. The void it left was deep. Though separated for long periods, they had shared a home on and off throughout their lives, bound together not just as kin but as kindred spirits: worldly in outlook, spiritually at home in Japan, deeply devoted to their work. Because neither of them married, Eliza was now the last of their immediate family.

George's passing had implications beyond the grief Eliza felt. He represented her oldest and most durable ties to Japan. After four decades, one of the most defining chapters of her life was about to end.

***

Eliza went on to spend her own final years in Europe. Some accounts after her death would say she left America as a protest against the country's racist 1924 Immigration Law.[32] Adopted amid anti-immigration fervor after the war, the law enforced a quota system that severely limited the number of people from certain nationalities. Japanese and other Asians were banned altogether.[33]

Prejudice against Asians had been escalating for years. In 1913, the same year some of the cherry trees from Japan bloomed in Washington for the first time, California enacted a law that prevented Japanese residents from owning or leasing land in the state. The move inspired similar measures in other parts of the country, paving the way for Congress's passage of the restrictive immigration law.

Over the years, Eliza never hesitated to express her outrage over the anti-Japanese hostility. In Washington, people regarded her activism on Japan's behalf as part of who she was. Though her bias was evident, one reporter wrote, Miss Scidmore was no "dreamy idealist," but "a straightforward, logical-minded open champion [well] equipped for a doughty debate on the merits of the Flowery Kingdom and her people with any anti-Japanese."[34]

Eliza no doubt felt disillusioned by America's failure to live up to its democratic ideals. As for why she decided to leave the country and move abroad, circumstances point to a much deeper and more personal state of anguish. And in fact, even before the bill leading to the 1924 Immigration Law was introduced, she had already made her intentions fully clear.

***

After her brother's death, Eliza inherited his house and other property in Yokohama. Early in 1923, she went to Japan to settle the affairs of his estate. Traveling with a friend, she sailed home in July via Seattle, aboard the *President Jackson*. The *Evening Star* reported on Miss Scidmore's return from abroad, adding that she was expected to spend the winter in Washington.[35]

She had barely resettled when, on September 2, newspapers in America began reporting an unfathomable horror in Japan: a powerful earthquake had leveled much of Tokyo and Yokohama. Fires were raging in both cities. After twenty-four hours nearly all of Yokohama was destroyed.[36]

The catastrophe had arrived with little warning on the morning of September 1. Residents awoke to thundershowers, though by 10 o'clock the weather was clearing, and light breezes broke the oppressive heat. Then, at two minutes before noon, "the smiles vanished, and for an appreciable instant everyone stood transfixed at the sound of unearthly thunder," one survivor said.[37]

A rupture in the rocky continental plates beneath the Bay of Tokyo had set off a violent shaking of the earth. The waves of death and destruction that followed seemed apocalyptic. Buildings in Yokohama collapsed like dominoes; tsunami waves drowned fishermen at sea and slammed steamships anchored in the harbor. Guests sitting down to lunch at the Grand Hotel were pinned beneath collapsed timber. Many people who survived the initial destruction raced through the streets and into the hills, only to be consumed by an inferno of advancing flames. As the shuddering of the ground advanced across the Great Kanto plain toward Tokyo, it tossed trains from their tracks, triggered mudslides, and toppled entire villages.

It was the worst seismic disaster the modern world had ever seen. By the end of the day, the region lay in ruins, with 145,000 people dead and nearly two million left homeless. The victims, Eliza learned, included her brother George's successor in Yokohama, the consul general Max D. Kirjassoff, and his wife.

Reading the press accounts, Eliza could hardly bear it. Her heart ached at the enormity of the lives lost and the destruction of so much she loved. Early in childhood she had been uprooted from home in Madison, Wisconsin. Now, in the twilight of her life, she once again faced a profound sense of displacement that left her feeling untethered from ties to any one place. The twin tragedies of her brother's death and the disaster in Japan made her think seriously about her own mortality, and how she wished to spend her remaining years. While details of the horror in Japan were still unfolding, Eliza told her cousin Mary Oakley Hawley that she hoped to make a permanent home in the south of France.[38]

Eliza's engagements in Washington that fall included a lecture on Japan at the Twentieth Century Club's first meeting of the season, scheduled for October 4 at the Cosmos Club. The program announcement had already gone out when she canceled.[39] In late October, the *Washington Post* noted that Miss Scidmore had closed up her apartment and would sail shortly for Europe, "where she will remain for several years."[40]

Her destination was Nice, on the French Riviera. During the prewar Belle Époque era, Nice and neighboring towns of the Côte d'Azur had attracted high-profile winter crowds of European royalty and aristocrats. By the 1920s Nice had grown somewhat tawdry—less fashionable, but popular with a broad mix of people. Along the busy palm-lined Promenade des Anglais, dandies in white flannel suits mingled with diamond- and pearl-studded old ladies leading their dogs on leashes.[41]

Summer tourism was catching on in Nice, but Eliza felt drawn there in winter, when the mellow sunshine and gentle rhythm of the sea offered the soothing environment she craved. One day in November 1923, not long after she arrived, she visited the offices of the U.S. Consulate to have a letter notarized. Addressed to Everett Frazar, her late brother's friend and legal adviser, it authorized him to act as her agent for the sale of the property on the Bluff in Yokohama where her brother George's house had stood.[42]

Three months later, on February 4, 1924, Eliza sat down at her desk in Nice and typed up her own last will and testament.[43]

***

On Sunday, January 4, 1925, the *New York Times* heralded an important auction of Oriental art, set for the following Saturday. "Miss Scidmore's Collection on Display at Anderson Galleries," the *Times* informed the public. The objects could be viewed at the firm's showrooms, at Park Avenue and 59th Street. Besides the empress dowager's throne chair that Eliza had acquired in Peking and her prized imperial yellow Kangxi bowl, the 232 pieces on offer included:

A pair of seventeenth-century lacquered Japanese *fusuma* (paper sliding doors), decorated in gold on a black background, originally from the Momoyaya castle in Kyoto

A Ming period ovoid-shaped dark-blue porcelain jar incised with phoenixes and clouds

A brocade seat cover and an embroidered melon-shaped pillow from the Summer Palace in Peking

A set of five "chicken-blood red" Yung Ching porcelain plates

A set of "liver-red" porcelain plates of the Kuang Sun period

A set of twenty-two blue and white eighteenth-century Japanese Hirado ware porcelain plates, with delicately scalloped edges and a royal crest

Japanese, Korean, and Chinese silver and bronze pieces

Pottery vessels used in Japanese tea ceremonies, some now rather rare after so many were destroyed in the earthquake.[44]

The collection netted Eliza $6,480.[45] She had already arranged the sale, through Sloan's auction house, of the Asian objects from her brother's estate.[46] Those proceeds and a year's salary from George's government job—a payment of $8,000 authorized by Congress for her relief—gave Eliza the means to plan for her retirement.[47] In preparation for the move abroad, she sent lantern slides and other items to the Smithsonian, and put a few special possessions in storage in Washington.[48]

Eliza started the search for a new home by scouting out a place to live in Nice. Unfortunately, the prices turned out to be too dear, though she hoped for better prospects after the summer real estate boom petered out.[49] She went instead to Geneva, Switzerland, where her friends the Nitobes had a home in the suburbs, not far from the League of Nations.

By the spring of 1926, Eliza had settled into an apartment at 31 Quai du Mont Blanc, on a main boulevard overlooking Lake Geneva. As she told family back in America, springtime in the city was exquisite. Heavy rains made everything so green and lush that "the leaf buds swell while you look at them." The season brought "huge, pale red strawberries" and "dark purple lilacs that are almost black." The pleasures included "wonderful ruby sunsets on Mt. Blanc," and dramatic atmospheric effects as the clouds that brought showers and rainbows parted to reveal snow on the hilltops. She wrote, in what sounded like a note of contentment and relief: "I am here for keeps now."[50]

All her life Eliza had been self-possessed. So it remained as she adjusted to her new life abroad with petty complaints, but not self-pity. She had long accepted as a matter of fact the personal circumstances of her existence. The feeling of oneness she felt with herself and her surroundings was rooted in her character and enhanced no doubt by Buddhist ideas she absorbed in Japan.

Like the beauty of the cherry blossoms she cherished, every day brought constant reminders that life was transient; pleasure, as well as sorrow, was fleeting.

The fact was, the world was changing, and there could be no turning of the tides. In letters to friends and family, Eliza lamented the passing of the "glory days" in the Far East. "Just now, China crowds everything out of my mind," she wrote to her cousin Mary in the spring of 1927, amid reports of internal turmoil and violent clashes between nationalists and communists. "I jump for the papers like in war time."[51] In Japan, the previous winter had brought to the throne a new emperor, Hirohito, whose long reign would be marked by upheaval and militarism leading up to World War II. Meanwhile, mail from America all too often brought heartbreaking news of yet another friend's death. "That is the trouble with growing old," Eliza told Mary. "Soon no one left on the U.S. side."

Living in Geneva provided stimulation enough. The League of Nations offered interesting meetings to follow, and a parade of personalities ripe for wry commentary. Eliza entertained Americans passing through town, as she and her mother had done for years in Japan.[52] Having never lost her wanderlust, she made periodic jaunts around Europe—to Paris and Athens; to Como and Milan during Holy Week. Now and then, as an antidote to her insomnia, she spent restful periods in Nice, where a week or two at the Grand Hotel O'Connor, on the Rue du Congrès, was so blissful she slept "like a baby."[53]

One fall, after a few days in the French spa town of Aix-les-Bain, Eliza headed off to the Alps with her car and driver to follow in the footsteps of Hannibal and Napoleon across the mountain passes. When a fever and bronchitis forced her to abort the trip, she returned to Geneva and lay low for several days, to let everyone think she was still out of town. Confined to bed, she caught up with her reading, including Inazo Nitobe's latest book, and worked on a piece of needlepoint, a new hobby.[54] With her dog Meah Goo—the "the sweetest, softest little ball"—camped atop her, she had all the company she needed for the moment. The dog was so badly spoiled, Eliza herself admitted, that she had to ban it from her bedroom at night to get what little rest she could.[55]

***

Though Eliza was accustomed to bouts of solitude, loneliness could wear at times. The family ties that had once been such a central part of her life had shrunk. Her cousin Mary Oakley Hawley remained like a sister and her most faithful correspondent. While living in Geneva, Eliza pined for visits by Mary's brother Horace Oakley, whom she now looked to as the head of the family. A wealthy lawyer and philanthropist in Chicago, and a lifelong bachelor, Horace traveled regularly to Europe as a trustee of the American School of Classical Studies in Athens and because he owned a villa near Florence, Italy.[56]

She and Horace shared many scholarly and humanitarian interests, including peace activism and support for the American Red Cross.[57] Enticing him to visit, Eliza described the attractions of the local golf club, which had recently added nine holes. She herself was a member—not to play, but to take tea in the open air and go for walks on the grassy turf. Though she kept a room ready for him, Horace was an extremely busy man and eluded her frequent invitations. They met up occasionally, but he found it impossible to give his cousin all the time she wanted.

Unexpectedly, another relative arrived to fill the void. In the fall of 1925, soon after Eliza moved to Geneva, Mary Atwood, the great-granddaughter of Eliza's uncle David Atwood, turned up for a visit with her "Cousin Lillie."[58] A recent graduate of the University of Wisconsin, Mary was fleeing an unhappy situation at home. Though never a mother herself, Eliza could see that Mary needed guidance; she was obviously a bright young woman, but difficult, and immature for her age. With the help of Eliza and her friends, Mary found a French teacher and enrolled in local classes. She stayed through the winter, marking the start of a relationship that would become surprisingly tender for both of them, despite the fifty-year difference in their ages.

After teaching for a while back in the States, Mary returned to Geneva in the summers of 1927 and 1928. Eliza was impressed by how much Mary blossomed during that time. Now more confident and socially poised, she had started reading major newspapers at home to keep up with European politics and world affairs; on one occasion she even lectured on the League of Nations to a civic group in Detroit. That Mary seemed determined to continue her education in Europe and expand her horizons pleased Eliza immensely. "Mary is going to do us credit," Eliza wrote happily to Horace—that is, if she didn't lose "that good-looking head" of hers and "make a fool marriage."[59]

Mary remained in Europe in the fall of 1928 to travel with her elderly cousin to Italy. They planned to stay at Horace's villa, La Badia, to enjoy the silky autumn weather. Eliza had purchased a new Buick a year earlier, at the expense of any new clothes. She had her driver, Emile, lined up and ready to go.

Late in the day on October 4, however, Eliza experienced acute pains that forced them to call in a doctor. He arrived for a consultation, and within minutes, he whisked Eliza and Mary off to the hospital in his car. Tests confirmed appendicitis. An hour later, Eliza was wheeled out of the operating room. "She was a brick about it all," Mary told Horace, keeping him updated in telegrams and letters. Twenty-four hours after the surgery and morphine, Eliza awoke feeling wobbly but in little pain, and with no complications. As Mary reported to Horace: "It is very difficult for her to have to be still, because she is always so active."[60]

During the two-week hospitalization, while she worked to regain her strength, Eliza entertained a few visitors, including General William Crozier and his wife, Mary. Old friends from Washington, the couple had called on Miss Scidmore during their ten days in Geneva. Now they stopped by to wish her well before they left town. They all looked forward to meeting up again soon in Florence.[61]

Eliza and Mary Atwood still hoped to start for Italy in time to reach Horace's villa by the first of November. In the third week of October, however, Eliza took a turn for the worse when she developed heart and lung complications. The situation spiraled downward over the next week, as she experienced hallucinations and stubbornly resisted food and remedies. A specialist who was brought in eased her discomfort but concluded there was little more he could do.

Mary Atwood stood close by throughout the ordeal. As she told Horace later, she was near at hand when Eliza called out to her at about 1 o'clock in the morning on November 3. She was holding their elderly cousin's hand and head when Eliza died at 2:30, at the age of seventy-two. For a week before her death, Mary told Horace, Lillie had come to realize she was not going to get better. Still, she lived her final days to the fullest, "with her newspapers, and making her plans, and being certain I was doing it all right." In the end, Mary noted, it was the heart that finally took their indomitable cousin. "It just was worn out and could not be mended."[62]

Epilogue: At Home in the World

"It is probable that no American woman had a more cosmopolitan assembly of friends or more varied interests of work than Miss Scidmore has enjoyed," an obituary observed after Eliza's death.[1] William and Mary Crozier, having seen her so chatty and positive at the hospital, expressed the shock all her friends felt at the news of her passing. As Mary Crozier wrote in condolence to Horace Oakley: "I cannot think of that forceful, cheerful and loyal personality, as gone from our Earth."[2]

Eliza had requested cremation, and there was no funeral. But a few days after her death, on November 7, the Croziers and a few others gathered to memorialize her at the American Church in Geneva. A hymn, prayers, and some readings from Scripture, followed by personal remarks; it was all "very simple—as she would have wished it to be," Mary Crozier reported to Horace. The American and Japanese ministers both motored down from Bern to attend. The Japanese minister, Isaburo Yoshida, who had known Eliza twenty years, spoke movingly of how Japan had lost in her "one of its best and most staunch friends."[3]

With Horace still in America, Mary Atwood agreed to stay on in Geneva through the winter to close down the apartment and wrap up their cousin's affairs. Fortunately, most of the legal matters were taken care of. Eliza had left clear instructions with the American consul general in Geneva, along with a $1,000 Liberty bond to pay her debts.[4] In her will, she bequeathed a few personal possessions. A trunkful of Asian art and textiles she earmarked for a friend in New York; her silver, lace, and jewelry she left to her cousin Mary Oakley Hawley. Any remaining property of value, including items on loan to the National Museum in Washington, was to be sold and the proceeds donated to fund scholarships at two schools in Japan: the Doshisha Girls School in Kyoto, where the Scidmores' cousin Mary Denton taught for many years, and the Girls School of the Convent of the Infant Jesus, located on the Bluff in Yokohama.[5]

To Mary Atwood, Eliza left the sum of $3,000, "to be used for a year of study and travel in France." It was a vote of confidence in Mary, whose growing independence and desire to continue her studies in Europe seemed an endorsement of the kind of choices Eliza herself had made as a young woman. On top of Eliza's bequest, Mary got the surprise of her life when Horace wrote to say he was making her their Cousin Lillie's official heir, instead of him.[6] He relinquished his entire interest in Eliza's property, with the sole exception of a pink Oriental rug from the Geneva apartment, possibly of museum quality, that he thought would look good in his Italian villa.[7]

As Mary sorted through closets and drawers, it became clear that for all of Eliza's love of beautiful objects, she had been quite thrifty in her personal habits. As Mary told Horace, she found mostly old hats, shoes, dresses, and undergarments, "which are not in very good condition, but might keep some needy souls warm."[8] A sapphire ring and some amber beads appeared to be the only jewelry of any value. The biggest heartache came in making final arrangements for Eliza's dog, Meah Goo, who was sweet but badly spoiled and hard to handle. Mary knew that keeping the dog would be impossible once she started her new job at the Students' International Union in Geneva and moved into a small flat nearby. In the spring of 1929, she had Meah Goo euthanized. The caretaker at Eliza's apartment building buried the dog's body in the woods.[9]

Meanwhile, the task of going through Eliza's papers dragged on for months. It seemed she had never thrown anything out, Mary lamented to Horace.[10] One day during the process Mary received a visit from Eliza's longtime friend Mary Nitobe, whose husband was now one of the highest-ranking officials in the League of Nations Secretariat. Because she and Miss Scidmore had shared many confidences, Mrs. Nitobe explained to the younger Mary, she asked that all the letters between them be destroyed. In the end, Mary Atwood disposed of nearly all of Eliza's papers, telling Horace later that she did so "knowing it to be Cousin Lillie's wishes that all her personal letters be done away with."[11] She held in reserve only two small pocket travel diaries and the 400-page typewritten manuscript of an unpublished novel, titled "I, Anastasia." Mary sent the surviving papers off to Horace, along with the pink rug.

A few months after Eliza's death, a handful of her friends—both Japanese and American—began writing to Horace and Mary about the final disposition of her ashes. Ever pragmatic and unsentimental, Eliza had stated in her will that she wished her ashes to be "dispersed in some seemly way at the time and place of the cremation." She added, as though to stress her point: "I do not ask to have them preserved or deposited in any special place."[12] Wisconsin, Washington, Japan, and Geneva—none of the places where she had lived claimed her eternal allegiance. She preferred to make her everlasting home in the larger world by having her ashes scattered to the winds.

Horace and Mary were inclined to honor Eliza's wishes. But her friends implored the family to allow at least a portion of the ashes to be carried to Japan for burial in the plot that held the remains of her mother and brother. As Everett Frazar, the good friend of Eliza's late brother George, pointed out, having Miss Scidmore's name and remains added to the family's grave site would provide a lasting record of her life and death.[13]

The drumbeat of appeals worked. In September 1929, a Japanese official in the health section at the League of Nations carried an urn containing Eliza's ashes with him when he returned home. Everett Frazar collected the remains when they arrived. On the afternoon of November 30, about a hundred people gathered at the Foreign General Cemetery in Yokohama to bury Eliza's ashes in the family plot. Eliza's good friend Inazo Nitobe delivered the memorial address in English.[14]

***

And so, for several decades, Eliza's remains lay in the cemetery in Yokohama, while she grew all but forgotten in her native America. Articles about Washington's cherry trees often mentioned her as the earliest champion of the idea. Yet no one seemed to realize she was buried in Japan—the place that had inspired her vision—or to have much knowledge about her.[15]

A crack opened in her story in April 1962, when the Japanese marked the fiftieth anniversary of the cherry tree planting in Washington. To honor Eliza Scidmore's role in U.S.–Japan friendship, a local foundation wanted to present a memorial gift to her closest living relative. But no one could figure out who that was. When the U.S. Embassy in Tokyo tried to trace Scidmore's family, it came up empty-handed. Trying another lead, the Japanese sought the help of Walter Tobriner, president of the Board of Commissioners in Washington, who had sent a congratulatory letter for the cherry tree commemoration in Tokyo.[16]

Tobriner located a copy of Eliza's will in D.C. records. It mentioned her "young cousin, Mary Atwood," of Madison, Wisconsin. So Tobriner contacted the mayor of Madison, who in turn ordered the local police department to look into the matter as a "missing person" case. The search led, in the spring of 1965, to the discovery of Mary—then in her early sixties—living in Montreal. She had moved there with her husband, Henri T. P. Binet, a Canadian attorney whom she met while he was working in the International Labor Office of the League of Nations. Mary Atwood Binet still had in her possession a special keepsake of her "Cousin Lillie": the autograph book from Eliza's childhood.[17]

Twenty-five years later, Eliza's story again stirred interest in Japan with the 1987 publication of a Japanese translation of her book *Jinrikisha Days in Japan*.[18] Upon realizing she was buried in Yokohama, a group of local citizens visited her grave site during cherry blossom season to pay tribute to her as a friend of Japan. The visit led to the annual memorial service organized by the Eliza Scidmore Sakura-no-kai (Cherry Blossom Society).

Today, a prominent kiosk about Scidmore stands in the Yokohama neighborhood of Motomachi, near a path leading uphill to the cemetery where she is buried. A cameo of her also appears, alongside an image of the Japanese chemist Jokichi Takamine, on plaques at about 120 sites around Japan where so-called home-coming cherry trees were planted to memorialize the centennial of Washington's cherry trees. The saplings were propagated from surviving 1912 trees around the Tidal Basin and sent to Japan as part of ongoing efforts to preserve the trees' historic legacy and genetic lineage.[19]

During my research trip to Japan in 2013, it surprised me to discover how many people there knew of Scidmore. As one young woman told me: "Eliza Scidmore understood that *sakura* is the soul of the Japanese people." Her association with cherry blossoms intertwines her legacy with the nation's most powerful cultural symbol.

Nothing unites the Japanese more than *sakura*. Having arrived in Japan in the spring, I followed from my Yokohama hotel room TV news reports that tracked—in graphic clouds of bright pink—the *sakura zensen*, or cherry blossom front, as it moved across the archipelago. Because of variations in climate, tree varieties, and other factors, the country's cherry blossoms bloom in waves. The annual spectacle begins as early as January in the far south, around Okinawa, then moves northward to reach Kyoto, Tokyo, and Yokohama in late March or early April. The fanfare ends in May a thousand miles north in the higher latitudes of Hokkaido.

During the period of peak blooming, parks and temple grounds in Japan assume the air of a vast festival as people engage in the ancient tradition of *hanami*. As I walked in Eliza Scidmore's footsteps beneath the cherry blossoms at Mukojima in Tokyo, I felt moved by the spirit that led her, a century ago, to seek a similar cherry tree park in Washington. Having learned something of her personal story, I came away with an understanding of how it was not just the beauty of the trees that inspired her, but *hanami* itself that she wanted to re-create on the banks of the Potomac; that Japanese ritual of people from all walks of life coming together in a great mingling of humanity. Anyone who has felt the transcendent joy of strolling beneath the cherry blossoms ringing the Tidal Basin in Washington realizes just how well her vision was finally realized.

***

In her life of extraordinary achievement, Eliza Scidmore gives us an inspiring model of female ambition and success. Her story is also significant because she serves as a representational figure at a time of great transition in American history, when women's roles were being shaped by influences beyond their position in the family. Social attitudes were changing, individual expectations were expanding, options were growing. While it may be tempting to think of Scidmore as a woman who set out to smash gender barriers, it seems closer to the truth to view her as someone who found cracks in the wall and pushed her way through.

Scidmore possessed traits typical of successful women. She was smart, enterprising, hard-working, and forceful, and she clearly mastered the art of networking. Her success rested as well on the international power structure of the day in which she traveled with great advantages in many places as a well-connected American woman. And yet, in years of teasing out the details of her life, I kept feeling there was something more individual and particular that helped explained her success story; something deep at the core that led her to move through the world in the way she did.

An answer came when I revisited a book-length study the psychiatry expert Kay Redfield Jamison published years ago in which she explored the personality trait she calls "exuberance."[20] Suddenly, I could see in Eliza Scidmore an embodiment of many of the ideas Jamison discussed. Jamison views exuberance as the emotional wellspring of people who live with passion. "Their love of life and of adventure is palpable," she wrote, describing Theodore Roosevelt and John Muir among her examples. "Exuberant people," she says, "take in the world and act upon it differently than those who are less lively and less energetically engaged."[21]

Exuberance, Jamison goes on to explain, carries people to places they would otherwise not go, as irrepressible curiosity fuels an inner drive. The concept enabled me to see afresh the high-spirited young woman traveling in remote Alaska, whose joy and enthusiasm was so infectious it inspired the captain and pilot of her steamer to name an island for her in Glacier Bay. Scidmore's passion for discovery helped me appreciate her frustration, while traveling in Shanghai and other coastal cities of China, at meeting incurious Westerners who had no desire to learn anything about the local Chinese among whom they dwelled.

A characteristic of many exuberant people, Jamison observes, is that they are, by temperament, "incapable of being indifferent."[22] So I was reminded of Scidmore's activism for preservation of the Alaska wilderness; of her call for urgent action by the scientific community to save the crumbling temple of Borobudur on Java. Once the vision of Japanese cherry blossoms in Washington came to her, she was incapable of abandoning the idea.

While a desire for adventure was no doubt a motive in Scidmore's travels, her life as a whole seems best explained as an intellectual journey. Geography became the lens through which she came to understand the world in its absorbing diversity. Her early work as a newspaper journalist, reporting on little-known places and people, gave her a sense of purpose and direction, and provided a template for how to live her life. The expansiveness of thought she derived from her ever-widening range of experience kept her on course. Her tendency to live at something of a remove from the everyday world, even as she enjoyed its pleasures, gave her the space she needed to pursue life on her own terms.

Among the principles that guided Scidmore was a deep faith in human progress, whether of individuals or of nations. That attitude explains her admiration for the Japanese, whom she saw as enlightened in seeking the best for their country without losing sight of traditional values. It was that same impulse toward human progress that made her exasperated by cultural forces that impeded modernization in China and led her to see British colonialism as a positive force for development in India.

As I embarked on this study of Eliza Scidmore, I found her a compelling subject of female ambition because of questions about her life that resonated with my own experience and that of other women I know. Questions of how to make a life when one's instincts don't accord with family or societal expectations; how to reconcile a hunger for creative expression and freedom with the need to make a living; how to say true to oneself in a world that values conformity. As with most people, her life story was about finding her place in the world—in her case, literally as well as figuratively.

I came away from this study convinced that one of the biggest factors in Scidmore's success was a brazen embrace of her own individuality. Driven by inner values and a strong sense of her own worth, she acted with a self-assurance that translated as confidence. That and her basic cheerfulness and optimism won people over.

The gift of imagination, on top of her passion and talent, also helps explain Eliza Scidmore's legacy of success. In the spirit of all artists, visionaries, and other creative thinkers, she saw things in ways that were not typical of most people. The cherry trees in Washington are one product of that imagination. Another is the nature of her own life and identity. She figured out how to live a rich and fulfilling life outside the expectations of what a woman should do or be, in an expression of the most creative human act of all: that of self-invention. In the end, her story shows us what it means to follow your mind and heart.

Acknowledgments

Any book of historical nonfiction like this one builds on a deep foundation of knowledge and scholarship. I am indebted to the many historians, authors, archivists, librarians, and other experts whose work made it possible for me to tell this story.

For early encouragement and support, I am grateful, first of all, to Daniel and Jennie Sidmore, who welcomed me into their home and generously shared their research on the family. Fellow authors and Eliza Scidmore "fan girls" Ann McClellan and Andrea Zimmerman provided many useful leads. Former National Geographic News colleague David Braun gave me an important early forum for telling Scidmore's story, as did Miki Ebara of Japan's NHK television.

My book-writing group, still going strong after a decade, is awesome. Thank you, Bonny Miller, Cheryl LaRoche, Judi Moore Latta, Kenneth Ackerman, Michael Kirkland, Michael Scadron, Nancy Derr, Sonja Williams, and Tara McKelvey. Several workshops helped me find my way, and I want to acknowledge the instructors: Kyoko Mori at George Mason University; research guru Thomas Mann, Library of Congress; Kenneth Ackerman and David O. Stewart, Writer's Center in Bethesda, Maryland; and Rae Bryant, Johns Hopkins University. The support of David Everett at Johns Hopkins meant a lot to me as a writer, and I regret that his tragically premature death prevents me from sharing this book with him.

As a first-time biographer, I benefited greatly from a period of mentoring with James McGrath Morris under a Mayborn Fellowship in Biography. I want to thank Jamie and his wife, Patty, for sponsoring the award and for their warm hospitality during a residency in Santa Fe. I am also grateful to Biographers International Organization for giving me its 2017 Hazel Rowley Prize. BIO's Cathy Curtis proved the perfect coach at a critical stage of this book.

I spent several years researching this book at the Library of Congress, where my interactions with the excellent staff across many divisions made working there a pleasure as well as immensely productive. I got lucky the first week when I met another independent scholar, Asian studies specialist Peg Spitzer Christoff; sharing a study room with her was a joy. Research specialists Tomoko Steen, Mari Nakahara, Katherine Blood, and Betty Culpepper were consistently helpful, as were colleagues in the Women's History and Gender Studies Discussion Group. Hugs also to my fellow docents in the library's public tour program for their friendship and continued interest in this book.

For archival access at the National Geographic Society, I am indebted to Susan Fifer Canby, Cathy Hunter, Renee Braden, Greg McGruder, Sara Manco, Chelsey Perry, and retired photo archivist Bill Bonner. At the New York Public Library, Thomas Lannon and Tal Nadan went the extra mile to make a critical records collection available. Other individuals to whom I owe thanks include Anne McDonough at the D.C. History Center's Kiplinger Library; Deborah Shapiro, Caitlin Haynes, Gina Rappaport, and Daisy Njoku, Smithsonian Institution; Amy Morgan, U.S. Agricultural Library; Deirdre Hester and Sister Joanne Gonter, Georgetown Visitation; Simone Munson and Lisa Marine, Wisconsin Historical Society; Ken Grossi, Oberlin College Archives; Crystal Hurd and Kimberley Springle, D.C. Public Schools; Cynthia Sternau, Japan Society in New York; and Harold Wood of the Sierra Club's John Muir website. Historian Jay Sherwood of British Columbia provided important leads on Eliza Scidmore's travels in the Pacific Northwest.

For other research support, I am grateful to staff at the Alaska State Library and Archives; Boston Public Library; George Washington University's Library and Special Collections; Morgan Library in New York; New South Wales Library; New York Society Library; St. Louis Mercantile Library; Schlesinger Library at the Harvard Radcliffe Institute; Stark County Public Library in Canton, Ohio; University of British Columbia; U.S. National Archives; and U.S. National Park Service.

My travel for research was extra special in taking me to both Japan and Alaska for the first time. The kindness and hospitality I received in Japan was extraordinary. My deep thanks to Kaoru Onji, Eliza Scidmore Cherry Blossom Society; Mina Ozawa, Yamate Museum; Kunihiko Nakada, Japan–America Society; Kanami Nakatake Hori, Yokohama Archives of History; Hakuei Wikaya, NYK Maritime Museum; and Masakai Hasebe, Jokichi Takamine Research Foundation. Also, Mikado Fusako, Hiromitsu Nomura, Yoshitaka Watanabe, Mitsuo Isada, Toshie Komatsu, Keiko Akiyama, Akira and Yoshiko Yamamoto, and Rumi Umino. U.S. State Department veteran and good friend Laurie Trost kindly provided logistical advice. Michiko Okubo and Kyoko Tanitsu were perfect companions in retracing Eliza Scidmore's footsteps under the cherry blossoms at Mukojima. I am hugely grateful to Miki Ebara, Wakako Hisaeda, and their NHK colleagues for allowing me to share Scidmore's story with millions of TV viewers in Japan. In a remarkable story of cross-cultural connection, I am deeply indebted to Ichiro Fudai, an amateur historian in Hanamaki who sought me out online after my TV appearance in Japan, then generously spent many hours tracking down and translating records that led to some important findings in this book.

At Glacier Bay National Park in Alaska, Park Ranger Caitlin Campbell showed such enthusiasm for Eliza Scidmore's story that hanging out with Caiti was a delight. I want to thank her for contacting me and for arranging, with the help of Emma Johnson and Jacob McLaughlin, several presentations I made at Glacier Bay. I also want to thank Natasha Newman and her Boundless Productions film crew for including me in BBC2's "Great American Railroad Journeys" episode in Juneau, in which the host Michael Portillo interviewed me about Scidmore's 1893 guidebook on Alaska.

For friendship, hospitality, and other forms of support during this long book-writing project, others I want to thank include Susan Burnore, Tom and Carol Stephens, Tom Gliatto, Maria Rudensky Silver, Blair Gately, Dan and Kendall Silberstein, Suzanne Bowler, Kathleen Stoner, Paula Tarnapol Whitacre, Sara Fitzgerald, Yoko Moskowitz, and the late Vertrees Malherbe in Cape Town. Also, Joyce Finn and all the other "WACS" who made our writers' retreats in the woods energizing and fun. Adventures in Southeast Asia with Helen Leitch, Sharmini Blok, and others brought me to this book; lunches with former *National Geographic* colleague Betty Clayman-DeAtley reminded me why I stuck with it.

I feel fortunate, especially as a first-time author, to have found the perfect publisher in Oxford University Press. A chance meeting on an airplane opened the right doors when fellow D.C. science writer Rich Blaustein referred me to his editors, Ania Wronski and Sonke Adlung, who commissioned this book. Project editor Giulia Lipparini won my deep admiration for her grace and competence in ushering the book through the myriad steps of publication. My thanks also to everyone involved in production and to Stephen Aucutt for his help on contract issues.

A huge shout-out goes to my large family in Ohio, who listened for years to my stories about Eliza Scidmore and never stopped believing I could write this book. Finally, I want to express my deepest thanks to my infinitely patient and supportive spouse, Bruce Parsell. No one could make a better companion, in travel or in life, than this man who first won my heart through his witty way with words and his ability to make me laugh. That hasn't changed.

Notes

Prologue

1 Translation by Toshie Komatsu, in conversation with the author, March 2013.

2 Sabin, *A Historical Guide to Yokohama*, 230–2.

3 As part of internal power struggles at the time, an antiforeigner movement arose in Japan "to revere the Emperor and repel the barbarians." Hane and Perez, *Modern Japan*, 68.

4 Patricia McCabe, *Gaijin Bochi: The Foreigners' Cemetery* (London: British Association for Cemeteries in South Asia, 1994), ix–x.

5 Eliza Ruhamah Scidmore [ERS], *Jinrikisha Days in Japan*, 1.

6 In his highly influential book *Orientalism* (Pantheon, 1978), Edward Said writes that knowledge about the East in the nineteenth-century was largely a false, "Orientalized" picture based on Euro-centric prejudices that presented peoples and cultures of Asia and the Middle East as alien and inferior. Such assumptions provided a basis for the colonial and imperialist policies of Western powers that saw themselves as bringing "enlightenment" to uncivilized people.

7 Lloyd Griscom, quoted in Hammersmith, *Spoilsmen in a "Flowery Fairy-land,"* vii.

8 ERS, *Jinrikisha Days in Japan*, 1–2.

9 For general background on Japan and its history in the late-nineteenth and early-twentieth-century span of this book, see Reis-chauer, *Japan: The Story of a Nation*; also Hane and Perez, *Modern Japan: A Historical Survey*, 1–199.

10 Colta Ives, "Japonisme," in *Heilbrunn Timeline of Art History* (New York: Metropolitan Museum of Art, 2000), http://www.metmuseum.org/toah/hd/jpon/hd_jpon.htm (October 2004). Elements of Japanese art that influenced the Impressionists and other artists included tightly cropped composition, lack of perspective, blocks of vibrant color, and decorative treatment of simple, everyday subjects. The approach marked a great departure from the formal academic conventions of Greco–Roman art.

11 Alice Fahs, *Out on Assignment* (Chapel Hill: University of North Carolina Press, 2011).

12 In light of colonialist and other connotations raised by scholar Edward Said and others, the word *Oriental* is losing favor as a general

reference to people and cultures of Asia and the Middle East. The term *Oriental art* is used throughout this text as appropriate to the nineteenth century.

13 McCarry, *From the Field*, 4–5.
14 Clifford, *A Truthful Impression of the Country*, 177.
15 ERS, *Winter India*, 84; *St. Louis Globe–Democrat*, May 12, 1881.
16 Clifford, *A Truthful Impression of the Country*, 30.

Chapter 1

1 This explains why many biographical accounts of Eliza Ruhamah Scidmore give Madison, Wisconsin, as her birthplace.
2 Warren Skidmore provides the most extensive genealogy of the family in *Thomas Skidmore (Scudamore)*, . . . (6th ed., 2010). He calls an earlier history by Emily C. Hawley, *A Genealogical and Biographical Record of the Pioneer Thomas Skidmore* (1911), useful but not fully accurate for relying on "hearsay evidence" as well as records. Family background also comes from Sidmore, "Eliza Ruhamah Scidmore: More Than a Footnote" (2000).
3 ERS, "I, Anastasia," Chapter 22, 1–3, Mss. QM, Oakley–Hawley Papers, Wisconsin Historical Society.
4 Erika Janik, *A Short History of Wisconsin* (Madison: Wisconsin Historical Society, 2010), 37.
5 Mollenhoff, *Madison: A History of the Formative Years*, 46.
6 *Proceedings of Wisconsin Editorial Association*, June 1870 (Madison, WI: Atwood & Culver, 1871), 42.
7 Foner, *Free Soil, Free Labor, Free Men*, xx–xvii.
8 Sidmore, "More Than a Footnote," 8.
9 George Bolles Scidmore was the son of Solomon Scidmore, of Saratoga Springs, New York, and the former Ruhamah Bolles Woodward, a Maryland native and reportedly the daughter of a French military officer who accompanied General Lafayette to America when he arrived to fight with George Washington in the Revolutionary War. Hawley genealogy, 69.
10 Connor Sweeney's second wife, the maternal grandmother of Eliza Ruhamah Scidmore, was Susan Schreiver Meyers, a widow of Pennsylvania Dutch ancestry. The family's shop and home was at Tuscarawas Street West, now a major intersection in Canton, Ohio. John Danner, *Old Landmarks of Canton and Stark County, Ohio*, vol. 1 (Logansport, IN: B. F. Bowen, 1904), 180, 189–90.
11 Heppen and Otterstrom, *Geography, History, and the American Political Economy*, 185.

12 *Evening Repository* (Canton, OH), July 11, 1927.
13 Eliza C.'s only brother, John L. Sweeney, may have led the family's move to Wisconsin. Identified in records as a miner and furniture maker, he later went west to Montana on the trail of the pioneer settler John Bozeman. Joaquin Miller, *An Illustrated History of the State of Montana* (Chicago: Lewis Publishing, 1894), 431.
14 The quote appears in Thwaites, *The Story of Madison*, 14.
15 Butterfield, *History of Dane County, Wisconsin*, 592–3; *St. Lawrence* (NY) *Plain Dealer*, April 28, 1935.
16 Degler, *At Odds*, 168–9.
17 The divorce is described in *Trial of Impeachment of Levi Hubbell, Judge of the Second Judicial Circuit, by the Senate of the State of Wisconsin, vol. 15, June 1853* (Madison: Argus & Democrat Steam Press, 1853), 4–15, 604–6.
18 *Acts and Resolves Passed by the Legislature of Wisconsin in the Year 1850* (Madison: Dave Dickson, State Printer, 1850), 24.
19 The U.S. Census of 1850 lists her at two different addresses.
20 Sidmore, "More Than a Footnote," 9. Land records at Huntertown Historical Society in Allen County, Indiana, show that in January 1852, George Scidmore sold forty acres of land for $160.
21 Matthew George Easton, *Easton's Bible Dictionary* (1897).
22 Detail on Frances Lavicy Brooks from typewritten family tree of "Connor Murray Sweeny" in Oakley–Hawley Papers.
23 In their respective genealogies, Skidmore and Hawley differ on the exact lines of the family's origins in Britain, but both date ancestors to at least the eleventh century, in Herefordshire, England.
24 Sidmore, "More Than a Footnote," 14.
25 A. P. De Milt, *Story of an Old Town, With Reminiscences of Early Nebraska and Biographies of Pioneers* (Omaha: Douglas Publishing, 1902), 34, 132.
26 The U.S. Census of 1860 recorded what appeared to be three other couples living with the Scidmores, as well as a half-dozen young people who likely worked as hired hands.
27 Barbara Welter, "The Cult of True Womanhood: 1800–1860," *American Quarterly* 18 (Summer 1966): 151–74; DuBois, *Feminism and Suffrage*, 16–18.
28 Throughout his life, Atwood held nearly every civic and political office available to a man of his day, from justice of the peace to a member of the U.S. Congress. Atwood obituary, *Wisconsin State Journal*, December 12, 1889; Reuben Thwaites, "General David Atwood," *Magazine of Western History* 5 (Cleveland: 1886): 549–64. On the historic Atwood home, see Gary Tipler, *Mansion Hill: Glimpses of Madison's Silk Stocking District* (Madison: 1981).

29 *Japan Gazette*, November 29, 1922.
30 *National Tribune* (Washington, DC), June 20, 1895.
31 Rufus R. Dawes, *Service with the Sixth Wisconsin Volunteers* (Marietta, OH: A. D. Alderman, 1890), 148.
32 Hurn, *Wisconsin Women in the War*, 6–7; Dawes, *Service with the Sixth Wisconsin Volunteers.*
33 American Civil War Soldiers, Ancestry.com.
34 *Wisconsin Daily Patriot*, July 15, 1861.
35 Silber, *Daughters of the Union*, 20.
36 Letter of Edward P. Brooks, October 20, 1861, published in *Wisconsin State Journal*; photocopy in Oakley–Hawley Papers.
37 Military pension records for George B. Scidmore, National Archives, Washington, DC.
38 Skidmore genealogy, 621–3.
39 Adage cited in Degler, *At Odds*, 47. A distant cousin, Daniel Sidmore, speculated in interviews with the author that poor health related to chronic sinusitis may have factored into George B. Scidmore's desire to live in the far West, where the dry climate would have promised some relief.
40 *Rocky Mountain News*, August 22, 1862.
41 Military pension records for George B. Scidmore, National Archives, Washington, DC.
42 George B. Scidmore married Melissa A. Ferris in Ogle County, Illinois, in 1868; by 1870 they had separated. Sidmore, "More Than a Footnote," 3.
43 Homestead claim no. 1499, Pueblo County, Colorado Territory.
44 Newspapers and public records show that George B. Scidmore worked as an insurance executive in St. Louis and as a carpenter, land agent, seller of farm equipment, and assistant postmaster in other places.
45 Sidmore, "More Than a Footnote," 3–4.
46 Silber, *Daughters of the Union*, 206; Wagner, *Library of Congress Civil War Desk Reference*, 701.
47 "Diary of George W. Stoner, 1862," *Wisconsin Magazine of History* 21, no. 2 (December 1937): 210–12.
48 It's not clear whether George H. Scidmore accompanied his mother and sister to Washington or continued his schooling in Madison for a while and joined them later. They were all together in Washington by 1864.

Chapter 2

1 Winkle, *Lincoln's Citadel*, xiv.

2 Details here and throughout on Civil War Washington come chiefly from Winkle and from Green, *Washington: A History of the Capital, 1800–1950*; Janke, *A Guide to Civil War Washington, D.C.*; Dickey, *Empire of Mud*; and Ziparo, *This Grand Experiment*.

3 Quoted in Ziparo, *This Grand Experiment*, 102.

4 *Daily Morning Chronicle*, March 5, 1867.

5 See Berg, *Grand Avenues*.

6 Twain and Warner, *The Gilded Age*, 221–2.

7 Before the early 1900s, the White House was commonly referred to as the *President's House* or *Executive Mansion*. This book uses the more familiar term *White House* for continuity.

8 Cornelius W. Heine, "The Washington City Canal," *Records of the Columbian Historical Society, Washington, D.C.* 53/56 (1953/1956): 23.

9 Winkle, *Lincoln's Citadel*, 358–60.

10 Janke, *A Guide to Civil War Washington*, 67–8.

11 "Prospectus," *Catalogue of Pupils of the Georgetown Academy of the Visitation, B.V.M., for the Academic Year 1862–63* (Baltimore: Kelly, Hedian & Piet, 1863), 3. Georgetown Visitation operates today as a prep school, at P and 35th Streets NW.

12 See Mary A. Mitchell, "An Intimate Journey Through Georgetown in April 1863," *Records of the Columbia Historical Society, Washington, D.C.* 60/61 (1960/1962): 84–102. Georgetown was a separate town until it became part of Washington in 1871.

13 Janke, *A Guide to Civil War Washington*, 66.

14 Details on the school come mostly from Eleanore C. Sullivan and Susan Hannan, *Georgetown Visitation, Since 1799*, 2nd ed. (Washington, DC: Georgetown Visitation, 2004), xi–xiii, 118–31. The author also benefited from site visits arranged by Visitation alumna Deirdre Hester and Sister Joanne Gonter, who provided access to the school's archives.

15 Georgetown Visitation website, http://www.visi.org/about-us/history/index.aspx.

16 *Catalogue of Pupils, 1862–63*, 7–11. Many of Visitation's archives were destroyed in a fire, but surviving registries show "Lilly Scidmore, [from] Wisconsin" enrolled from at least 1862 to 1864. She may have completed her schooling elsewhere, as a *Washington Post* obituary of November 4, 1928, states that she was educated at Visitation "and in private schools."

17 *Catalogue of Pupils, 1862–63*, 17–18.

18 On American geography education in the nineteenth century, see Schulten, *Mapping the Nation,* 11–40, and *The Geographical Imagination in America*, 11–38; also "19th Century Maps by Children," David Rumsey Map Collection, https://www.davidrumsey.com.

19 While Scidmore approached her foreign reporting with the aim of promoting cultural understanding and sympathetic human connection, she also perpetuated images and ideas that Western countries drew on to justify colonialism, including the view of some countries as "inferior." She found the "picturesqueness" of India's diverse groups appealing, but also admitted revulsion toward the native people in general for what she called their lack of attractiveness and charm. She endorsed British imperialism as an instrument for stability and progress in a sprawling nation where "race hatreds, jealousies, animosities, and antipathies would never permit a native leader to be acceptable to all the native malcontents." *India, the Long-Lived Empire*, xiv. For a related critique, see Kase Sease, "Cherry Blossoms, Travels Logs, and Colonial Connections: Eliza Scidmore's Contributions to the Smithsonian," August 18, 2020, Smithsonian Institution Archives blog, https://siarchives.si.edu.

20 *Daily Inter Ocean* (Chicago), October 18, 1890, and August 3, 1895. Georgetown Visitation's "Prospectus" shows that special attention was given in geography classes to the use of globes. *Catalogue of Pupils*, 3. Education historian Kim Tolley writes in *The Science Education of Girls: A Historical Perspective* (New York: Routledge Falmer, 2003) that the study of geography, more than any other discipline, opened the door to science for girls and women in the nineteenth century.

21 The Scidmores' residences over the years are listed in various editions of *Boyd's Washington and Georgetown City Directory.*

22 *Washington Post*, March 9, 1913.

23 Record of Clerks in the Office of the Commissioner of Internal Revenue 1862–1869, RG 58, National Archives, Washington, DC; *Federal Register of Employees, 1865*, 33.

24 On the hiring of female federal workers during the Civil War, see Ziparo, *This Grand Experiment*. The Treasury Department became a receptive environment for female employees after Treasurer Francis E. Spinner found women more dexterous than men at trimming and sorting bank notes.

25 Boardinghouse work offered one of the few respectable occupations available to a white middle-class woman. See Wendy Gamber, *The Boardinghouse in Nineteenth-Century America* (Baltimore, MD: Johns

Hopkins University, 2007). As at many boardinghouses, many of Mrs. Scidmore's patrons came from referrals by family and friends. Amasa Cobb of Wisconsin boarded with her at 23 4½ Street West while serving in the thirty-ninth Congress. *National Intelligencer*, December 8, 1865. One of her regular guests during the war was the Rev. John Pierpont, a Unitarian minister and impassioned abolitionist from Boston, and the grandfather of the American financier J. P. Morgan. Pierpont was a widely published poet, and during a stay with Mrs. Scidmore he dedicated one of his poems to young Eliza. John Pierpont to Mrs. Eliza C. Scidmore, September 25, 1865; Mrs. Eliza C. Scidmore to J. Pierpont Morgan, March 5, 1909, John Pierpont Morgan Papers, Morgan Library and Museum, New York.

26 In the 1880 Census, Eliza C. Scidmore gave her marital status as "widowed." Her October 20, 1916, obituaries in the *Evening Star* and *Madison Democrat* both describe her as having been "a widow" when she arrived in Washington early in the war.

27 Wagner, *Civil War Desk Reference*, 312–13.

28 Obituary of Mrs. Scidmore, *Madison Democrat*, October 20, 1916.

29 Records of the Office of the Secretary of War, 1791–1948, RG 107, National Archives, Washington, D.C.

30 Brooks was reportedly captured again that summer with a group of his men in an ill-fated incident around Lee's army known as "Brooks Raid." He escaped from the Confederate Salisbury Prison in North Carolina and made his way up the coast to Washington. E. P. Brooks obituary, *Wisconsin State Journal*, May 6, 1893.

31 Mrs. Lathrop E. Smith, "My Recollections of Civil War Days," *Wisconsin Magazine of History* 2, no. 1 (September 1918): 26–39.

32 Martin G. Murray, "Traveling with the Wounded: Walt Whitman and Washington's Civil War Hospitals," *Washington History: Magazine of the Historical Society of Washington, D.C.* 8 (Fall/Winter 1996–97); online at www.whitmanarchive.org.

33 Amanda Akin Stearns, *The Lady Nurse of Ward E* (New York: Baker & Taylor, 1909), 8.

34 Horace Oakley, a cousin of Eliza and George, described the incident when he donated George's autograph book to the Newberry Library in Chicago. Report of the Trustees, Newberry Library (Chicago, 1922), 19. Eliza's album, containing the signatures of seventy-seven eminent men and women, was acquired by an unidentified buyer at an auction in New York on June 22, 2012. Christie's Sale 2572, Lot 250.

35 "The Route of Abraham Lincoln's Funeral Train," http://rogerjnorton.com/Lincoln51.html.
36 City directories show the Scidmores living in 1870 at 1335 F Street NW, a "boardinghouse," and for several years at 1407 F Street NW. The 1870 Census noted that Eliza C. Scidmore, "a 46 yo white female from Ohio keeps house worth $2K."
37 See Kenneth R. Bowling, "From Federal Town to National Capital: Ulysses S. Grant and the Reconstruction of Washington, D.C.," *Washington History* 14, no. 1 (Spring/Summer, 2002): 8–25.
38 Miller, *Washington in Maps, 1606–2000*, 94–9; Caemmerer, *Historic Washington*, 58–9; "The New Washington," *The Century*, March 1884, 643–59.
39 Caemmerer, *Historic Washington*, 59.
40 *Evening Star*, July 21, 1870; *Federal Register of Employees, 1871*, 26. An 1879 federal registry of civil-service employees shows that Mrs. Scidmore earned $900 a year as "matron" at the Treasury Department. In an odd coincidence, the U.S. Treasurer for part of the decade she worked there was Albert U. Wyman, her stepson during her brief marriage to William Wyman. There is no evidence they interacted during their overlapping service.
41 *New York Times*, October 3, 1865; Matthew Harper, "Living in God's Time: African-American Faith and Politics in Post-Emancipation North Carolina" (PhD diss., University of North Carolina–Chapel Hill, 2009), 42.
42 D.C. directories and other sources show that Edward P. Brooks worked as a correspondent for the *New York Times, New Orleans Republican,* and *Philadelphia Press.*
43 Ritchie, *Press Gallery*, 73–9. The term *lobbyist* emerged from the practice of hanging out in the bar at the Willard Hotel to request favors from politicians.
44 *Janesville* (Wisconsin) *Gazette*, February 10, 1876.
45 *Evening Star*, October 3, 1859.
46 *Evening Star*, October 30, 1871; *Wisconsin Sentinel*, April 9, 1874; *National Republican*, January 12, 1875.
47 *Washington Post*, March 11, 1913. In response to an inquiry from the author in 2012, Marti Scheel of Luther Place Memorial discovered, in the church's archives, a brass pew plate, No. 74, bearing the name "Scidmore." Such plates were made to acknowledge individual donations of fifty dollars to support the church's founding.
48 *New York Times*, July 6, 1873; Sidmore, "More Than a Footnote," 10. A legation secretary in London clearly annoyed by Mrs. Scidmore's demands on his office derided her as "a vulgarian to whom such a

letter should never have been given." The Papers of Ulysses S. Grant, vol. 24: 1873, John Y. Simon, ed. (Southern Illinois University Press, 2000), 183–4.

49 Butterfield, *History of Dane County*, 548. Charles Atwood no doubt obtained his appointment through his father's influence, as the U.S. consul in Liverpool was Lucius Fairchild, a former governor of Wisconsin.

50 William Benning Webb et al., *Centennial History of the City of Washington, D.C.* (Dayton, OH: H. W. Crew, 1892), 509–11; *Report of the Commissioner of Education for the Year 1873* (Washington: U.S. Government Printing Office, 1873), 440. National University grew out of a desire, dating from President George Washington's day, to establish a "flagship" university in the nation's capital. At its opening in 1869, the school offered mainly legal studies. National University later merged with George Washington University.

51 Mrs. Scidmore's older son, Edward P. Brooks, also received a consular appointment, serving as U.S. consul to Cork, Ireland, from 1880–1882. E. P. Brooks obituary, *Wisconsin State Journal*, May 6, 1893.

52 According to a federal report, about 11,000 young women were pursuing higher education in 1879. They attended ladies' seminaries, land-grant universities, coed colleges, and several new all-female colleges. Horowitz, *Alma Mater*, 56.

53 John Barnard, *From Evangelicalism to Progressivism at Oberlin College, 1866–1917* (Columbus: Ohio State University, 1969), 3.

54 *Pictorial Memories of Oberlin* (Oberlin, OH: Rotary Club of Oberlin, 1976).

55 Oberlin College Archives, RG 2811, Alumni Records (nongraduates), 225. A note in the files indicates that the Preparatory Department was comparable to a modern high school.

56 Barnard, *From Evangelism to Progressivism at Oberlin College*, 18–21; *Catalogue of the Officers and Students of Oberlin College for the College Year 1873–74* (Cleveland: Fairbanks, Benedict and Co., Herald Office, 1873).

57 Ronald W. Hogeland, "Coeducation of the Sexes at Oberlin College: A Study of Social Ideas in Mid-Nineteenth-Century America," *Journal of Social History* 6, no. 2 (Winter 1972–3): 160–76; see also Degler, *At Odds: Women and Family in America*, 310. Many of Oberlin College's female graduates at the time became missionaries or the wives of missionaries.

58 *Catalogue of the Officers and Students of Oberlin College for the College Year 1874–75* (Oberlin, OH: Pratt and Brattle, 1874).

59 *Oberlin Review*, October 10, 1875, 192–93.

60 Horowitz, *Alma Mater*, 11–90; Louise Stevenson, *The Victorian Homefront*, 134.

Chapter 3

1 Faith K. Pizor, "Preparations for the Centennial Exhibition of 1876," *Pennsylvania Magazine of History and Biography* 94, no. 2 (April 1970): 231–2.
2 "The World's Festival," *National Republican*, May 12, 1876, by "Ruhamah."
3 *The United States International Exhibition* (Philadelphia: Times Printing House, 1875), 17.
4 William Dean Howells, "A Sennight of the Centennial," *Atlantic Monthly*, July 1876, 92–107.
5 Greenhalgh, *Ephemeral Vistas*, 2–15; "United States Centennial Exhibition," Centennial Exhibition Digital Collection, Free Library of Philadelphia, https://libwww.freelibrary.org/collections/centennial/exhibition.
6 "The United States International Exhibition: The Organization, the Work Proposed, the Work Already Done" (Philadelphia, 1876), n.p.
7 Pizor, "Preparations for the Centennial Exhibition of 1876"; *Visitors' Guide to the Centennial Exhibition and Philadelphia* (J. B. Lippincott, 1876); "Tours," Centennial Exhibition Digital Collection, Free Library of Philadelphia, https://libwww.freelibrary.org/digital/feature/centennial/tours.
8 "Uncle Sam's Show," *National Republican*, May 12, 1876.
9 Mary Frances Cordato, "Toward a New Century: Women and the Philadelphia Centennial Exhibition, 1876," *Pennsylvania Magazine of History and Biography* 107, no. 1 (January 1983): 113–35; "Women Working, 1800–1930," Harvard University Library Open Collections, http://ocp.hul.harvard.edu/ww/centennial.html.
10 Ruhamah [ERS], *St. Louis Globe-Democrat*, September 3, 1876.
11 Ruhamah [ERS], *National Republican*, May 12, 1876. All of Scidmore's details and quotes on the Women's Pavilion are from this source.
12 Harper, *Women During the Civil War*, 311–13.
13 Editors used the term *letters* for general freelance dispatches.
14 Ruhamah [ERS], *National Republican*, May 12, 1876.
15 Pizor, "Preparations for the Centennial Exhibition of 1876," 231–2.
16 Gray, *Reluctant Genius*, 132.
17 Ruhamah [ERS], *National Republican*, May 13, 1876.
18 Schlereth, *Victorian America*, 1.
19 Cashman, *America in the Gilded Age*, 12–23.

20 Cited in "Tours," Centennial Exhibition Digital Collection, Free Library of Philadelphia, https://libwww.freelibrary.org/collections/centennial/tours.
21 Ruhamah [ERS], *National Republican*, October 20, 1876.
22 Elizabeth Mitchell, *Liberty's Torch: The Great Adventure to Build the Statue of Liberty* (New York: Atlantic Monthly Press, 2014), 113–22.
23 Greenhalgh, *Ephemeral Vistas*, 105.
24 *New York Times*, May 23, 1876.
25 Ruhamah [ERS], *National Republican*, May 13, 1876.
26 Anna Jackson, "Imagining Japan: The Victorian Perception and Acquisition of Japanese Culture," *Journal of Design History* 5, no. 4 (1992): 245–56.
27 Bonita Billman, "Looking East: Japonisme in American Art," Smithsonian Associates Seminar, Washington, DC, September 29, 2012.
28 Ruhamah [ERS], *National Republican*, May 12, 1876.

Chapter 4

1 *St. Louis Globe–Democrat*, December 18, 1876. Eliza's columns from the capital bore the label "Washington Gossip" and were signed with the pen name "Ruhamah."
2 Formed from an 1875 merger, the *St. Louis Globe–Democrat* became the region's leading newspaper until overtaken by Joseph Pulitzer's *St. Louis Post–Dispatch* at the turn of the century. Jim Alee Hart, "The Other Newspaper in St. Louis," *Journalism Quarterly* 39, no. 3 (September 1962): 324; Charles Clayton, *Little Mack: Joseph B. McCullagh of the St. Louis Globe–Democrat* (Carbondale: Southern Illinois University, 1969), 77–86.
3 Edward Brooks and Joseph McCullagh were among the two dozen charter members of the short-lived Washington Correspondents Club, organized in 1867 to professionalize the capital's press corps by weeding out lobbyists and office-seekers. Bryan, *History of the National Capital*, 586.
4 Fahs, *Out on Assignment*, 242.
5 *St. Louis Globe–Democrat*, December 25, 1876.
6 Scidmore's details and quotes on the event are from *St. Louis Globe–Democrat*, January 7, 1877.
7 Anthony, *Nellie Taft*, 31–2.
8 Anthony, *Nellie Taft*, 32.

9 Maurine Hoffman Beasley, *The First Women Washington Correspondents*, Washington Studies No. 4 (Washington, DC: George Washington University, 1976), 25.
10 Ritchie, *Press Gallery*, 145–57.
11 By 1879, nineteen women were accredited to the congressional press galleries. A year later, women all but disappeared from the roster, after a rules change restricted privileges to reporters who wrote for dailies and filed their dispatches by telegraph. Because most women, as well as black reporters, were freelancers who submitted their stories by mail or wrote for weeklies, the move effectively barred them from eligibility. Beasley, *First Women Washington Correspondents*, 25; Ritchie, *Press Gallery*, 145–6, 260n8.
12 An 1879 painting of the Senate chamber by Cornelia Fassett, now hanging in the U.S. Capitol, shows women in the press gallery among the male reporters. United States Senate, "Art & History," https://www.senate.gov/artandhistory/art/common/image/Painting_33_00006.htm.
13 Jacob, *Capital Elites*, 64.
14 Emily Edson Briggs, *The Olivia Letters*, January 1869; quoted in Jacob, *Capital Elites*, 64.
15 Twain and Warner, *The Gilded Age*, 220.
16 *Wisconsin State Journal*, January 4, 1883; *Washington Hatchet*, July 2, 1886.
17 *St. Louis Globe–Democrat*, March 11, 1877.
18 Fedler, *Lessons from the Past*, 18.
19 As a reporter in Washington, Joseph McCullagh popularized the now-common use of direct quoting of sources when he employed the technique in an exclusive interview with President Andrew Johnson during his impeachment. Many editorial and marketing practices McCullagh implemented in twenty years as editor of the *St. Louis Globe–Democrat* were later adopted, often to more sensationalized effect, in the New Journalism of publishers such as Joseph Pulitzer and William Randolph Hearst. Jim Allee Hart, *A History of the* St. Louis Globe–Democrat (Columbia: University of Missouri, 1961), 136–56; Clayton, *Little Mack*, 86–7.
20 Hart, "The Other Newspaper in St. Louis," 329.
21 Pay rates from Clayton (97), Fahs (34), and Ritchie (152–3). A reporter or editor typically earned about thirty-five dollars a week, equivalent to eight hundred dollars today (based on a calculator at https://www.measuringworth.com/). Some writers with a strong following earned as much as thirty or forty dollars a column.

22 Scidmore's quotes and details on the journey are from *St. Louis Globe–Democrat*, August 12, 1877.

23 The American writer Bret Harte became well known for his stories featuring miners, gamblers, and other colorful figures of the California Gold Rush.

24 *St. Louis Globe–Democrat*, August 26, 1877.

25 This and related quotes from *St. Louis Globe–Democrat*, December 2, 1877.

26 "A Lesson from the Nez Perces," *New York Times*, October 15, 1877.

27 Orlando S. Goff set up studios at Bismarck and Fort Lincoln in the 1870s. He took the last-known portrait of General George Custer not long before his death and traveled around the Great Plains in a wagon studio photographing Native Americans. See Louis N. Hafermehl, "Chasing an Enigma: Frontier Photographer Orlando S. Goff," *North Dakota History* 81, no. 2 (Spring 2016): 3–26, https://www.history.nd.gov/publications/goff-article.pdf.

28 *St. Louis Globe–Democrat*, September 23, 1879.

29 *National Republican*, May 21, 1881.

30 "Newspaper Work, How and By Whom It Is Done at Washington," *St. Paul Daily Globe*, February 18, 1884.

31 Jacob, *Capital Elites*, 118.

32 *St. Louis Globe–Democrat*, February 5, 1882.

33 O'Toole, *The Five of Hearts*, 94–5, 381.

34 Passes for trial attendance, Miscellany, 1881–82, o.1, c.f., Oakley–Hawley Papers.

35 *St. Louis Globe–Democrat*, July 2, 1882.

36 *St. Louis Globe–Democrat*, July 13 and 14, 1882.

37 *St. Louis Globe–Democrat*, July 30 and August 1, 1882.

38 *St. Louis Globe–Democrat*, August 8, 13, and 17, 1882.

39 *St. Louis Globe–Democrat*, June 1, 1882.

40 *St. Louis Globe–Democrat*, August 20 and 22, 1882.

41 Schlereth, *Victorian America*, 217–18; Fahs, *Out on Assignment*, 68. Kate Field, America's best-known female journalist at the time, helped popularize outdoor adventures in the Adirondacks by writing about her own experiences in the woods, inspired by the nature writings of Henry David Thoreau. Field's interest in the area included efforts to preserve the abolitionist John Brown's former home as a landmark open to visitors. Gary Scharnhorst, *Kate Field: The Many Lives of a Nineteenth–Century American Journalist* (Syracuse, NY: Syracuse University Press, 2008), 68–72.

42 *St. Louis Globe–Democrat*, September 22, 1882.

43 *National Republican*, May 21, 1881.
44 "Newspaper Work, How and By Whom It Is Done at Washington," *St. Paul Daily Globe*, February 18, 1884.
45 Eve LaPlante, *Marmee and Louisa* (New York: Simon & Schuster Paperbacks), 281.
46 ERS, "I, Anastasia," Chapter 22, p. 1, Wis Mss. QM, Oakley–Hawley Papers.
47 Fahs, *Out on Assignment*, 235–6.
48 Scharnhorst, *Kate Field*, xi.
49 In one example, Scidmore wrote to an acquaintance in Boston in 1885: "I quite count upon having Miss Field in our midst next winter." ERS to Lilian Whiting, Ms. KF.1650, Special Collections and Rare Manuscripts, Boston Public Library.

Chapter 5

1 ERS, *Alaska: Its Southern Coast and the Sitkan Archipelago*, v. Because the book merges Scidmore's reporting from two Alaska journeys, in 1883 and 1884, her actual itineraries may have been somewhat different from the published narrative.
2 See Ted C. Hinckley, "The Inside Passage: A Popular Gilded Age Tour," *Pacific Northwest Quarterly* 56, no. 2 (April 1965): 67–74.
3 Scidmore (*Alaska*, 2) and others used the analogy in writing about Alaska, a reference that reflected public interest in adventures such as the newspaperman Henry Stanley's 1871 quest to track down the British explorer David Livingstone, who went missing in Africa.
4 Borneman, *Alaska: Saga of a Bold Land*, 106–12.
5 ERS, *Alaska*, 1.
6 Borneman, *Alaska*, 101–5; W. P. Woodring, "William Healey Dall" (Washington, DC: National Academy of Sciences, 1958), 95–8; Geological Survey Professional Paper, vol. 567, U.S. Geological Survey (Washington, DC: Government Printing Office, 1967), 39.
7 The William Healey Dall Papers at the Smithsonian Institution include Scidmore's letters to Dall requesting copies of his *Pacific Coast Pilot* to carry with her on her second Alaska trip. ERS to William H. Dall, June 17 and July 12, 1884. She cited her reliance on Dall's expertise, including his seminal work, *The Resources of Alaska*, in the preface to her own book.
8 Borneman, *Alaska*, 109. Population figure from 1890 Census.
9 Hinckley, "The Inside Passage," 67–8.
10 Muir quote from Hinckley, "The Inside Passage," 68.

11 John Muir, "The Discovery of Glacier Bay," *Century Illustrated Magazine*, June 1895, 235; ERS, "The Muir Glacier in Alaska," *Harper's Weekly*, July 23, 1892, 711.

12 Rick Kurtz, *Glacier Bay National Park and Preserve Historic Resources Study* (Anchorage, AK: National Park Service, 1995).

13 Muir, "The Discovery of Glacier Bay," 243; Gifford, *John Muir: His Life and Letters and Other Writings*; Robert Engberg and Bruce Merrell, eds., *John Muir: Letters from Alaska* (Madison: University of Wisconsin, 1993). For many years John Muir was credited with being the first white man to "discover" Glacier Bay, an assertion Scidmore made in some of her own writings. Historical records indicate that others explored parts of the bay earlier, some in the company of native Alaskans, though access was limited by ice blockages.

14 Engberg and Merrell, *John Muir: Letters from Alaska*; xi, xvi–ii. In what was likely the first organized excursion to Alaska along the Inside Passage, the railroad magnate Henry Villard and Army general Nelson Miles hosted a sightseeing part of eighty people in 1881. Campbell, *In Darkest Alaska*, 94.

15 Richard Engeman and Chet Orloff, "City of Portland Civic Planning, Development & Public Works: A Historic Context," Bureau of Planning and Sustainability, Portland, OR, 2009.

16 ERS, *Alaska*, 5.

17 E. W. Wright, ed., *Lewis and Dryden's Marine History of the Pacific Northwest* (Portland, OR: Lewis and Dryden, 1895), 150; Will Lawson, *Pacific Steamers* (Glasgow: Brown, Son and Ferguson, 1927), 198.

18 Borneman, *Alaska*, 145–7.

19 ERS, *Alaska*, 28–30.

20 In Scidmore's repeated travels to Alaska, her journalistic work included ethnographic articles on native customs and crafts, including baskets, bracelets, ornamental spoons, and textiles such as Chilkat tribal blankets woven from the fine hair of mountain goats.

21 John Muir, *San Francisco Daily Evening Bulletin*, September 6, 1879; quoted in Engberg and Merrell, 11.

22 ERS, *Alaska*, 83.

23 In *Alaska* (112), Scidmore related the story of Davidson's stay in the home village of the Chilkat chief Kloh-Kutz in the summer of 1869, during an expedition to view an eclipse. Under Davidson's influence, she wrote, the chief had the name "Seward" tattooed on his arm, in honor of "the great *Tyee*" who had bought Alaska from the Russians.

Scidmore may have heard the story from Seward's son Frederick, whom she acknowledged in the preface of the book.

24 Scidmore later wrote about the Klondike gold rush in "The Stikene River in 1898," published in *National Geographic* in January 1899.

25 ERS, *Alaska*, 105.

26 Willoughby also achieved fame as a teller of tall tales. In the most outrageous one, he produced a photograph in 1889 of what he said was a phantom city suspended above Muir Glacier, which had appeared to him in a mirage. He sold thousands of copies to tourists for fifty cents each, and newspapers around the world carried reports of the "Silent City." The hoax was finally revealed when a reader identified the city in the picture as Bristol, England. ERS, *Appletons' Guide-Book to Alaska*, 106.

27 ERS, *Alaska*, 128.

28 "Reports of Captain L. A. Beardslee, U. S. Navy, relative to affairs in Alaska, and the operations of the U.S.S. *Jamestown* under his command, while in the waters of that territory," S. Exec. Doc. No. 71, 47th Cong., 1st Sess. (1882), 70–1, 93–4. The name Glacier Bay was officially adopted in 1890.

29 ERS, *Alaska*, 131.

30 According to Scidmore, "her" island lay at latitude 58° 29′ north and longitude 135° 52′ west from Greenwich. *Alaska*, 131–2. Several late-nineteenth-century travelers referred to Scidmore Island in published accounts of their own visits to Glacier Bay. The name of the island was later changed. GIS specialist Whitney Rapp at Glacier Bay National Park speculates that it may have been today's Strawberry Island or Young Island. Email correspondence with the author, July 31, 2018.

31 ERS, *Alaska*, 132–3.

32 "Willoughby Island," https://alaska.guide/island/willoughby-island.

33 ERS, *Alaska*, 134.

34 ERS, *Alaska*, 135, 143.

35 ERS, *Alaska*, 139.

36 ERS, "The Discovery of Glacier Bay," *National Geographic*, April 1896, 144.

37 ERS, *Alaska*, 154.

38 ERS, *Alaska*, 293–4.

39 ERS, *Alaska*, 289.

40 ERS, *Alaska*, 16.

41 ERS, *Alaska*, 21.

Chapter 6

1 *St. Louis Globe–Democrat*, November 4, 1883.
2 *St. Louis Globe–Democrat*, November 22, 1883.
3 "The River Front Improvement," *Washington Star*, November 16, 1883. Scidmore almost certainly read the *Star* article because her own later article in the November 25, 1883, *St. Louis Globe–Democrat* used phrasing similar to that of Hains.
4 Scott, *Capital Engineers*, 79.
5 Weather Conditions, *Washington Star*, November 21, 1883.
6 Cowdrey, *A City for the Nation*, 31–2; Scott, *Capital Engineers*, 125.
7 *New York Times*, January 27, 1882.
8 Cowdrey, *A City for the Nation*, 32.
9 Scott, *Capital Engineers*, 127.
10 Scott, *Capital Engineers*, 126–7; Cowdrey, *A City for the Nation*, 31–2.
11 *Washington Star*, November 16, 1883.
12 Jacob, *Capital Elites*, 167–76.
13 "Jackson Hole and the President Arthur Yellowstone Expedition of 1883," https://jacksonholehistory.org/jackson-hole-the-president-arthur-yellowstone-expedition-of-1883.
14 *St. Louis Globe–Democrat*, November 18, 1883.
15 *Cleveland Leader*, September 30, 1883; quoted in Jacob, *Capital Elites*, 169.
16 *St. Louis Globe–Democrat*, March 21, 1877.
17 "The New Washington," *The Century*, March 1884, 643–59.
18 George J. Olszewski, *A History of the Washington Monument, 1844–1968* (Washington, DC: National Park Service, 1971), 28–32.
19 *St. Louis Globe–Democrat*, November 25, 1883.
20 The height record was eclipsed four years later with the building of the Eiffel Tower in Paris.
21 *St. Louis Globe–Democrat*, November 25, 1883. Scidmore's critical comments may have been inspired by the unfavorable views of some people in the architectural community who deemed the monument "one of the blankest, meanest, ugliest and most unbecoming piles that ever encumbered the globe." Quoted in Cowdrey, *A City for the Nation*, 28.
22 Monument details from Olszewski, *A History of the Washington Monument*, 27–49. Also, Thomas Allen, *The Washington Monument: It Stands for All* (New York: Discovery Books, 2000); Chalmers M. Roberts, *The Washington Monument: The Story of a National Shrine* (Washington, DC: Litho Process Co., 1948).

23 Details and quotes on Scidmore's reporting that day from *St. Louis Globe–Democrat*, November 25, 1883.

24 The Tidal Basin began as four ponds, or tidal reservoirs, with intake and outflow gates that flushed out the Washington Channel at each tide to keep the water from stagnating and silting up. Scott, *Capital Engineers*, 124, 126.

25 ERS, "The Wonders of Alaska," *New York Times*, October 27, 1884; "The Indians of Alaska," *New York Times*, November 23, 1884.

26 ERS, *Alaska*, 7.

27 ERS, *Alaska*, 8.

28 ERS, *Alaska*, 296.

29 ERS, *Alaska*, iv–v.

30 Scidmore's 1885 Alaska book is now considered a classic of travel literature. Isabel C. McLean, "Eliza Ruhamah Scidmore," *Alaska Journal* 7, no. 4 (Autumn 1977): 238.

31 ERS to Lilian Whiting, July 8, 1885, Ms. KF. 1650, Special Collections and Rare Manuscripts, Boston Public Library.

32 "The Evening Lamp," *Christian Union*, June 4, 1885.

33 *Literary World*, April 18, 1885.

34 ERS to Lilian Whiting, July 8, 1885, Boston Public Library.

Chapter 7

1 Wharf details from Robert A. Weinstein, "Northwest from Panama, West to the Orient: The Pacific Mail Steamship Company," *California History* 57, no. 1 (Spring 1978): 51.

2 *New York Times*, May 14, 1874.

3 "SS *City of Tokio*," https://en.wikipedia.org/wiki/SS_City_of_Tokio.

4 ERS, *St. Louis Globe–Democrat*, August 15, 1885.

5 ERS, *Jinrikisha Days*, 2.

6 Cited in Hammersmith, *Spoilsmen in a "Flowery Fairyland,"* 141, 310n50.

7 Benfey, *The Great Wave*, xvii–iii.

8 ERS, *Jinrikisha Days*, 3.

9 *Japan Weekly Mail*, June 27, 1885; *San Francisco Chronicle*, June 26, 1885; E. Mowbrey Tate, *Transpacific Steam* (New York: Cornwall Books, 1986), 32.

10 *St. Louis Globe–Democrat*, August 18, 1885.

11 ERS, *Jinrikisha Days*, 6–7.

12 ERS, *Jinrikisha Days*, 4.

13 At one point during her stay Scidmore tested what it felt like to pull one of the carriages. She described the methods of the "jinrikisha coolies" that enabled them to "trot along as regularly as a horse." ERS, *Jinrikisha Days*, 8–9.
14 Benfey, *The Great Wave*, xii–iv; Richard Hofstadtler, *American Political Tradition* (New York: Vintage, 1938), 206.
15 See Allen Hockley, "Globetrotters' Japan: Foreigners on the Tourist Circuit in Meiji Japan," *MIT Visualizing Cultures*, https://visualizingcultures.mit.edu.
16 On Isabella Stewart Gardner's travels in Japan and China, see the website of the museum that bears her name, at https://www.gardnermuseum.org/.
17 ERS, *Jinrikisha Days*, 16.
18 ERS to Lilian Whiting, July 8, 1885, Special Collections and Rare Manuscripts, Ms. KF. 1650, Boston Public Library.
19 *St. Louis Globe–Democrat*, September 20, 1885.
20 "The Tea Trade," *Jinrikisha Days*, 350–58.
21 *St. Louis Globe–Democrat*, September 20, 1885.
22 U.S. State Department Foreign Service List, vol. 1901.
23 *Japan Weekly Mail*, June 18, 1887; George Hawthorne Scidmore, *A Digest of Leading Cases Decided in the United States Consular Court at Kanagawa, Japan, . . .* (R. Meiklejohn & Company, 1882).
24 *St. Louis Globe–Democrat*, November 11, 1885.
25 *St. Louis Globe–Democrat*, December 26, 1885. Some restaurants today still offer *ikizukuri*—fish or other organisms served and eaten alive—as a traditional Japanese delicacy, though the practice is waning out of concerns about animal cruelty.
26 ERS, *Jinrikisha Days*, 24–5.
27 "About *The Hatchet*," Chronicling America, https://chroniclingamerica.loc.gov/lccn/sn82014159/. The paper cheekily adopted the slogan "I can't tell a lie," drawn from the apocryphal story of George Washington chopping down a cherry tree.
28 *St. Louis Globe–Democrat*, July 22, 1886.
29 *Japan Weekly Mail*, November 27, 1886; *Japan Weekly Mail*, January 22, 1887.
30 Washington *Evening Star*, March 3, 1886.
31 Eliza Catherine Scidmore to Minnie Oakley, August 1, 1887, Oakley–Hawley Papers, Wisconsin Historical Society.
32 *St. Louis Globe–Democrat*, September 11, 1887.
33 ERS, "'Chin, Huang Ta-Ta!'" *Wide Awake*, March 1892, 342–53.

34 *St. Louis Globe–Democrat*, September 7, November 25, and December 2, 1887.
35 ERS, *Jinrikisha Days*, 174.
36 ERS, *Jinrikisha Days*, 173.
37 ERS, *Jinrikisha Days*, 177.
38 ERS, *Jinrikisha Days*, 180.
39 ERS, *Jinrikisha Days*, 186–7.
40 ERS, *Jinrikisha Days*, 188.

Chapter 8

1 Newspaper datelines indicate that Scidmore wrote for the *St. Louis Globe–Democrat* from 1876 to the end of 1887.
2 See Frank Luther Mott, "The Magazine Revolution and Popular Ideas in the Nineties," *Proceedings of the American Antiquarian Society* 64, Part 1 (April 1954): 195–214. In 1870 about 1,200 magazine titles were being produced in the United States; by 1890 the number had grown to 4,500.
3 On the popularity of Japanese goods and other foreign imports among middle-class consumers at the turn of the twentieth century, see Hoganson, *Consumers' Imperium: The Global Production of American Domesticity, 1865–1920*, and Yoshihara, *Embracing the East: White Women and American Orientalism*.
4 *Science*, February 15, 1889, 120; ERS, "Korean Women," *Harper's Bazaar*, March 9, 1889, 171.
5 The Women's Anthropological Society was founded in Washington in 1885 as the first American organization to promote scientific investigation by women. Its members included women who did pioneering work in areas such as ethnological studies of Native American cultures. "The Women's Anthropological Society of America," *Science*, March 29, 1889, 240–2.
6 ERS, "Yoshi Hito, Haru no Miya, the Child of Modern Japan," *St. Nicholas*, July 1889, 671.
7 ERS, "Haruko of Japan," *Frank Leslie's Popular Monthly*, October 1890, 470.
8 Junior League of Washington, *The City of Washington*, 244, 252–68.
9 Olszewski, *A History of the Washington Monument*, 1844–1968, 56–9.
10 "Making the First Trip," *Washington Post*, October 10, 1888.
11 ERS, "The Cherry Blossoms of Japan," *The Century*, March 1910, 653.
12 ERS, "The Cherry Blossoms of Japan," 650.

13 ERS, "The Cherry Blossoms of Japan," 645.
14 ERS, *Jinrikisha Days in Japan*, 65.
15 Mari Nakahara and Katherine Blood, *Cherry Blossoms: Collections from the Library of Congress* (Washington, DC: Smithsonian Books, 2020), 21.
16 ERS, "The Cherry Blossoms of Japan," 648.
17 A horticultural magazine reported during the Civil War that an American scientist had returned from Japan with "fifteen new double flowering cherries," one with blossoms "as large as a rose." Jefferson and Fusonie, *The Japanese Flowering Cherry Trees of Washington, D.C.*, 2–3; John L. Creech, "Highlights of Ornamental Plant Introduction in the United States," Longwood Program Seminars, No. 6 (Kennett Square, PA: Longwood Botanic Gardens, 1974), 21–5.
18 ERS, "Capital Cherry Blossoms Gift of Japanese Chemist," Washington *Sunday Star*, April 11, 1926.
19 ERS, "The Cherry Blossoms of Japan," 653.
20 On cherry tree cultivation in America, see Samuels, *Enduring Roots.*
21 Among the most famous images of Uyeno Park in cherry blossom season are hand-colored photographs from *Japan: Described and Illustrated by the Japanese*, a multivolume, limited-edition collection edited by Captain Francis Brinkley and published in 1897–8.
22 ERS, "The Cherry Blossoms of Japan," 653.
23 ERS, "The Cherry Blossoms of Japan," 653. Abe describes in *The Sakura Obsession* (204–6, n345) how domesticated Yoshino cherry trees, with their soft pink blooms, became the dominant variety in Japan during the Meiji era, making up the majority of cherry trees in public parks like Mukojima. About sixty percent of the cherry tree saplings sent to Washington in 1912 were Yoshinos.
24 E. W. Clement, "The Japanese Floral Calendar," *Open Court*, April 1904, 215.
25 ERS, *Jinrikisha Days*, 74.
26 ERS, "Capital's Cherry Blossoms Gift of Japanese Chemist."
27 ERS, "Twelve Varieties of Japanese Cherry Trees Are Blooming in Potomac Park," *Washington Star*, March 27, 1921; "Capital's Cherry Blossoms Gift of Japanese Chemist," *Sunday Star*, April 11, 1926.
28 ERS, "Family Life in the White House," *Harper's Bazaar*, June 1, 1889, 404. A similar article by Scidmore appeared April 20, 1890, in the *New York Herald.*
29 Annette Scherber, "'Underrated' First Lady Caroline Scott Harrison: Advocate for the Arts, Women's Interests, and Preservation of the White House," Indiana History Blog, https://blog.history.in.gov/.
30 ERS, San Francisco *Morning Call*, November 29, 1891.

31 Society items in the *Evening Star* and *Washington Post* show that Eliza Scidmore and Mary Harrison McKee gathered periodically with friends until at least 1913.

32 *Washington Post*, January 26, 1890.

33 On at least one occasion, Scidmore socialized with Colonel Ernst and his daughters at a musical program the Harrisons hosted at the White House. *Evening Star*, April 26, 1890.

34 See Scott, *Capital Engineers* (72), on the responsibilities associated with the position of superintendent of the Office of Public Buildings and Grounds.

35 ERS, "Capital's Cherry Blossoms Gift of Japanese Chemist."

36 Apart from her complaints in several published articles, see Scidmore's letter to Robert U. Johnson, August 20, 1909, Century Collection, New York Public Library.

37 Frederic V. Abbot, "Fiftieth Annual Report of the Graduates of the United States Military Academy" (Saginaw, MI: Seeman and Peters, 1919); "John Moulder Wilson (1837–1919)," 5, compiled by Jerry Olson, Olson Engineering, http://www.olsonengr.com/download/globios/wilsonjohnmbio.pdf.

38 ERS, "Capital's Cherry Blossoms Gift of Japanese Chemist."

39 ERS, "Capital's Cherry Blossoms Gift of Japanese Chemist."

40 William Seale, *The Imperial Season*, 11.

41 "A Day at Waukesha—Bethesda's Birthday," *Wisconsin State Journal*, August 16, 1889.

42 At nearly fifty, living alone in Washington, Scidmore wrote to her elderly Aunt Jane in Wisconsin imploring her to visit "for as much of the month of April as you can give me." ERS to Jane Sweeney Oakley, March 1903, Oakley–Hawley Papers, Wisconsin Historical Society.

43 Quoted by Jane Sweeney Oakley to Walter Oakley and family, January 7, 1891, Oakley–Hawley Papers.

44 Lilian Whiting, "Brainy Boston Women," *Bismarck Daily Tribune*, April 29, 1890. Scidmore and Whiting likely first met in Washington at one of the frequent soirées that correspondent Mary Clemmer Ames held at her home for local and out-of-town women journalists. See *Milwaukee Sentinel*, February 18, 1883; also, ERS letter to Lilian Whiting, July 8, 1885, Special Collections and Rare Manuscripts, Ms. KF. 1650, Boston Public Library.

45 Elizabeth A. Thompkins, "Literary Washington," *Cosmopolitan*, December 1889, 191.

46 *American Law Review*, January/February 1888. Today the school is Chuo University.
47 Obituary of George Scidmore, November 28, 1922, *Japan Advertiser*.
48 *Philadelphia Inquirer*, June 10, 1890; Award from Imperial Japan to G. H. Scidmore, U.S. Vice Consul General at Yokosuka, July 4, 1889, translated from Japanese in email correspondence to Daniel Sidmore in December 2006.
49 Quote from ERS, "I, Anastasia," Part 3, p. 11, Wisc Mss. QM, Oakley–Hawley Papers.
50 ERS, "The First District of Alaska from Prince Frederick Sound to Yakutat Bay," *Report on Population and Resources of Alaska at the Eleventh Census: 1890* (Washington, DC: U.S. Government Printing Office, 1893), 42–53.
51 Mary Louise Atwood to Mary Oakley, July 27, 1890, Oakley–Hawley Papers.

Chapter 9

1 Hinckley, "The Inside Passage," 71. According to U.S. Census figures cited in Hinckley, summer tourists traveling the Inside Passage increased from 1,650 passengers in 1884 to 5,007 in 1890.
2 *Milwaukee Sentinel*, September 9, 1890.
3 Victoria Wyatt, *Images from the Inside Passage: An Alaskan Portrait by Winter and Pond* (Seattle: University of Washington Press; Juneau: Alaska State Library, 1989), 23.
4 "All About Alaska," Pacific Coast Steamship Company (1890), 3–4, 15.
5 Campbell, *In Darkest Alaska*, 19.
6 Heacox, *John Muir and the Ice That Started a Fire*, 97.
7 M. Wood, diary entry August 6, 1887; cited in Campbell, 62.
8 See Shaffer, *See America First: Tourism and National Identity, 1880–1940.*
9 "Are Nervous Diseases Increasing?" *The Century*, May 1896, 146
10 *Milwaukee Sentinel*, September 9, 1890.
11 Scidmore included John Muir in the acknowledgments of her 1885 book *Alaska.* Soon after its publication she sent him a copy with a note attached: "One of these raw democratic officials here casually asked me the other day if I 'knew anything about Alaska.' I shall be anxious to know what your opinion may be on that subject." ERS to John Muir, December 27, 1885, John Muir Papers, Holt–Atherton Special Collections, University of the Pacific.

12 Scidmore may have had an introduction to the Muirs through Jeanne Carr. A May 20, 1876, letter by Carr in the Muir Papers indicates that Mrs. Carr and Mrs. Scidmore had a mutual friend in Madison, Wisconsin. Scidmore's own letters in the collection make it clear that she visited the Muirs, Carrs, and Hoopers over the years.
13 ERS to Richard Watson Gilder, August 10, 1890, Century Collection.
14 Noonan, *Reading* The Century *Illustrated Monthly Magazine*, xi–iii.
15 Johnson, *Remembered Yesterdays*, 88–95, 112.
16 Noonan, *Reading* The Century, xiii, 17; Johnson, *Remembered Yesterdays*, 99–100. Photocopied pages in files of the Eliza R. Scidmore Collection of the Smithsonian Institution's National Anthropological Archives (NAA Photo Lot 139) show examples of illustrations *The Century* made from some of Scidmore's later photographs from her travels in Asia.
17 Camera details from West, *Kodak and the Lens of Nostalgia*, 23–4, 50–1. Scidmore was clearly an early adopter of Kodak photography, as only 3,250 of the No. 1 cameras were manufactured by 1889, and 7,000 No. 2 Kodaks in 1890. The twenty-five-dollar price made the cameras unaffordable to most households.
18 Noonan, *Reading* The Century, xviii, 33–4, 199n40. Scidmore faced strong odds in approaching *The Century*. Noonan notes that while women writers were well represented in an "evangelized and feminized" phase of *The Century* in the 1870s, by the 1880s men dominated the main content, with contributions from women limited mostly to poetry.
19 Lilian Whiting, "Boston Life: A Chatty Interview with Miss Scidmore, the Traveler and Correspondent," *Daily Inter Ocean* (Chicago), October 10, 1890.
20 *Milwaukee Sentinel*, September 21, 1890.
21 Johnson, *Remembered Yesterdays*, 283.
22 Worster, *A Passion for Nature*, 310–16; Fox, *The American Conservation Movement*, 1985, 98–9.
23 Heacox, *John Muir and the Ice That Started a Fire*, 72–3. The school is now Case–Western Reserve University.
24 *Milwaukee Sentinel*, September 21, 1890.
25 ERS, *Appletons' Guide-Book to Alaska*, 96.
26 ERS to Louisa Strentzel Muir, December 31, 1890, Muir Papers.
27 ERS to John Muir, October 22, 1890, Muir Papers.
28 ERS, "The Alaska Boundary Question," *The Century*, May 1896, 145.

29 Because the Cosmos Club banned women from using the front door, female guests had to enter through a separate entrance around the corner on H Street NW.
30 Bryan, *The National Geographic Society*, 24.
31 For example, John Wesley Powell's botanist sister, Ellen Powell Thompson, accompanied her husband and brother on field expeditions. William Dall's mother, Caroline Healey Dall, was an outspoken feminist and social reformer. Anita McGee, the daughter of astronomer Simon Newcomb and the wife of geologist W J McGee, practiced medicine in Washington, D.C.
32 Jenkins, *National Geographic: 125 Years*, 34.
33 San Francisco *Morning Call*, December 7, 1890.
34 ERS to John Muir, October 22, 1890, Muir Papers.
35 Heacox, *John Muir and the Ice That Started a Fire*, 63–4.
36 John Muir to Robert U. Johnson, October 16, 1890, Century Collection.
37 Alice W. Rollins to Louisa Wanda Muir, n.d., Muir Papers.
38 *Evening Star*, November 14, 1890. Scidmore visited Georgina Jones, the wife of silver-mining magnate and Republican senator John P. Jones, a cofounder of Santa Monica. The Joneses had just built a seventeen-bedroom oceanfront villa they named Miramar.
39 Worster, *A Passion for Nature*, 314–20.
40 ERS to John Muir, October 22, 1890, Muir Papers.
41 "The National Geographic's Society's First Expedition Leader," National Geographic Blog, January 5, 2018. The 1890 expedition team named some peaks and glaciers, including Mount Hubbard for the National Geographic Society's president and chief patron, Gardiner G. Hubbard. During a follow-up expedition a year later, Israel Russell and his party named Mount Ruhamah for Eliza Scidmore. In 1937, Scidmore Glacier and Bay, in the upper western flank of Glacier Bay, were also named for her.
42 See Patrick D. Sylvestre, "The Art and Science of Natural Discovery: Israel Cook Russell and the Emergence of Modern Environmental Exploration" (master's thesis, Colorado State University, 2008).
43 ERS to John Muir, October 22, 1890, Muir Papers.
44 "Among Vast Glaciers," *Daily Inter Ocean* (Chicago), October 20, 1890. The same article appeared in many newspapers under different headlines.
45 ERS to John Muir, October 22, 1890, Muir Papers.
46 John Muir to Robert U. Johnson, October 16, 1890, Muir Papers. John Muir and Mark Kerr were among the founding directors of

the Sierra Club, established in 1892 to provide public stewardship of Yosemite and other wilderness areas of Northern California.

47 ERS to Louisa Strentzel Muir, December 31, 1890, Muir Papers.

48 *National Geographic*, April 1891, 293; *Morning Call*, December 7, 1890.

49 *National Geographic*, April 1891, 293.

50 ERS to Louisa Strentzel Muir, December 31, 1890, Muir Papers.

51 *Evening Star*, November 12, 1890; D.C. Real Estate Atlas, Permit 0432, August 22, 1889, microfilm box 129, Historical Society of Washington, DC.

52 Lilian Whiting, "Miss Eliza Ruhamah Scidmore: The Globe-Trotter Who Has Written of Japan, China, and Alaska," syndicated in *Springfield Republican*, September 19, 1902.

53 Jane Sweeney Oakley to Walter Oakley family, January 7, 1891, Oakley–Hawley Papers.

Chapter 10

1 ERS, "The Disputed Boundary Between Alaska and British Columbia," *The Century*, July 1891, 473–5.

2 *Literary World*, August 29, 1891.

3 *San Francisco Chronicle*, June 21, 1891; *New York Times*, June 22, 1891; *Washington Post*, June 23, 1891; and *The Critic*, August 22, 1891.

4 See, for example, ERS, "What Has the United States Done with Alaska?," *The Century*, March 1895.

5 ERS to John Muir, October 22, 1890, Muir Papers.

6 ERS to Louisa Strentzel Muir, December 31, 1890, Muir Papers.

7 ERS to Robert U. Johnson, [n.d.], Century Collection.

8 Details and quotes from ERS, "The Disputed Boundary Between Alaska and British Columbia," *The Century,* July 1891, 473–5; also a later article, "The Alaska Boundary Question," *The Century*, May 1896, 143–6. Scidmore's handwritten manuscript of the latter article is archived in the Wickersham Collection at the Alaska State Library in Juneau.

9 ERS to John Muir, September 12, 1891, Muir Papers.

10 ERS, "Goat-Hunting at Glacier Bay, Alaska," *Californian Illustrated Magazine*, April 1894, 537–44; "The Muir Glacier in Alaska," *Harper's Weekly*, July 23, 1892, 711–12.

11 Isabel C. McLean identified Eliza's companions in "Eliza Ruhamah Scidmore," *The Alaska Journal* 7, no. 4 (Autumn 1977): 240–1.

12 ERS, "Goat-Hunting," 539–40.

13 The first half of the article ran in April 1894, with the second part promised to readers in the next issue. But the magazine apparently ceased publication and the sequel never appeared. Scidmore never got paid for the piece, despite repeated attempts to collect the money.
14 *The* (Sitka) *Alaskan*, August 1, 1891.
15 ERS, "The Muir Glacier," 711–12. Harry Fielding Reid acknowledged Scidmore's photographs and descriptions in an assessment of ice-front changes he reported in "Studies of Muir Glacier, Alaska," *National Geographic*, March 21, 1892, 42.
16 ERS, "Goat-Hunting, 543.
17 ERS, "Goat-Hunting," 544.
18 ERS to John Muir, September 12, 1891, Muir Papers.
19 *Current Literature*, March 1892, 476.
20 Washington *Evening Star*, July 9, 1892.
21 *National Geographic*, April 1891, 316–34.
22 Donald J. Orth, *Dictionary of Alaska Place Names* (Washington, DC: U.S. Government Printing Office, 1967), 820.
23 See, for example, Gilbert M. Grosvenor, "The Society and the Discipline," *The Professional Geographer* 36, no. 4 (November 1984): 414.
24 Martin, *American Geography and Geographers*, 463.
25 See Flack, *Desideratum in Washington.*
26 McLean, "Eliza Ruhamah Scidmore," 239.
27 Nancy Pagh, "Imagining Native Women: Feminine Discourse and Four Women Travelling the Northwest Coast," in *Telling Tales: Essays in Women's Western History*, Catherine A. Cavanaugh and Randi R. Warne, eds. (Vancouver: University of British Columbia, 2001), 86.
28 ERS to Robert U. Johnson, February 21, 1893, Century Collection.
29 See Homer E. Socolofsky, "Benjamin Harrison and the American West," *Great Plains Quarterly* 5 (Fall 1985): 249–58. A clause in the 1891 Land Revision Act gave the president the authority to set aside tracts of forested federal land until their future use could be decided. By the time President Harrison left office on March 4, 1893, he had designated 22 million acres of protected timberlands that laid the foundation of a national forest system.
30 ERS, *Jinrikisha Days in Japan*, 189.
31 ERS to Robert U. Johnson, March 2, 1893, Century Collection.
32 Scidmore's first book on Alaska went out of print by the end of 1890. That fall, she wrote to John Muir: "I have acquired my Alaska [printing] plates myself and I shall no doubt take your advice and write

an entirely new book." ERS to John Muir, October 22, 1890, Muir Papers.

33 John W. Noble to Robert U. Johnson, March 2, 1893, Muir Papers; ERS to Robert U. Johnson, March 21, 1893, Century Collection. The fifteen new "national forest reserves" were all in western states and Alaska, as most forests in the East and Midwest had already come under private control.

34 ERS, "Our New National Forest Reserves," *The Century*, September 1893, 793.

35 See Theodore Catton, "The Campaign to Establish Mount Rainier National Park, 1893–1899," *Pacific Northwest Quarterly* 88, no. 2 (Spring 1997): 70–81.

36 Fox, *The American Conservation Movement*, 110; John Muir to Louie Muir, September 30, 1892, Muir Papers.

37 ERS to John Muir, January 7, 1894, Muir Papers.

38 *Tacoma Daily News*, June 14, 1897.

39 "Abstract of Minutes," *National Geographic Magazine* 5, 1893, xi.

40 *Idaho Statesman*, May 14, 1891.

41 *Washington Post*, March 25, 1893.

42 Scidmore's remark echoed prevalent notions of her day. In *The Critic* of April 29, 1893, for example, a reviewer commenting on a new book published in English by a Japanese author wrote (270): "It is almost a commonplace that the white brain cannot understand the yellow brain." In *Becoming Yellow: A Short History of Racial Thinking* (Princeton: Princeton University, 2011), Michael Keevak explains that "yellow" in reference to East Asians emerged in the late-eighteenth century when scientists such as Carl Linnaeus classified people by four continental types and assigned each group a color; Europeans, Africans, and Native Americans were referred to, respectively, as "white," "black," and "red." The categories evolved into a hierarchy of racial groups defined by physical and cultural features and presumed levels of civilization. The term *yellow races* gained currency in the United States, as Chinese and Japanese immigration increased in the nineteenth century.

43 *Washington Post*, February 4, 1894.

44 *Washington Post*, February 4, 1894.

45 *New York Tribune*, July 2, 1893; *Chautauqua Assembly Herald* 18, June 1983, 1–2; Schlereth, *Victorian America*, 253–7.

46 Norman Bolotin and Christine Laing, *The World's Columbian Exposition* (Champaign: University of Illinois, 2002), 5; *Photographs of the World's Fair* (Chicago: Warner Company, 1894).

47 Gardiner G. Hubbard to ERS, May 16, [1893], National Geographic Society Library and Archives.
48 Baron of Marajó to ERS, July 24, 1893, NGS Library and Archives.
49 Martin, *American Geography and Geographers,* 454.
50 ERS to Richard Watson Gilder, March 6, 1894, Century Collection.
51 *Chicago Daily Tribune*, July 28, 1893; ERS, "Recent Explorations in Alaska," *National Geographic*, January 31, 1894, 110, 173–9.
52 Frederick Jackson Turner, "The Significance of the Frontier in American History," Annual Report of the American Historical Association, 1893, 197–227.
53 "Two Visions of the Frontier," Newberry Library essay, http://www.newberry.org/.
54 Rydell, *All the World's a Fair*, 31–65; Schlereth, *Victorian America*, 172–3.
55 Rydell, *All the World's a Fair*, 48–51.
56 *Milwaukee Sentinel*, February 21, 1893; Gozo Tateno and Augustus O. Bourn, "Foreign Nations at the World's Fair," *North American Review* 156, no. 434 (January 1893): 34–5, 42.
57 ERS, "Chicago's Japan," *Harper's Bazaar*, August 19, 1893.
58 On the women's library at the Chicago World's Fair, see Sarah Wadsworth and Wayne Wiegand, *Right Here I See My Own Books* (Amherst: University of Massachusetts, 2012).
59 *Milwaukee Sentinel*, September 13, 1893.
60 ERS, "The Discovery of Glacier Bay, Alaska," *National Geographic,* April 1896, 145.
61 McLean says in "Eliza Ruhamah Scidmore" (241) that although Scidmore published an article in *National Geographic* titled "Northwest Passes to the Yukon," she did not trace the gold trails herself.
62 ERS, "The Stikine River in 1898," *National Geographic*, January 1899, 105–12.
63 ERS to Robert U. Johnson, July 25, 1898, Century Collection.
64 Tours of Glacier Bay resumed in the 1950s, with only small boats allowed. Today's huge cruise ship industry arose in the 1970s. Because of the severe retreat of Muir Glacier, most large ships now sail instead into the western arm of the bay. Karen Jettmar, *Alaska's Glacier Bay: A Traveler's Guide* (Anchorage: Alaska Northwest Books, 1997), 73.

Chapter 11

1 Canadian Pacific details from Shannon, *Finding Japan*, 77–94.
2 Shannon, *Finding Japan*, 89.

3 Shannon writes in *Finding Japan* (90) that Van Horne financed the trip that led to Hearn's first article on Japan, which appeared in the November 1890 *Harper's Bazaar*. Hearn, who lived in Japan to the end of his life, gained a wide readership for writings that presented glimpses of an older, more mystical Japan before the country's rapid modernization.
4 ERS, *Westward to the Far East*, 1891 ed., 6.
5 Robert Gardiner describes a typical journey by way of the Canadian Pacific in *Japan as We Saw It* (Boston: Rand Avery, 1892).
6 Shannon, *Finding Japan*, 79.
7 See "The Safest Ships Afloat," *Chap-Book*, September 1, 1895.
8 ERS, *Westward*, 1891 ed., 9–10.
9 ERS, *Westward*, 1902 ed., 17.
10 ERS, *Westward*, 1891 ed., 12–13.
11 ERS, *Westward*, 1902 ed., 9.
12 Scidmore reported on Gardner Inlet in her 1893 *Appletons' Guide-Book to Alaska and the Northwest Coast*, 28–9. The author is grateful to Jay Sherwood for providing information about Scidmore's British Columbia travels in a series of email exchanges in 2014.
13 *British Colonist*, September 2, 1892; *Washington Post*, September 2, 1892, and January 1, 1893.
14 "Her Remarkable Career," *Biloxi Daily Herald*, May 9, 1901. The syndicated article states that Scidmore made a tour of the world "in the interests of a railway company."
15 ERS, *Westward*, 1891 ed., 5–6.
16 "At Home Around the World," https://twain.lib.virginia.edu/onstage/world.html.
17 Scidmore's correspondence with Goode and other museum personnel is archived in Smithsonian Record Units 7050, 7073, and 189. The letters show that she wrote regularly asking for rolls of film, which she had sent to her at U.S. Consulates via the State Department's private dispatch bag. After Goode's untimely death in 1896, some Smithsonian officials balked at continuing to provide film but apparently deferred out of respect for Goode's arrangement.
18 The photographs are archived in the Smithsonian's National Anthropological Archives as Photo Lot 139, Eliza Ruhamah Scidmore Photographs Relating to Japan and China, c. 1914–1916; also in Photo Lot 97, Division of Ethnology Photographs.
19 For examples of Scidmore's photos at the Smithsonian that correlate with illustrations in her published articles, see "An Island

Without Death," *The Century*, August 1896, and "The Porcelain Artists of Japan," *Harper's Weekly*, January 22, 1898.

20 Acquisition records from the Smithsonian show that Scidmore donated items such as a pigeon whistle from Peking, a serpentine-shaped harpoon blade from Alaska, a U.S. Army regulation shoe from the Civil War, and the brass ammunition case for a gun fired from the U.S.S. *Olympia* on May 1, 1899, during the battle of Manila Bay. Smithsonian Archives, RU 0305.

21 Goode enjoyed a reputation as a brilliant museum administrator with a talent for innovative exhibits that engaged the public. Paul E. Oeshser, "George Brown Goode (1851–1896)," *Scientific Monthly* 66, no. 3 (March 1948): 195. Originally housed in the Smithsonian Castle, the United States National Museum expanded greatly in 1881 with the construction of a new building on the Mall, to showcase exhibits from the Centennial Exhibition. The present-day National Museum of Natural History opened in 1910 as a successor to the National Museum.

22 ERS, *Java, the Garden of the East*, 173.

23 ERS, *Westward*, 1891 ed., 34. Scidmore's companionship during this period of travel is unclear. In a letter of October 1894 describing her plans to visit Java she mentions "my friend and I." Her later book on Java suggests she traveled the island with two other American women.

24 "Along an Inland Sea," *Chicago Daily Tribune*, January 15, 1895.

25 ERS, "An Island Without Death," 494.

26 ERS, *Java*, 12.

27 Alfred Russel Wallace to his mother, July 20, 1861, Wallace Letters Online, Natural History Museum, Oxford University, UK, http://www.nhm.ac.uk/research-curation/scientific-resources/collections/library-collections/wallace-letters-online/375/5918/S/details.html.

28 ERS to Robert U. Johnson, February 25, 1897, Century Collection. Most English-language books on Java at the time were scientific or scholarly in nature, and not generally useful for tourists.

29 ERS, *Java*, 30.

30 For Scidmore's discussion of the "culture system," see *Java*, 94–125.

31 ERS, *Java*, 29.

32 ERS, *Java*, 25.

33 ERS, *Java*, 29–30.

34 ERS, *Java*, 30.

35 ERS, *Java*, 67–8.

36 ERS to Clarence Buel, October 7, 1897, Century Collection.
37 ERS, *Java*, 80.
38 ERS, *Java*, 80–91.
39 ERS to George Brown Goode, December 8, 1894, Smithsonian Archives, RU 189.
40 ERS, *Java*, 171–2.
41 ERS, *Java*, 175.
42 ERS, *Java*, 179.
43 Mary Russell discusses the interactive aspects of women's travel writing in *The Blessings of a Good Thick Skirt*. In a review of Scidmore's Java book, a *New York Times* reviewer wrote that Scidmore's sympathetic depiction of the Javanese people showed "a feeling that approaches affection." "The Garden of the Far East," *New York Times*, December 18, 1897.
44 ERS, *Java*, 204.
45 ERS to George Brown Goode, December 8, 1894, Smithsonian Archives, RU 189.
46 ERS to George Brown Goode, March 5, 1895, Smithsonian Archives, RU 189.
47 ERS to John Muir, September 16, 1895, Muir Papers.
48 "Sixth International Geographical Congress, London, 1895," *Journal of the American Geographical Society of New York* 27, no. 2 (1895): 237.
49 "Womankind Abroad," *Daily Inter Ocean* (Chicago), August 19, 1895; also "Wise Men Are There," *Daily Inter Ocean*, July 27, 1895; "Labors of Geographers," *New York Times*, July 27, 1895.
50 *Chicago Daily Tribune*, August 4, 1895.
51 *Daily Inter Ocean*, August 19, 1895.
52 Mary Leiter's father was a wealthy Chicago merchant who moved the family to Washington in the 1880s. As one of the many American heiresses who married into English nobility at the turn of the century, Mary Leiter Curzon was among the models scriptwriter Julian Fellowes had in mind when he created the character of Cora in the popular TV series "Downton Abbey." See Dave Itzkof, "Julian Fellowes Discusses a Season of Comings and Goings at 'Downton Abbey,'" *New York Times*, February 18, 2013.
53 Ghose, *Women Travelers in Colonial India*, xi.

Chapter 12

1 Records show the Scidmores living at different times at No. 3 and No. 6 on the Bund, on either side of the Club Hotel. Street details

from Sabin, *A Historical Guide to Yokohama*, 48–51; ERS, *Jinrikisha Days*, 4–5; Shannon, *Finding Japan*, 143; and ads in *Japan Weekly Mail*.

2 Eliza Catherine Scidmore obituary, Washington *Evening Star*, October 20, 1916.

3 Sidmore, "More Than a Footnote," 79.

4 "Her Remarkable Career," *Biloxi Daily Herald*, May 9, 1901.

5 ERS to Robert U. Johnson, August 20, 1896, Century Collection.

6 "Her Remarkable Career," *Biloxi Daily Herald*, May 9, 1901.

7 ERS, "The Wonderful Morning-Glories of Japan," *The Century*, December 1897, 281–9.

8 ERS, "Asagao (*Ipomea purpurea*), the Morning Flower of Japan," *Transactions and Proceedings of the Japan Society, London* 5–6 (London: Japan Society, 1902): 198–217.

9 Present-day scientists have estimated the 1896 Japanese earthquake's force of 8.5 magnitude. U.S. Geological Survey, Earthquake Hazards Program, https://earthquake.usgs.gov/.

10 "Sanriku Coast of Iwate, Japan," https://sanriku2019.jp/en/sanriku-coast-iwate.html.

11 ERS to Robert U. Johnson, July 3, 1896, Century Collection.

12 *Japan Weekly Mail*, June 27, 1896. The first person on the scene was reportedly an American missionary, the Rev. Rothesay Miller, who pedaled his bicycle over the mountains. Local newspaper accounts suggest Frederick Eastlake, an American teacher in Japan, may also have been a key source of information after visiting the affected region on behalf of the U.S. ministry in Tokyo. Email correspondence to the author from Ichiro Fudai, November 16, 2015.

13 ERS, "The Recent Earthquake Wave on the Coast of Japan," *National Geographic*, September 1896, 285–9.

14 Various reports have asserted that the Japanese word *tsunami* first appeared in English in Scidmore's *National Geographic* article of September 1896. But the word had already been used in the American press weeks earlier in news dispatches, as in "Victims of the Sea," *Monroeville* (Indiana) *Breeze*, July 23, 1896, and "The Japanese Tidal Wave," *Pomeroy* (Iowa) *Herald*, July 30, 1896. Those news accounts, which described victims fleeing to cries of "tsunami! tsunami!," cited as their source a report from U.S. officials in Tokyo.

15 Historical records of *Asahi Shimbun* show that the photographs published in *National Geographic* appeared originally in that newspaper, which acquired its images from four Japanese contributors. The author is grateful to Ichiro Fudai for his research assistance in Iwate Prefecture, Japan. His examination of visitor logs and other local

records offers no evidence that Scidmore reported directly from the scene of the tragedy. Passages in her article suggest she drew heavily on the *Japan Weekly Mail*'s coverage from June 20 to July 4, 1896.

16 Julyan H. E. Cartwright and Hisami Nakamura, "Tsunami: A History of the Term and of Scientific Understanding of the Phenomenon in Japanese and Western Culture," *Notes and Records of the Royal Society of London* 62, no. 2 (June 20, 2008): 152–5.

17 The scene Scidmore described prefigured events of a century later when, on March 11, 2011, the most powerful earthquake ever recorded in Japan erupted in the same region. Registered at a magnitude of 9.1, it triggered a tsunami that killed more than 19,000 people and caused explosions in three nuclear reactors at the Fukushima power plant.

18 ERS to George Goode, August 21, 1896, Smithsonian Archives, RU 189.

19 ERS, "Reports of the Sailing Schooners Cruising the Neighborhood of the Tuscarora Deep in May and June 1896," *National Geographic*, September 1896, 310–12.

20 ERS to Robert U. Johnson, March 10, 1897, Century Collection.

21 Review in *Chap-Book*, December 15, 1897; ERS, *Java, the Garden of the East*, 109.

22 ERS to Robert U. Johnson, February 25, 1897, Century Collection.

23 ERS, "Down to Java," *The Century*, August 1897, 527–46; ERS, "Prisoners of State at Boro Boedor," *The Century*, September 1897, 655–70.

24 ERS in note of explanation to "The Editor of the Century Magazine," n.d. [fall 1897], Century Collection.

25 ERS to Clarence Buel, n.d. [fall 1897], Century Collection.

26 ERS to Clarence Buel, October 15, 1897, Century Collection.

27 ERS to Clarence Buel, October 7, 1897, Century Collection.

28 ERS to Robert U. Johnson, May 8, 1898, Century Collection.

29 ERS to Robert U. Johnson, July 25, 1898, Century Collection.

30 Dean Worcester's prolific writing on the Philippines at the turn of the century greatly influenced public opinion as readers sought to understand the "white man's burden" in America's first-time role as an imperial power. As a member of the U.S. Philippine Commission, Worcester took thousands of photographs of indigenous people, many of which appeared in *National Geographic*. Mark Rice, "Dean Worcester's Photographs, American National Identity, and *National Geographic Magazine*," *Australasian Journal of American Studies* 31, no. 2 (December 2012): 42–56.

31 ERS to Robert U. Johnson, November 4, 1898, Century Collection.

32 ERS, "An Intrepid American: Some Personal Reminiscences of Admiral Dewey," *Outlook*, January 31, 1917, 190–1.

33 ERS to Robert U. Johnson, February 9, 1899, Century Collection.

34 Hand-written reply on Western Union telegraph cable from Scidmore in Manila to *The Century* magazine in New York, February 10, 1899, Century Collection.

35 ERS to Robert U. Johnson, February 28, 1899, Century Collection.

36 General James Rusling, "Interview with President William McKinley," *Christian Advocate*, January 22, 1903, 17.

37 ERS to Robert U. Johnson, February 28, 1899, Century Collection.

38 ERS to Robert U. Johnson, November 4, 1898, Century Collection.

39 ERS to Robert U. Johnson, April 14, [n.d., 1899?], Century Collection.

Chapter 13

1 ERS, *China: The Long-Lived Empire*, 60–1.

2 ERS to Gardiner G. Hubbard, October 31, 1896, NGS Library and Archives.

3 ERS, *China*, 8.

4 ERS, *China*, 4.

5 ERS to Robert U. Johnson, September 3, 1896, Century Collection.

6 ERS to Robert U. Johnson, November 24, 1896, Century Collection.

7 ERS, *China*, 66.

8 ERS, *China*, 73.

9 ERS, *Westward to the Far East*, 1891 ed., 42.

10 ERS, *China*, 73–92; Preston, *The Boxer Rebellion*, 5–6.

11 ERS, *China*, 81.

12 ERS, *China*, 77. Scidmore wrote several times about the distinctive headdress, or *liangbatou*, of the Manchu women. See examples in *China*, 75 and 129; also "Mukden, the Manchu Home, and Its Great Art Museum," *National Geographic*, April 1910, 300.

13 ERS, *China*, 179.

14 ERS, "The Greatest Wonder in the Chinese World: The Marvelous Bore of Hang-Chau," *The Century*, April 1900, 852–9.

15 ERS, "The River of Tea," *The Century*, August 1899, 547–9, and "Cruising Up the Yangtze," *The Century*, September 1899, 668–79.

16 In her 2013 biography *Empress Dowager Cixi*, Jung Chang challenges long-held views of the empress dowager as a cruel despot, and credits her with instigating the reforms of 1898.

17 ERS to Robert U. Johnson, October 1, 1898, Century Collection.

18 Scidmore family records in the Oakley–Hawley Papers at the Wisconsin Historical Society include a photo of the adolescent "Lizzie" Yu.
19 Grant Hayter-Menzies, *Imperial Masquerade: The Legend of Princess Der Ling* (Hong Kong: Hong Kong University Press, 2008), 70, 75.
20 ERS to Robert U. Johnson, February 9, 1899, Century Collection.
21 The rate of $125 per article comes from a February 16, [1910], letter by Scidmore to *The Century's* editors.
22 Editor's note, S.J./Q/287–88 [1898?], Century Collection.
23 ERS to Robert U. Johnson, February 9, 1899, Century Collection.
24 Goodwin, *The Bully Pulpit*, 265.
25 Cited in Anthony, *Nellie Taft*, 33.
26 Taft, *Recollections of Full Years*, 33, 66.
27 Preston, *The Boxer Rebellion*, 40–46.
28 Taft, *Recollections of Full Years*, 55–6.
29 Writing on the eve of William Taft's election as president, Mrs. Scidmore told relatives that "we have seen much of them ever since he was Gov. in Manila." Taft was named governor-general of the Philippines in 1901. Eliza Catherine Scidmore to Walter Oakley and family, October 10, 1908, Oakley–Hawley Papers.
30 Taft, *Recollections of Full Years*, 74–6.
31 Preston, *The Boxer Rebellion*, x–xiv, 84.
32 "The Crisis in China," *Outlook*, July 21, 1900. Not everyone liked the book. After reading a copy donated to the National Geographic's library by the journalist Ida Tarbell, Alexander Graham Bell's wife judged *China: The Long-Lived Empire* "not a bit interesting." Mabel Hubbard Bell to Alexander Graham Bell, September 14, 1900, Alexander Graham Bell Papers, Library of Congress.
33 ERS to Mr. Scott [book division editor], October 5, 1900, Century Collection.
34 In *A Truthful Impression of the Country*, Clifford calls Scidmore "an unwitting postmodernist before her time" (52) in questioning whether a truly enlightened China had ever really existed except in the Western mind. She wrote in her book *China* that the Western world began to discover the "actual China" (4) only at the onset of its 1894–5 war with Japan.
35 ERS, *China*, 1–7
36 ERS, *China*, 5.
37 ERS, *China*, 459.

38 See Thurin, *Victorian Travelers and the Opening of China*: 1842–1907; also, Clifford, *A Truthful Impression of the Country*. Clifford calls Scidmore "one of the pioneers of Western tourism in East Asia" (19).
39 Julia Kuehn "China of the Tourists: Women and the Grand Tour of the Middle Kingdom, 1878–1923," in Steve Clark and Paul Smethurst, eds., *Asia Crossings: Travel Writing on China, Japan and Southeast Asia* (Hong Kong University Press, 2008), 115, 122–6.
40 ERS, *China*, 10.
41 Kuehn, in "China of the Tourists" (122), notes that the itinerary Scidmore recommended in *Westward to the Far East* closely paralleled Cummings's route.
42 ERS, "Mrs. Bishop's 'The Yangtze Valley and Beyond,'" *National Geographic*, September 1900, 366–8. Isabella Bird was also known as Mrs. Bishop after a late-in-life marriage.
43 *Evening Star*, September 11, 1899; "Twelfth International Congress of Orientalists in Rome, 1899," *The Journal of the Royal Asiatic Society of Great Britain and Ireland* (January 1900): 181–6.
44 *Evening Star*, April 17, 1903.
45 On the craze for Chinese porcelains, see Meyer and Brysac, *The China Collectors*, 141–56.
46 *Sydney Morning Herald*, quoted in Meyer and Brysac, *The China Collectors*, 26.
47 Preston, *The Boxer Rebellion*, 283–93. The wave of looting and destruction was reminiscent of an earlier incident, in 1860, when Anglo-French armies laid waste to the Old Summer Palace—a Versailles-like compound northwest of Peking—during the final phase of the second Opium War.
48 Meyer and Brysac, *The China Collectors*, 26–9, 238–9.
49 See *Chinese & Japanese Porcelains & Potteries & and Other Far Eastern Objects of Art*, The Collection of Miss Eliza Ruhamah Scidmore, Washington, D.C., Sale No. 1903, Anderson Galleries, New York, January 10, 1925.
50 "Introduction," in 1925 Anderson Galleries catalog of Scidmore's Far East collection.
51 ERS to Emily Eames MacVeagh, March 18 and July 28, 1904, Franklin MacVeagh Papers, Library of Congress; ERS to George Goode, August 21, 1896, Smithsonian Archives, RU 189. The Smithsonian Institution exhibited Hippisley's Chinese porcelains for many years until the collection was sold at auction. "Royal Porcelains of China on View," *New York Times*, January 25, 1925.

52 In May 1920, many of Scidmore's art books sold at auction in New York as part of a private collection. *Sale of Library of Amos Warner, of Minnesota, including collections from library of Eliza Ruhamah Scidmore, of Washington,* D.C., Anderson Galleries, 1920.
53 Amelia Gere Mason, *Memories of a Friend* (Chicago: Laurence. C. Woodworth, 1918), 17. Scidmore describes meeting Emily MacVeagh in Calcutta in notes of a pocket diary from November 1900 to August 1901; Wisc. Mss. Qm., Oakley–Hawley Papers.
54 Scidmore's letters to Emily MacVeagh are archived in the Franklin MacVeagh Papers at the Library of Congress.
55 ERS to Emily E. MacVeagh, November 30, [1902?], Franklin MacVeagh Papers.
56 ERS to Emily E. MacVeagh, September 10, 1909, Franklin MacVeagh Papers.
57 ERS to Mr. Scott, October 5, 1900, Century Collection.
58 See Nigel Nicolson, *Mary Curzon* (New York: Harper Collins, 1977).
59 ERS, *Winter India*, 94, 96.
60 Entry of December 28, 1900, Scidmore pocket diary, November 1900 to August 1901; Wisc. Mss. Qm., Oakley–Hawley Papers.
61 ERS, *Winter India*, xi-ii.
62 ERS, *Winter India*, xi–v.
63 A companion of Scidmore during one of her trips to India may have been Mrs. Caroline Tousey Burkam, a New York resident who lived at the Plaza Hotel. Scidmore dedicated her 1903 book, *Winter India*, to Burkam.

Chapter 14

1 ERS to Robert U. Johnson, December 9, 1903, Century Collection.
2 ERS to Emily E. MacVeagh, March 28, 1904, Franklin MacVeagh Papers.
3 *Chicago Daily Tribune*, December 22, 1903.
4 *Chicago Daily Tribune*, December 23, 1903.
5 *Chicago Daily Tribune*, December 23 and December 24, 1903.
6 John W. Dower, "Throwing Off Asia III," *MIT Visualizing Cultures*, https://visualizingcultures.mit.edu/.
7 *Chicago Daily Tribune*, December 25, 1903.
8 ERS, *Westward to the Far East*, 1902 ed., 44–5.
9 Xunling, Der Ling's brother, photographed Cixi and her court after taking up photography as a hobby. See David Hogge, "The Empress

Dowager and the Camera: Photographing Cixi, 1903–1904," *MIT Visualizing Cultures*, https://visualizingcultures.mit.edu/.

10 *Chicago Daily Tribune*, January 3, 1904.

11 *Chicago Daily Tribune*, January 13 to February 12, 1904.

12 *Evergreen* is a traditional newspaper term for non-time-sensitive material, such as features and human-interest stories, that can be published any time.

13 The series ran in the *Chicago Daily Tribune* from December 21, 1903, to August 25, 1904.

14 ERS to Gilbert H. Grosvenor, February 5, 1914, NGS Library and Archives.

15 ERS to Emily E. MacVeagh, March 18 and July 24, 1904, Franklin MacVeagh Papers.

16 *Chicago Daily Tribune*, December 19 and December 21, 1903.

17 ERS to Emily E. MacVeagh, March 18, 1904, Franklin MacVeagh Papers.

18 "Oriental Art Sale on Next Saturday," *New York Times*, January 4, 1925; Sale No. 1903, *Chinese & Japanese Porcelains & Potteries & and Other Far Eastern Objects of Art, The Collection of Miss Eliza Ruhamah Scidmore, Washington, D.C.*, Anderson Galleries, New York, January 10, 1925. Scidmore's acquisition of the imperial chair and other Asian objects is representative of the power imbalance that made it possible for Westerners like her to not just benefit from favorable exchange rates in buying curios from the East but also profit from the objects they carried home.

19 ERS to Robert U. Johnson, July 29, 1904, Century Collection.

20 Katharine A. Carl, *With the Empress Dowager* (New York: Century Company, 1905).

21 ERS to Robert U. Johnson, July 29, 1904, Century Collection.

22 See Ekai Kawaguchi, "The Latest News from Lhasa: A Narrative of Personal Adventure," *The Century*, January 1904; Count Kosui Otani, "The Japanese Pilgrimage to the Buddhist Holy Land: A Personal Narrative of the Hongwani Expedition of 1902–03," *The Century*, October 1906.

23 ERS to Emily E. MacVeagh, June 22, [1904], Franklin MacVeagh Papers.

24 ERS to Robert U. Johnson, July 29, 1904, Century Collection.

25 ERS to Robert U. Johnson, January 23, 1905, Century Collection.

26 Anita Newcomb McGee, a medical doctor, was among twenty-nine American nurses and war correspondents who received the Order

of the Precious Crown from the emperor of Japan in 1907 for their service in the Russo–Japanese War. *Evening Star*, November 3, 1912.

27 ERS to Robert U. Johnson, July 29, 1904, Century Collection.

28 Jane Oakley to Walter Oakley family, February 14, 1905, Oakley–Hawley Papers.

29 ERS to Robert U. Johnson, February 9, 1905, Century Collection.

30 The Hague 1899, Convention with Respect to the Laws and Customs of War on Land, Annex: Section 1—Belligerents, Chapter II—Prisoners of War; Article VII.

31 Discussions with Komura himself may have influenced the story idea Scidmore sent to her editors at *The Century* on February 9, 1905. She wrote that "the subject has been suggested to me and the official permits for visits offered me by a 'high personage.'" The army minister who handled the directive was Masatake Terauchi, whom Scidmore later profiled in her last article in *The Century*, after he had become prime minister. ERS, "Marshall Count Terauchi, the New Premier of Japan," *The Century*, August 1917.

32 Japan Center for Asian Historical Records, http://www.jacar.go.jp.

33 ERS, *As the Hague Ordains*, 1914 ed., vii–ix.

34 ERS to Robert U. Johnson, March 31, 1905, Century Collection; "Miss Frances Parmelee and Matsuyama," *Mission Studies: Woman's Work in Foreign Lands*, vols. 25–26 (Chicago: Congregational Church, 1907), 337–41.

35 Sidney Gulick, *The White Peril in the Far East* (New York: F. H. Revell Company, 1905), 99.

36 Eric Johnston, "Civility Shown to Russo–Japanese War POWs Lives on as Matsuyama's Legacy," *Japan Times* (online), August 22, 2016.

37 ERS to Robert U. Johnson, March 31, 1905, Century Collection.

38 Goodwin, *The Bully Pulpit*, 432–3.

39 Andrew Gordon, "Social Protest in Imperial Japan: The Hibaya Riot of 1905," *MIT Visualizing Cultures*, http://visualizingcultures.mit.edu/.

Chapter 15

1 Susan Schulten, "The Making of National Geographic: Science, Culture, and Expansionism," *American Studies* 41, no. 1 (Spring 2000): 5.

2 Gilbert Hovey Grosvenor in *National Geographic Magazine*, April 1914, 455; quoted in McCarry, *From the Field*, 12.

3 ERS to Gardiner G. Hubbard, October 31, 1896, NGS Library and Archives.

4 John Hyde, "Introductory," *National Geographic*, January 1896, 2.
5 Scidmore ended her tenure as corresponding secretary of the National Geographic Society in 1896.
6 "History of the National Geographic Library," National Geographic Library and Archives, https://nglibrary.ngs.org/public_home.
7 McCarry, *From the Field*, 10.
8 Gray, *Reluctant Genius*, 353–63.
9 Quoted in McCarry, *From the Field*, 11.
10 Poole, *Explorers House*, 32.
11 ERS to Gardiner G. Hubbard, December 8, 1896, NGS Library and Archives.
12 Gray, *Reluctant Genius*, 335–6.
13 Alexander Graham Bell to Mrs. Scidmore from Grand Hotel in Yokohama, October 31, 1898; letter now in private hands.
14 ERS to Mabel Hubbard Bell, August 14, 1899, NGS Library and Archives.
15 Mabel Hubbard Bell to Alexander Graham Bell, May 17, 1899, Alexander Graham Bell Papers, Library of Congress. Mabel Bell refers to her social interactions with Scidmore in several letters to her husband.
16 Gray, *Reluctant Genius*, 350–51.
17 Gray, *Reluctant Genius*, 351.
18 Alexander Graham Bell to Gilbert H. Grosvenor, March 5, 1900; quoted in Jenkins, *National Geographic: 125 Years*, 54.
19 Poole, *Explorers House*, 42.
20 Priit Juho Vesilind "National Geographic and Color Photography" (M.A. thesis, Syracuse University, 1977), 11. Ida Tarbell, who gained fame as one of the "muckraking" investigative reporters at *McClure's*, served for a while on *National Geographic*'s editorial board.
21 Alexander Graham Bell to Gilbert H. Grosvenor, March 5, 1900; quoted in Leah Bendavid-Val, *Stories on Paper and Glass*, 11.
22 The use of such images over the years and a strong emphasis on "exotic" human subjects led the magazine to apologize for its past racist coverage of people around the world. "National Geographic admits 'racist' past," *BBC News*, March 13, 2018.
23 Gilbert H. Grosvenor served as editor of *National Geographic* for fifty-five years, the first of three generations of Grosvenors in the position. Since 2019 the National Geographic Society has operated in partnership with the Walt Disney Company.
24 Poole, *Explorers House*, 66.

25 Bryan, *The National Geographic Society*, 121.
26 Poole, *Explorers House*, 67.
27 The halftone process entails the printing of fine dots that blend optically to appear like the original photograph. At the beginning of the twentieth century, a halftone image cost twenty dollars to make, compared with three hundred dollars for a labor-intensive wood engraving. "History of the Magazine Industry," https://www.encyclopedia.com/media/.
28 Poole, *Explorers House*, 66.
29 ERS to Gilbert H. Grosvenor, November 21, 1912, NGS Library and Archives.
30 ERS to Gilbert H. Grosvenor, June 18, 1907, NGS Library and Archives.
31 *National Geographic*, December 1906, 673.
32 See Newman, *Women Photographers at National Geographic*.
33 Bill Bonner, former photo archivist of the National Geographic Society, concluded that the elephant hunt photos were submitted by James Howard Gore, one of the society's founding members. Personal interview with the author at National Geographic Society on April 5, 2016.
34 In one example of Scidmore's reliance on other sources, she took a photographer from Kyoto with her to Ise, Japan, in the fall of 1912 to report on women pearl divers. The women were elusive, and the photos she and her companion took were not very interesting, she told Grosvenor, so the images she submitted included pictures from an out-of-date publication for which she had acquired permission to reproduce. ERS to Gilbert H. Grosvenor, October 26 and December 30, 1912, NGS Library and Archives.
35 See ERS, "Archaeology in the Air," *National Geographic*, March 1907.
36 ERS to Gilbert H. Grosvenor, November 25, 1907, NGS Library and Archives.
37 Gilbert H. Grosvenor to ERS, December 2, 1907, NGS Library and Archives.
38 ERS to Gilbert H. Grosvenor, December 2, 1907, NGS Library and Archives.
39 Gilbert H. Grosvenor to ERS, October 12, 1912, ERS to Gilbert H. Grosvenor, January 16, 1913, NGS Library and Archives.
40 ERS to Gilbert H. Grosvenor, June 24, 1909, NGS Library and Archives.
41 Gilbert H. Grosvenor to ERS, November 29, 1910, NGS Library and Archives.

42 ERS to Gilbert H. Grosvenor, June 24, 1909, NGS Library and Archives.

43 In the early years of the twentieth century, magazines were beginning to adopt color printing for uses beyond advertising.

44 Autochrome, the first viable form of color photography, was introduced in 1907. *National Geographic* printed its first autochrome in 1914 and became a pioneer in the use of them.

45 See Mio Wakita, "Sites of 'Disconnectedness': The Port City of Yokohama, Souvenir Photography, and Its Audience," *Transcultural Studies* 4, no. 2 (2013): 77–129. By the 1880s, Japanese photo studios were selling millions of hand-tinted stock photographs to travelers and mail-order customers around the world. Prints and negatives of popular images circulated widely, often with no identifying information about the original photographers. In an interview with the author on April 5, 2016, Bill Bonner, former photo archivist at the National Geographic Society, noted that the organization's Eliza Scidmore Collection includes "some of the best hand-tinted photos in the [society's] entire image collection." Some were likely the product of studios or professional photographers, as suggested by the highly stylized poses, painted backdrops, and superb technical quality.

46 In her correspondence with Gilbert H. Grosvenor, Scidmore made several references to her use of colorists in Japan, as in a letter of November 21, 1912, NGS Library and Archives.

47 Gilbert H. Grosvenor to ERS, August 13, 1909, NGS Library and Archives.

48 Jenkins, *National Geographic: 125 Years*, 76.

49 Gilbert H. Grosvenor to ERS, May 31, 1912, NGS Library and Archives.

50 ERS to Gilbert H. Grosvenor, September 18, 1912, NGS Library and Archives. The images likely included some of the photographs published with Scidmore's final article for *National Geographic*, "Young Japan," in July 1914.

51 Gilbert H. Grosvenor to ERS, October 24, 1912, NGS Library and Archives.

52 ERS to Gilbert H. Grosvenor, November 21, 1912, NGS Library and Archives.

53 ERS to Gilbert H. Grosvenor, November 16, 1912, NGS Library and Archives.

54 Gilbert H. Grosvenor to ERS, December 30, 1912, NGS Library and Archives. The amount of $450 was about what a schoolteacher made

in a year, according to the article "How Other People Live" in the September 1912 *Ladies Home Journal*.

Chapter 16

1 The original city plan of the capital provided for a profusion of trees, in accordance with the wishes of George Washington and Thomas Jefferson. By the late-nineteenth century the nation's capital had become well known for its park-like character and the variety and abundance of its trees. Jonnes, *Urban Forests*, 26–8.
2 *Washington Star*, March 27, 1908.
3 Scidmore refers in her letters and travel diaries to her chronic insomnia and seeking treatment for her aching feet.
4 Rutkow, *American Canopy*, 130–2; Jonnes, *Urban Forests*, 18–34.
5 President Theodore Roosevelt, "To the School Children of the United States," April 15, 1907, Library of Congress, https://www.loc.gov/resource/rbpe.24001100/?sp=1.
6 "Arbor Day Observed," *Washington Evening Star*, March 27, 1908. Scidmore mentioned in a private letter a year later that Colonel Bromwell was among the park superintendents she approached—unsuccessfully—in her various attempts to have Japanese cherry trees planted in Potomac Park. ERS to Robert U. Johnson, August 20, 1909, Century Collection.
7 Franklin School, D.C., https://www.franklinschooldc.org/.
8 Gray, *Reluctant Genius*, 204–5, 209–13; Robert V. Bruce, *Bell: Alexander Bell and the Conquest of Solitude* (Ithaca, NY: Cornell University Press, 1990), 338.
9 Fairchild's grandfather was one of the founders of Oberlin College, and his father and uncles became presidents of colleges in the Midwest.
10 Stone, *The Food Explorer*, 14–16.
11 Fairchild, *The World Was My Garden*, 61–2. In the 1920s Fairchild and his wife built an estate and gardens at Biscayne Bay in Coconut Grove, Florida, and named it "the Kampong," a native word on Java for a family compound.
12 Rutkow, *American Canopy*, 203–4; Fairchild, *The World Was My Garden*, 84.
13 Samuels, *Enduring Roots*, 77.
14 See Fairchild, *The World Was My Garden*, and Daniel Stone, *The Food Explorer*.
15 David Fairchild to W. A. Taylor, June 24, 1910, National Agricultural Library, U.S. Department of Agriculture.

16 McClellan, *The Cherry Blossom Festival*, 29.
17 David Fairchild, "The Cherry Blossoms of Japan," n.d., n.p., unpublished manuscript, Item 2009.2093.46, Chevy Chase Historical Society.
18 Samuels, *Enduring Roots*, 77; Abe, *The Sakura Obsession*, 118–21.
19 Fairchild, *The World Was My Garden*, 254.
20 Fairchild, "The Cherry Blossoms of Japan." Fairchild noted that the tree on Massachusetts Avenue was likely from China and not Japan.
21 The Fairchilds eventually built a permanent home on the property that had many Japanese-inspired features, such as paving-stone paths leading directly into the outdoors. A separate studio on the banks of the creek gave Alexander Graham Bell, Marian's father, a quiet retreat outside the city. Thomas McAdam, "In the Woods," *Country Life in America*, October 1914.
22 Details here and as follows are from Fairchild's memoir, *The World Was My Garden*; Jefferson and Fusonie, *The Japanese Flowering Cherry Trees*, 4–8; and "Arbor Day Observed," *Washington Star*, March 27, 1908.
23 Jefferson and Fusonie, *Japanese Flowering Cherry Trees*, 7; Fairchild, "The Cherry Blossoms of Japan."
24 Fairchild said many years later in a public statement picked up by the press: "The idea of a field of cherries on the Speedway originated one afternoon during a visit to the cherry trees at 'In the Woods' of Miss E. R. Scidmore." "How United States and Japan Entered a League of Flowers," *National Geographic Bulletin*, n.d., No. 329–N.S; printed in *Trenton Evening Times*, October 31, 1920.
25 Gordon Chappell, "Historic Resource Study—West Potomac Park: A History," (Denver: National Park Service, U.S. Department of the Interior, 1973), 97.
26 "Arbor Day Observed," *Washington Star*, March 27, 1908.
27 Fairchild, *The World Was My Garden*, 412.
28 Fairchild, "The Cherry Blossoms of Japan."
29 ERS, "Twelve Varieties of Japanese Cherry Trees Are Blooming in Potomac Park," *Washington Star*, March 27, 1921.
30 Fairchild, "The Cherry Blossoms of Japan"; *The World Was My Garden*, 412.
31 "Authorship Solved," *Washington Star*, April 25, 1908.
32 ERS to Emily Eames MacVeagh, April 12, 1902, Franklin MacVeagh Papers. *Ground Arms*, first published in 1889 and reprinted in English as *Lay Down Your Arms*, sold a million copies, was translated into sixteen languages, and earned the author the Nobel Peace Prize in 1905, the first time a woman received the award.

33 ERS, *As the Hague Ordains*, 8.
34 ERS, *As the Hague Ordains*, 35.
35 See ERS, "The Moon's Birthday," *Asia Magazine*, March 1921.
36 *Washington Herald*, July 5, 1908.
37 *Washington Evening Star*, April 25, 1908.
38 Quote cited in Auslin, *Japan Society: 100 Years*, 11.
39 "Says Yellow Peril Means White Man Against the World," *New York Times*, December 17, 1908. Also see, for example, "War, Says Hobson," *New York Times*, July 8, 1907; Richmond P. Hobson, "America Is Defenseless in the Far East" and "Japan May Seize the Pacific Slope, Says Hobson," *Washington Times Magazine*, October 27 and November 3, 1907.
40 Mike McKinley, "Cruise of the Great White Fleet," Naval History and Heritage Command, https://www.history.navy.mil/.
41 *Japan Weekly Mail*, February 23, 1908.
42 An *Evening Star* article of November 3, 1912, reported that Scidmore was given the Order of the Rising Eastern Sun, though the version awarded to women at the time was the Order of the Precious Crown.
43 Records regarding the conferment on Eliza R. Scidmore; from Prime Minister and Decorating Bureau President, February 20, 1908, Japan Ministry of Foreign Affairs.
44 William Crozier, U.S. Army (Retired), Address at Services in Memory of Eliza Ruhamah Scidmore, held in Geneva, Switzerland, on 7 November 1928; in Oakley–Hawley Papers.

Chapter 17

1 ERS, "Capital Cherry Trees Gift of Japanese Chemist," *Sunday Star*, April 11, 1926.
2 ERS, "Capital Cherry Trees Gift of Japanese Chemist."
3 "Japanese Commissioners Here," *Evening Star*, April 5, 1909.
4 *Washington Post*, January 8, 1909.
5 ERS to Emily E. MacVeagh, February 26, [1909], Franklin MacVeagh Papers.
6 ERS, "Capital Cherry Trees Gift of Japanese Chemist."
7 *Baltimore Sun*, March 5, 1909.
8 Anthony, *Nellie Taft*, 229–31.
9 *New York Times*, November 15, 1909.
10 Butt, *Taft and Roosevelt*, 39–40.

11 The Luneta had also been a killing field during Spanish rule of the Philippines, as a site of public executions meant to deter Filipino insurgents. The execution there of Filipino national hero José Rizal in 1896 gave the Luneta its current name of Rizal Park. Nenette Arroyo, "First Lady Helen Taft's Luneta," *White House History* 34, no. 34 (Fall 2013), https://www.whitehousehistory.org/.

12 Taft, *Recollections of Full Years*, 361–2.

13 Butt described his White House experiences in frequent letters to his sister, published in two volumes in 1930. Butt died in the *Titanic* disaster in 1912.

14 Anthony, *Nellie Taft*, 240–1.

15 Anthony, *Nellie Taft*, 142–3.

16 Anthony, *Nellie Taft*, 243.

17 The "city beautiful" movement inspired civic improvement and beautification projects in many U.S. cities and towns at the turn of the century. Some people viewed the nation's capital as a model for implementing the ideas.

18 Protective of their executive authority over the city's public buildings and grounds, the Army Corps of Engineers became "chief antagonists" of the McMillan Plan, according to Kohler and Scott, *Designing the Nation's Capital*, 7. War Department correspondence of 1907–8 in National Archives RG 42 shows that William Taft not only favored the McMillan Plan's aims but solicited ideas for Potomac Park improvements from one of the key planners, landscape architect Frederick Law Olmstead Jr.—much to the annoyance of the Army engineers.

19 Chappell, "Historic Resource Study–West Potomac Park," 77–101 and Appendix B.

20 *Washington Post*, March 15, 1909; *Baltimore Sun*, April 11, 1909; "Army Engineers Ran the White House," U.S. Army Corps of Engineers, http://www.usace.army.mil/.

21 See also "Gayety on Speedway," *Washington Post*, April 2, 1909; "Proceeds with Plan," *Washington Star*, April 3, 1904; and "Busy with Esplanade," *Washington Post*, April 4, 1909.

22 "Esplanade in Park," *Washington Post*, April 3, 1909.

23 "Busy with Esplanade," *Washington Post*, April 4, 1909.

24 ERS, "Capital Cherry Trees Gift of Japanese Chemist."

25 Fairchild, "The Cherry Blossom Trees of Japan," Chevy Chase Historical Society.

26 The letter has been lost, but Scidmore described it in her 1921 and 1926 articles in the *Washington Star*.

27 Helen Taft, April 7, 1909, William H. Taft Papers, Library of Congress.

28 Notes by Spencer Cosby; First Indorsement, Office of Public Buildings and Grounds, National Archives RG 42. In response to Cosby's briefing on the first lady's request, the chief landscaper George Brown responded that Japanese cherries and other low-growing flowering trees planted "on the line of the roadway on the river side of Potomac park . . . would be appropriate and prove an attractive feature." Second Indorsement, Office of Public Buildings and Grounds, April 7, 1909, Office of Public Buildings and Grounds, National Archives RG 42.

29 Described in letter from Spencer Cosby to David Fairchild, April 13, 1909, Office of Public Buildings and Grounds, National Archives RG 42.

30 *Washington Star*, April 5 to April 9, 1909.

31 Joan W. Bennett, "Adrenalin and Cherry Trees," *Modern Drug Discovery* 4, no. 12 (December 2001): 47–8, 51. Takamine's research advances included a novel process for the distillation of alcohol, adapted from methods of sake production, and the first U.S. patent for a microbial enzyme, diastase, which he licensed to a pharmaceutical company as a digestive aid.

32 Auslin, *Japan Society: 100 Years*, 11–12. Scidmore was listed among the members beginning in 1912; email correspondence with Cynthia Sternau of the Japan Society, July 2016.

33 ERS quoting Takamine in "Capital Cherry Trees Gift of Japanese Chemist."

34 ERS, "Twelve Varieties of Japanese Cherry Trees Are Blooming in Potomac Park," *Washington Star*, March 27, 1921.

35 Ichiro Fujisaki, "Washington's Cherry Blossoms: A Century-old Connection Between U.S., Japan," *Washington Post*, January 20, 2012.

36 Also present at their meeting was a "Mr. Legardo"—apparently Benito Legarda, a friend of the Tafts serving as a resident commissioner to Congress from the Philippines. Anthony, *Nellie Taft*, 246.

37 A copy of Scidmore's letter appears (in English) in Mitsuo Isada, *Kindai Nihon no Sohzohshi* (Tokyo: Jokichi Takamine Research Foundation, 2011), 50. Though dated only "Monday," the letter was apparently written on April 12, 1909. The author is grateful to Mr. Isada for providing a copy of the publication during an interview in Japan in April 2013.

38 Spencer Cosby to Hoopes Bros. & Thomas Company, April 12, 1909, Office of Public Buildings and Grounds, National Archives RG 42.

39 David Fairchild to Spencer Cosby, April 4, 1909, Office of Public Buildings and Grounds, National Archives RG 42.
40 Spencer Cosby to David Fairchild, April 13, 1909, Office of Public Buildings and Grounds, National Archives RG 42.
41 ERS, "Twelve Varieties of Japanese Cherry Trees Are Blooming in Potomac Park."
42 Anthony, *Nellie Taft*, 247; "Mr. Taft to Attend," *Washington Star*, April 17, 1909; "Great Throng Out on Potomac Drive," *Washington Star*, April 18, 1909.
43 Taft, *Recollections of Full Years*, 362.
44 Anthony, *Nellie Taft*, 156; also, Tichie Carandang-Tiongson, "A D.C. Springtime Concert Born in Manila," *Positively Filipino Magazine*, http://www.positivelyfilipino.com/.
45 "Washington Drive Opened," *New York Times*, April 18, 1909.
46 Anthony, *Nellie Taft*, 258–60.

Chapter 18

1 ERS to Emily MacVeagh, September 10, 1909, Franklin MacVeagh Papers.
2 Details from unidentified newspaper clippings, probably from the *Japan Mail*, enclosed in ERS letter to Robert U. Johnson, August 20, 1909, Century Collection.
3 From *The Autobiography of Ozaki Yukio: The Struggle for Constitutional Government in Japan* (Princeton: Princeton University, 2001); quoted in Samuels, *Enduring Roots*, 78.
4 ERS to Robert U. Johnson, August 20, 1909, Century Collection.
5 ERS to Emily MacVeagh, September 10, 1909, Franklin MacVeagh Papers.
6 ERS to Robert U. Johnson, September 22, 1909, Century Collection.
7 ERS to Robert U. Johnson, October 26, 1909, Century Collection.
8 Scidmore published a piece on the empress dowager soon after her death. ERS, "Secrets of a Forbidden Palace," *Harper's Weekly*, February 6, 1909, 11–12, 32.
9 In 2008, a century after Emperor Guangxu's death, the state-run *China Daily* reported that forensic tests done on samples of hair and clothing retrieved from his tomb led experts to conclude he was a victim of acute arsenic poisoning. Lin Qi, "The Poisoned Palace—Mystery of Last Emperor's Death," *China Daily*, November 21, 2008.
10 ERS to Gilbert H. Grosvenor, September 1, 1909, NGS Library and Archives.

11 Frederic J. Haskin, "Our Far East Problem," *Washington Post*, October 11, 1909.
12 Some of the photographs and lantern slides Scidmore used in her lecture and articles are archived in Photo Lot 139 of the Smithsonian Institution's National Anthropological Archives.
13 "Bone of Contention," *Evening Star*, January 8, 1910; ERS, "Mukden, the Manchu Home, and Its Great Art Museum," *National Geographic*, April 1910, 289–320.
14 ERS to Emily MacVeagh, [Fall 1908], Franklin MacVeagh Papers.
15 *Proceedings of the United States National Museum* 36 (Washington, DC: Government Printing Office, 1909), 338–48. Scidmore wrote the catalogue notes on the museum's collection of Chinese and Japanese rosaries.
16 Scidmore's photo of the Chinese rosary appeared in her April 1910 *National Geographic* article on Mukden, on page 311.
17 *New York Times*, August 20, 1909.
18 Keishiro Matsui to Alvery Adee, August 30, 1909, General Records of the Department of State, National Archives RG 59.
19 "Bring on the Cherry Trees," *Washington Post*, August 29, 1909.
20 "Mikado Offers New York Trees," *Christian Science Monitor*, August 27, 1909.
21 "The Friendship of Japan," *Outlook*, September 11, 1909.
22 See "Shibusawa Eiichi and the 1909 Business Mission," Shibusawa Eiichi Memorial Foundation, https://www.shibusawa.or.jp/english/eiichi/1909/mission.html.
23 See, as one example, the *Salt Lake Tribune*, September 5, 1909.
24 "Taft Toasts Japan as Sincere Friend," *New York Times*, September 20, 1909.
25 Yukio Ozaki to Spencer Cosby, October 13, 1909, Office of Public Buildings and Grounds, National Archives RG 42.
26 Masanao Takahiro to Spencer Cosby, October 29, 1909, Office of Public Buildings and Grounds, National Archives RG 42.
27 James Wilson to Spencer Cosby, November 13, 1909, Office of Public Buildings and Grounds, National Archives RG 42.
28 Fairchild, *The World Was My Garden*, 412.
29 "Auto Ride on Potomac Drive," *Washington Star*, October 18, 1909.
30 "Medal for Mikado," *Washington Post*, November 3, 1909.
31 ERS to Emily MacVeagh, December 1909, Franklin MacVeagh Papers.
32 Eliza C. Scidmore to Walter Oakley family, October 10, 1908, Oakley–Hawley Papers.

33 ERS to Emily MacVeagh, September 10, 1909, Franklin MacVeagh Papers.
34 ERS to Robert U. Johnson, December 12, 1909, Century Collection.
35 ERS, "The Cherry-Blossoms of Japan," *The Century*, March 1910, 648.
36 Anthony, *Nellie Taft*, 261–7.
37 "Diplomats at Play," *Washington Post*, July 25, 1909.
38 ERS to Robert U. Johnson, December 12, 1909, Century Collection.
39 Explanatory notes from George Brown to Spencer Cosby, April 13, 1909, Office of Public Buildings and Grounds, National Archives RG 42.
40 Spencer Cosby to William Howard Taft, November 29, 1909, Office of Public Buildings and Grounds, National Archives RG 42.
41 ERS to Robert U. Johnson, December 12, 1909, Century Collection.
42 The *Star*'s editor at the time, Theodore Noyes, was a great booster of local civic improvements and backed the efforts to bring flowering cherries to Washington. Eliza was a good friend of the Noyes family.
43 ERS to Robert U. Johnson, February 16, 1910, Century Collection.
44 Anthony, *Nellie Taft*, 277.
45 *Chicago Daily Tribune*, January 2, 1910; *Washington Post*, January 7, 1910; *Washington Star*, January 7, 1910; *Washington Star*, January 8, 1910.
46 Under the headline "Tokio's Gift to Mrs. Taft," an *Evening Star* photo of January 7, 1910, showed a federal worker standing with a mummy-like bundle of the newly arrived trees.
47 Philip J. Pauly, "The Beauty and Menace of the Japanese Cherry Trees: Conflicting Visions of American Ecological Independence," *Isis* 87, no. 1 (March 1996): 51. See also Stone, *The Food Explorer*, 220–2.
48 Charles Marlatt to James Wilson, January 19, 1910, Office of Public Buildings and Grounds, National Archives RG 42.
49 A few of the trees apparently survived the pyres. A *Washington Star* account noted that "about a dozen" of the "buggiest" trees were reserved and planted in an experimental USDA plot so scientists could study them in the spring "to see what sort of insects and diseases the trees harbor." "Gift Is Destroyed," *Washington Star*, January 29, 1910. The National Park Service has speculated that a grove of old and twisted cherry trees growing today on Hains Point, at the tip of East Potomac Park, might be survivors of the 1910 shipment. "A Century-Old Mystery Blooms in Grove of D.C. Cherry Trees," *Washington Post*, April 2, 2009; "1910 Japanese Flowering Cherry Trees," Historic American Landscapes Survey, National Park Service, U.S. Department of the Interior, HALS No. DC-8.
50 "Gift Is Destroyed," *Washington Star*, January 29, 1910.

51 In the spring of 1910, Ozaki traveled with a Japanese delegation whose goodwill tour included a visit to Washington. It may have been then that Ozaki met with Cosby to discuss replacing the destroyed cherry trees. A meeting between the two men is referenced in a December 9, 1910, letter from clerk William McNeir at the State Department to C. E. Campbell of the Great Northern Railway Company's freight division. U.S. National Arboretum Collection Cherry Tree Files, National Agricultural Library. In *The Food Explorer* (234), Stone writes that the USDA botanist David Fairchild visited Ozaki at his hotel in Washington to explain why the cherry trees had to be destroyed and to apologize for the incident.
52 Yukio Ozaki to Spencer Cosby, February 2, 1912, Office of Public Buildings and Grounds, National Archives RG 42.
53 "Sees War in 10 Months," *Baltimore Sun*, February 21, 1911.
54 Fairchild, *The World Was My Garden*, 413.

Chapter 19

1 Banquet details from *National Geographic*, March 1912, 272–98; "Factor for Peace," *Evening Star*, January 27, 1912.
2 *Evening Star*, January 27, 1912.
3 McCarry, *From the Field*, 197; *Evening Star*, January 27, 1912.
4 ERS to Gilbert H. Grosvenor, April 27, 1911; Gilbert H. Grosvenor to ERS, May 18, 1911, NGS Library and Archives. At Scidmore's suggestion, Grosvenor also arranged for Nitobe to deliver a public lecture on Formosa (Taiwan) during his time in Washington. A trained agronomist, Nitobe helped oversee development work on the island after it became a colony of Japan at the end of the Sino–Japanese War. See Inazo Nitobe, "Japan as a Colonizer," *Journal of Race Development* 2, no. 4 (April 1912): 347–61.
5 ERS, *Westward to the Far East*, 1891 ed., 5. Scidmore quoted from Nitobe's *Intercourse Between the United States and Japan: An Historical Sketch* (Baltimore: Johns Hopkins, 1891).
6 Auslin, *Japan Society: 100 Years*, 20.
7 *National Geographic*, March 1912, 272–4.
8 Inazo Nitobe, *Bushido: The Soul of Japan* [1900] (Tokyo: Kodansha International, 2002).
9 *Bushido* has been criticized as a highly romanticized interpretation of samurai culture in pre-Meiji Japan. See, for example, Michiyo Nakamoto, "Bushido: The Book That Changed Japan's Image," October 21, 2020, *BBC Culture*, https://www.bbc.com/.

10 Samuel M. Snipes, "The Life of Japanese Quaker Inazo Nitobe," *Friends Journal*, August 1, 2011, http://www.friendsjournal.org/; John F. Howes, ed., *Nitobe Inazo: Japan's Bridge Across the Pacific* (Boulder, CO: Westview Press, 1995).
11 Nitobe quotes from *National Geographic*, March 1912, 290–2.
12 *Evening Star*, March 15, 1912.
13 "Baron Uchida, Ambassador from Japan, Sees Only Peace Ahead," *Walsenburg* (Colo.) *World*, June 11, 1911.
14 Amelia Gere Mason, *Memories of a Friend* (Chicago: Laurence C. Woodworth, 1918), 147.
15 Michael E. Ruane, "The Cherry Trees Were Tinged with Sorrow," *Washington Post*, March 26, 2010. In the 1870s, Chinda was one of several Japanese young men sent abroad to study at what is now DePauw University in Indiana.
16 "Japan's Debt to U.S. Heavy, Chinda Says," *New York Times*, March 17, 1912.
17 Auslin, *Japan Society: 100 Years*, 19.
18 *Evening Star*, March 30, 1912; *Washington Post*, March 30, 1912.
19 *Evening Star*, March 21, 1912.
20 *Japan Weekly Mail*, March 21, 1912.
21 Daniel J. Wakin, "The Heist, the Getaway and the Sawed-Off Leg," *New York Times*, August 26, 2007.
22 Jefferson and Fusonie, *The Japanese Flowering Cherry Trees of Washington, D.C.*, 17.
23 Yoshinao Kozai to L. O. Howard, January 29, 1912, Office of Public Buildings and Grounds, National Archives RG 42.
24 Yei Theodora Ozaki to Helen Taft, February 26, 1911 [1912], Taft Papers.
25 Yukio Ozaki to Spencer Cosby, February 19, 1912, Office of Public Buildings and Grounds, National Archives RG 42.
26 *Evening Star*, March 21, 1912; *Washington Post*, March 29, 1912.
27 *Evening Star*, March 26, 1912.
28 *Evening Star*, March 21, 1912.
29 *Evening Star*, March 23 and 24, 1912; *New York Times*, March 24, 1912.
30 *Evening Star*, March 25–27, 1912.
31 Anthony, *Nellie Taft*, 331.
32 Spencer Cosby to James Wilson, March 26, 1912, Office of Public Buildings and Grounds, National Archives RG 42.
33 James Wilson to Spencer Cosby, March 27, 1912, Office of Public Buildings and Grounds, National Archives RG 42.

34 Details based on the author's visit to the original planting site on February 22, 2012.
35 "History of the Cherry Trees," National Park Service, https://www.nps.gov/subjects/cherryblossom/history-of-the-cherry-trees.htm.
36 O'Toole, *Five of Hearts*, 99.
37 *Washington Post*, March 28, 1912.
38 *Evening Star*, March 28, 1912.
39 A 300-year-old Japanese stone lantern was added near the site in March 1954, a gift from Japan to commemorate the hundredth anniversary of the first treaty providing for peaceful exchange between Japan and the United States.
40 "The Vandalization of the Cherry Trees in 1914," National Park Service, https://www.nps.gov/articles/the-vandalization-of-the-cherry-trees-in-1941.htm.

Chapter 20

1 Details and quotes from ERS, "Japan's Platonic War with Germany," *Outlook*, December 23, 1914, 914–20.
2 ERS, "The Japanese Red Cross in the Tsingtau Campaign and the European War," *American Red Cross Magazine* 10, no. 2 (February 1915): 51–6.
3 For examples, see Honolulu *Star–Bulletin*, February 6, 1915; "Japan and the War," *Outlook*, December 23, 1914; "A Remarkable Letter on the Japanese Question," *Outlook*, January 13, 1915; and "No Menace Seen in Demands of Japan," *Philadelphia Inquirer*, March 10, 1915.
4 Noonan, *Reading* The Century, 178; also biographical notes from Robert Underwood Johnson Papers, Archives and Manuscripts, MssCol 1575, New York Public Library. After an April 1912 article on "The Famous Gardens of Kioto," Scidmore published only one more article in the magazine: "Marshall Count Terauchi, the New Premier of Japan," *The Century*, August 1917.
5 Gilbert H. Grosvenor to ERS, February 14, 1913, NGS Library and Archives.
6 ERS to Gilbert H. Grosvenor, February 5, 1914, NGS Library and Archives.
7 ERS, "Japan's Coronation Season," *Outlook*, January 26, 1916, 203–10.
8 Eliza Catherine Scidmore to Mary Oakley, July 27, 1916, Oakley–Hawley Papers.
9 ERS to Mary Oakley, October 10, 1916, Oakley–Hawley Papers.

10 The *Washington Star*, *Washington Post*, and *New York Times* all ran obituaries of Eliza Catherine Scidmore. The *Star* carried a full-length portrait of her taken on her ninety-second birthday.

11 *Japan Gazette*, October 6, 7, and 13, 1916.

12 ERS to Mary Oakley, October 10, 1916, Oakley–Hawley Papers.

13 Telegram from George H. Scidmore to Secretary of State, October 30, 1916, and follow-up letters of November 23 and December 14, 1916, Consular Bureau Records, Department of State.

14 Mary Florence Denton lived many years in Japan with the support of an American mission board and taught at Doshisha University in Kyoto.

15 *Washington Post*, December 10, 11, and 29, 1916.

16 ERS, "Stories of the Internés: I—Hospitable Switzerland," *Outlook*, December 5, 1917, 556–8; ERS, "In the Wind-Swept Marne Country," *Outlook*, June 26, 1918, 342–3.

17 *Washington Herald*, January 8, 1917; *Washington Post*, January 18, 1917.

18 Washington *Spur*, November 15, 1918. The term *cave dweller* became popular in Washington in the early 1900s, usually in reference to those with ties dating back to the city's oldest families. See Sarah Booth Conroy, "D.C.'s First Families," *Washington Post*, June 14, 1987.

19 *Washington Times*, October 27, 1918; Elizabeth Foxwell, ed., *In Their Own Words: American Women in World War I* (Waverly, TN: Oconee Spirit Press, 1915), 121.

20 *Evening Star*, December 5, 1914; "The Elegant Stoneleigh Court Apartments," *Streets of Washington*, http://www.streetsofwashington.com.

21 Carrie Walsh, a former schoolteacher from Wisconsin, owed her wealth to her husband's gold-mining bonanza in Colorado. The Walshes entertained lavishly in their elaborate mansion at Dupont Circle (the present-day Indonesian Embassy). Among their many interactions, Scidmore escorted Walsh and a friend on a tour of Japan in 1916.

22 George Washington University letter to ERS, May 5, 1919, Box 5, Oakley–Hawley Papers. At the time of the award, Scidmore's *National Geographic* editor, Gilbert H. Grosvenor, was serving on the trustees' committee to award honorary degrees, and several women of her acquaintance were members of the school's University Council. George Washington University Bulletin, June 1919, 9–12.

23 *Evening Star*, November 1, 1919. Scidmore may have taken the flight thanks to her National Geographic Society connections. She was on hand the evening Alexander Graham Bell and his family feted

the plane's developer, Glenn Curtiss, when he went to Washington in 1913 to receive the Smithsonian's Langley Medal for outstanding contributions in aeronautics. *Evening Star*, May 7, 1913.

24 *Washington Post*, July 23, 1922; *New York Times*, July 23 and 26, 1923.

25 William Crozier, U.S. Army (Retired), Address at Services in Memory of Eliza Ruhamah Scidmore, held in Geneva, Switzerland, on 7 November 1928; in Oakley–Hawley Papers.

26 Japan joined the League of Nations despite the indignity it suffered at the Paris Peace Conference when several of the participating powers, including the United States, rejected a "racial equality" clause the Japanese sought to include in the treaty. See Thomas W. Burkman, "Japan and the League of Nations: An Asian Power Encounters the 'European Club,'" *World Affairs* 158, no. 1 (Summer 1995): 45–57.

27 ERS to Mary Oakley, July 2, 1922, Oakley–Hawley Papers. The manor house, which now operates as a luxury hotel, is on the National Heritage List for England. When Scidmore visited in 1922, the estate had recently been sold at auction, after 600 years in the same family, and its future was up in the air. "Holme Lacy House History," Herefordshire Past, https://herefordshirepast.co.uk.

28 *Japan Weekly Chronicle*, November 30, 1922. Additional details on George Scidmore from *Japan Gazette*, November 27 and 29, and *Japan Advertiser*, November 28 and 30, 1922.

29 George Scidmore lived from 1920–2 in a house on plot No. 246 in the Yamate-cho district of Yokohama. *Japan Directory*, vol. 4 (Yokohama: Gazette, 1913), 111.

30 *Japan Gazette*, November 27, 1922. George Scidmore owned a succession of a yachts and a schooner and served many years as commodore of the Yokohama Yacht Club.

31 Telegrams and letters between consular personnel in Yokohama and Department of State, October 22 to November 30, 1922.

32 The claim apparently originated with a Japanese tribute after Scidmore's death and got picked up in other articles over the years. "Ashes of Miss Scidmore, Friend of Japanese, Brought for Burial," *Osaka Mainichi*, September 11, 1929.

33 "The Immigration Act of 1924 (The Johnson–Reed Act)," Office of the Historian, U.S. Department of State, http://history.state.gov/milestones/. The ban on Japanese immigration was not lifted until 1952.

34 Washington *Spur*, August 1, 1923.

35 *Sunday Star*, July 29, 1923.

36 Sample articles in *New York Times*, September 2 and 3, 1923; *Baltimore Sun*, September 3 and 6, 1923; *Washington Post*, September 5, 1923; *Boston Daily Globe*, September 6, 1923.
37 Earthquake details from Hammer, *Yokohama Burning*; quote (88) from eyewitness account by Ellis Zacharias, a U.S. Naval intelligence officer in Tokyo.
38 *Capital Times* (Madison, WI), September 4, 1923.
39 *Washington Post*, September 30 and October 3, 1923.
40 *Washington Post*, October 22, 1923.
41 Robert Kanigel, *High Season in Nice* (London: Abacus, 2003), 168.
42 ERS to Everett W. Frazar, November 13, 1923, Oakley–Hawley Papers; notarized by U.S. Consulate in Nice, France.
43 ERS Last Will and Testament, February 4, 1924; filed June 10, 1930, with Register of Wills, Clerk of Probate Court, Washington, DC.
44 *Chinese & Japanese Porcelains & Potteries & and Other Far Eastern Objects of Art, The Collection of Miss Eliza Ruhamah Scidmore, Washington, D.C.*, Sale No. 1903, Anderson Galleries, New York, January 10, 1925.
45 *New York Times*, January 11, 1925. The throne chair, purchased by "E. C. Cockcroft," sold for $350.
46 *Washington Post*, November 16, 1924.
47 H.R. Bill 5489, 68th Congress; Appropriation for payment to sister of George H. Scidmore, *United States Statutes at Large*, vol. 43, Part 2, Department of State, 1925.
48 ERS to Walter Hough, head curator, Smithsonian Institution, United States National Museum, with accession memorandum, January 16, 1925.
49 ERS to Mary Oakley, April 7, 1925, Oakley–Hawley Papers.
50 ERS to Mary Oakley, April 7, 1925; ERS to Horace Oakley, April 24, 1926, Oakley–Hawley Papers.
51 ERS to Mary Oakley, April 6, 1927, Oakley–Hawley Papers.
52 After Scidmore's death, an Associated Press obituary of November 3, 1928, reported, in a lofty description, that "her salon [in Geneva] was a meeting place of all representative American visitors." In fact, her letters point to a far more modest style of entertaining, usually an invitation to lunch or tea—sometimes arranged on short notice, to the annoyance of her maid.
53 ERS to Mary Oakley, May 1, 1928, Oakley–Hawley Papers.
54 ERS to Mary Oakley, September 19, 1927, Oakley–Hawley Papers.
55 ERS to Mary Oakley, March 29/April 1, [1928?], Oakley–Hawley Papers.

56 ERS to Horace Oakley, April 18, April 24, May 10, May 17, 1926; February 7, 1928; Oakley–Hawley Papers.
57 See Horace Sweeney Oakley Papers, Newberry Library, Chicago. Oakley was a strong patron of the arts, served on President Woodrow Wilson's peace committee, and in 1918 joined an American Red Cross Commission to Greece for relief efforts in Macedonia.
58 Details on Mary Elizabeth Atwood from ERS letters and typewritten notes in Oakley–Hawley Papers. Mary E. Atwood should not be confused with Mary Louise Atwood, Eliza Scidmore's first cousin, whom she had known growing up.
59 ERS to Horace Oakley, February 7, 1928, Oakley–Hawley Papers.
60 Mary Atwood to Horace Oakley, October 5, 1928, Oakley–Hawley Papers.
61 Mary Crozier to Horace Oakley, November 7, 1928, Oakley–Hawley Papers. William Crozier was a weapons and ammunition expert who served as chief of ordinance during his forty-seven-year career in the U.S. Army. He and his wife lived in Washington after his retirement in 1919.
62 Mary Atwood to Horace Oakley, November 18, 1928, Oakley–Hawley Papers.

Epilogue

1 "Noted Author, Who Was Known Here, Dies in Geneva," *Wisconsin State Journal*, November 4, 1928.
2 Mary Crozier to Horace Oakley, November 7, 1928, Oakley–Hawley Papers.
3 Yoshida remarks included in Mary E. Atwood to Horace Oakley, December 18, 1928, Oakley–Hawley Papers; also see William Crozier, U.S. Army (Retired), Address at Services in Memory of Eliza Ruhamah Scidmore, held in Geneva, Switzerland, on 7 November 1928, in Oakley–Hawley Papers. The family sent an announcement of Scidmore's death, along with a printed copy of the memorial service, to more than a hundred people from her address book, including the Tafts and many other prominent people. Sidmore, "More Than a Footnote," 80–8.
4 Last Will and Testament of Eliza R. Scidmore, signed and sealed on February 4, 1924, filed at Supreme Court of the District of Columbia on June 10, 1930; Gilson G. Blake Jr. to Horace Oakley, November 10, 1928, Oakley–Hawley Papers.

5 Both schools are still in operation today, the first as Doshisha Women's College, the second as Saint Maur International School.
6 Horace Oakley to Mary E. Atwood, November 20, 1928, Oakley–Hawley Papers.
7 Mary E. Atwood to Horace Oakley, December 9, 1928, Oakley–Hawley Papers.
8 Mary Atwood to Horace Oakley, November 15, 1928, Oakley–Hawley Papers.
9 Mary Atwood to Horace Oakley, March 26, 1929, Oakley–Hawley Papers.
10 Mary Atwood to Horace Oakley, January 29, 1929, Oakley–Hawley Papers.
11 Mary Atwood to Horace Oakley, October 19, 1929, Oakley–Hawley Papers.
12 Last Will and Testament of Eliza R. Scidmore.
13 E. W. Frazar to Horace Oakley, March 7, 1929, Oakley–Hawley Papers.
14 "Ashes of Miss Scidmore, Friend of Japanese, Brought for Burial," *Osaka Mianichi*, September 11, 1929.
15 Soon after Scidmore's death, a dozen of her women friends arranged to endow a bed in her name at St. Luke's Hospital in Tokyo, at a cost of $2,500. Sidmore, "More Than a Footnote," 78. According to a February 1829 note in the Oakley–Hawley Papers [source unknown], they also discussed having a plaque placed in Potomac Park explaining her role in the creation of Cherry Tree Drive, along with Mrs. Taft and Dr. Takamine. Apparently, no further action was taken at the time, and no such marker appeared until nearly a century later.
16 All details from Madison, Wisconsin, *Capital Times*, articles by Irvin Kreisman, "Washington's Cherry Blossoms a Madison Woman's Inspiration," April 22, 1965, and "Madison Police Make Possible Japanese Friendship Gesture," December 10, 1965.
17 The autograph book is now in private hands, after being sold at auction in New York on June 22, 2012, to an undisclosed buyer. Christie's Sale 2572, Lot 250.
18 ERS, *Nihon · Jinrikisha ryojō* ("Japan: Journeys by Rickshaw," a Japanese translation of Scidmore's *Jinrikisha Days in Japan*), trans. by Onchi Mitsuo (Yokohama: Yūrindō, 1987).
19 "History of the Cherry Trees," National Park Service, https://www.nps.gov/subjects/cherryblossom/history-of-the-cherry-trees.htm.

In April 2013, the author visited a site in Kanazawa, in Ishikawa Prefecture, to view one of the "homecoming" trees.

20 Kay Redfield Jamison, *Exuberance: The Passion for Life* (New York: Vintage, 2004).

21 Jamison, *Exuberance*, 6.

22 Jamison, *Exuberance*, 7.

Bibliography

Primary Works by Eliza R. Scidmore (in chronological order)

Scidmore, E. Ruhamah. *Alaska: Its Southern Coast and the Sitkan Archipelago.* Boston: D. Lothrop, 1885.

Scidmore, Eliza Ruhamah. *Jinrikisha Days in Japan.* New York: Harper and Brothers, 1891.

Scidmore, Eliza Ruhamah. *Westward to the Far East, a Guide to the Principal Cities of China and Japan,* 1st ed. Montreal: Canadian Pacific Railway, 1891. [*East to the West*; version for those traveling in the opposite direction]

Scidmore, Eliza Ruhamah. *Appletons' Guide-Book to Alaska and the Northwest Coast.* New York: D. Appleton, 1893.

Scidmore, Eliza Ruhamah. *Java, the Garden of the East.* Century Company, 1897. [Paperback reprint: Oxford University Press/Singapore, 1986.]

Scidmore, Eliza Ruhamah. *China, the Long-Lived Empire.* New York: Century Company, 1900.

Scidmore, Eliza Ruhamah. *Winter India.* New York: Century Company, 1903.

Scidmore, Eliza R. *As the Hague Ordains: Journal of a Prisoner's Wife in Japan.* New York: Century Company, 1914. [Originally published anonymously by Henry Holt, 1907.]

Besides her books, Scidmore published about 800 articles in newspapers, magazines, and journals. The primary ones as mentioned in this text—including the D.C. *National Republican, St. Louis Globe–Democrat, Harper's Weekly* and *Bazaar, The Century Illustrated Magazine, National Geographic, Chicago Daily Tribune,* San Francisco *Call,* Washington *Evening Star, New York Times,* and *The Outlook*—are only a portion of the periodicals to which she contributed.

Select List of Other Sources

Abe, Naoko. *The Sakura Obsession: The Incredible Story of the Plant Hunter Who Saved Japan's Cherry Blossoms.* New York: Alfred A. Knopf, 2019.

Anthony, Carl Sferrazza. *Nellie Taft: The Unconventional First Lady of the Ragtime Era.* New York: William Morrow, 2005.

Auslin, Michael R. *Pacific Cosmopolitans: A Cultural History of U.S.–Japan Relations*. Cambridge, MA: Harvard University, 2011.

Bendavid-Val, Leah. *Stories on Paper and Glass: Pioneering Photography at National Geographic*. Washington, DC: National Geographic, 2001.

Benfey, Christopher. *The Great Wave: Gilded Age Misfits, Japanese Eccentrics, and the Opening of Old Japan*. New York: Random House, 2003.

Berg, Scott. *Grand Avenues: The Story of Charles L'Enfant, the French Visionary Who Designed Washington, D.C.* New York: Vintage, 2007.

Borneman, Walter R. *Alaska: Saga of a Bold Land*. New York: Perennial/Harper Collins, 2003.

Boyd, Andrew. *Boyd's Washington and Georgetown Directory*. Washington, DC: Hudson Taylor, 1865. [Also, other editions.]

Brands, H. W. *The Reckless Decade: American in the 1890s*. Chicago: University of Chicago, 1995.

Bryan, C. D. B. *The National Geographic Society, 100 Years of Adventure and Discovery*. New York: Harry N. Abrams, 1987

Bryan, W. B. *History of the National Capital from Its Foundation through the Period of Its Adoption of the Organic Act, Vol II, 1815–1817*. New York: Macmillan, 1914–1916.

Butterfield, Willshire, ed. *History of Dane County, Wisconsin*. Chicago: Western Historical Company, 1880.

Butt, Archibald Willingham. *Taft and Roosevelt: The Intimate Letters of Archie Butt, Military Aide*. 2 vols. Garden City, NY: Doubleday, Doran, 1930. [Reprinted by Kennikat, 1971.]

Caemmerer, H. Paul. *Historic Washington: Capital of the Nation*. Washington, DC: Columbia Historical Society, 1948.

Campbell, Robert. *In Darkest Alaska: Travel and Empire Along the Inside Passage*. Philadelphia: University of Pennsylvania, 2007.

Cashman, Sean Dennis. *America in the Gilded Age*. New York: New York University, 1988.

Chang, Jung. *Empress Dowager Cixi: The Concubine Who Launched Modern China*. New York: Alfred A. Knopf, 2013.

Clark, Steve and Paul Smethurst, eds. *Asian Crossings: Travel Writing on China, Japan and Southeast Asia*. Hong Kong: Hong Kong University, 2008.

Clifford, Nicholas. *A Truthful Impression of the Country: British and American Travel Writing in China, 1880–1949*. Ann Arbor: University of Michigan, 2001.

Cott, Nancy. *The Grounding of Modern Feminism*. New Haven, CT: Yale University, 1987.

Cowdrey, Albert E. *A City for the Nation: The Army Corps of Engineers and the Building of Washington, D.C., 1790–1967.* Washington, DC: Historical Division, Office of the Chief of Engineers, 1979.

Daniels, Roger. *The Politics of Prejudice: The Anti-Japanese Movement in California and the Struggle for Japanese Exclusion.* Berkeley: University of California, 1962.

Degler, Carl. *At Odds: Women and Family in America.* New York and Oxford: Oxford University, 1980.

Dickey, J. B. *Empire of Mud: The Secret History of Washington, D.C.* Guilford, CT: Globe-Pequot, 2014.

Donnelly, Mabel Collins. *The American Victorian Woman: The Myth and the Reality.* New York: Greenwood, 1986.

Dower, John W. et al. Visualizing Cultures (website), Units on China and Japan, Massachusetts Institute of Technology, https://visualizingcultures.mit.edu/home/.

DuBois, Ellen Carol. *Feminism and Suffrage: The Emergence of an Independent Women's Movement in America, 1848–1869.* Ithaca, NY: Cornell University, 1978.

Fahs, Alice. *Out on Assignment: Newspaper Women and the Making of Modern Public Space.* Chapel Hill: University of North Carolina, 2011.

Fairchild, David. *The World Was My Garden: Travels of a Plant Explorer.* New York: Charles Scribner's, 1938.

Fedler, Fred. *Lessons from the Past: Journalists' Lives and Work, 1850–1950.* Prospect Heights, IL: Waveland, 2000.

Flack, J. Kirkpatrick. *Desideratum in Washington: The Intellectual Community in the Capital City, 1870–1900.* Cambridge, MA: Schenkman, 1975.

Foner, Eric. *Free Soil, Free Labor, Free Men: The Ideology of the Republican Party Before the Civil War.* New York: Oxford University, 1995.

Fox, Stephen. *The American Conservation Movement: John Muir and His Legacy.* Madison: University of Wisconsin, 1985.

Foxwell, Elizabeth, ed. *In Their Own Words: American Women in World War I.* Waverly, TN: Oconee Spirit Press, 2015.

Furgerson, Ernest B. *Freedom Rising: Washington in the Civil War.* New York: Vintage, 2004.

Ghose, Indira. *Women Travelers in Colonial India: The Power of the Female Gaze.* New York: Oxford University, 1998.

Gibbon, John Murray. *Steel of Empire: Romantic History of Canadian Pacific, the Northwest Passage of Today.* Indianapolis: Bobbs–Merrill, 1935.

Gifford, Terry, ed. *John Muir: His Life and Letters and Other Writings*. London: Bâton Wicks; Seattle: Mountaineers, 1996.

Ginger, Ray. *The Age of Excess: The United States From 1877 to 1914*. New York: MacMillan, 1965.

Gluck, Carol. *Japan's Modern Myths: Ideology in the Late Meiji Period*. Princeton, NJ: Princeton University, 1985.

Goodwin, Doris Kearns. *The Bully Pulpit: Theodore Roosevelt, William Howard Taft, and the Golden Age of Journalism*. New York: Simon & Schuster, 2013.

Gray, Charlotte. *Reluctant Genius: Alexander Graham Bell and the Passion for Invention*. New York: Arcade, 2006.

Green, Constance McLaughlin. *Washington: A History of the Capital, 1800–1950*. 2 vols. Princeton, NJ: Princeton University, 1962.

Greenhalgh, Paul. *Ephemeral Vistas: The Expositions Universelles, Great Expositions and World's Fairs, 1851–1939*. Manchester, UK: Manchester University, 1988.

Grosvenor, Edwin S. and Morgan Wesson. *Alexander Graham Bell: The Life and Times of the Man Who Invented the Telephone*. New York: Henry Abrams, 1997.

Gruen, J. Philip. *Manifest Destinations: Cities and Tourists in the Nineteenth-Century American West*. Norman: University of Oklahoma, 2014.

Hammer, Joshua. *Yokohama Burning: The Deadly 1923 Earthquake and Fire That Helped Forge the Path to World War II*. New York: Free Press, 2006.

Hammersmith, Jack. *Spoilsmen in a "Flowery Fairyland": The Development of the U.S. Legation in Japan, 1859–1906*. Kent, OH: Kent State University, 1998.

Hane, Mikiso and Louis G. Perez. *Modern Japan: A Historical Survey*. 5th ed. Boulder, CO: Westview, 2013.

Harper, Judith. *Women During the Civil War: An Encyclopedia*. New York and London: Routledge, 2004.

Hawkins, Sallie. *American Iconographic: National Geographic, Global Culture, and the Visual Imagination*. Charlottesville: University of Virginia, 2010.

Hawley, Emily Carrie. *A Genealogical and Biographical Record of the Pioneer Thomas Skidmore (Scudamore) of the Massachusetts and Connecticut Colonies in New England and of Huntington, Long Island, and of His Descendants. . . .* Brookfield Center, CT: Hawley, 1911.

Heacox, Kim. *John Muir and the Ice That Started a Fire: How a Visionary and the Glaciers of Alaska Changed America*. Guilford, CT: Lyons, 2014.

Heilbrun, Carolyn. *Writing a Woman's Life*. New York: Ballantine, 2002.

Heppen, John and Samuel Otterstrom. *Geography, History, and the American Political Economy*. Lanham, MD: Lexington, 2009.

Hodgson, Barbara. *Dreaming of the East: Western Women and the Exotic Lure of the Orient*. Vancouver, BC: Greystone, 2005.

Hoganson, Kristen. *Consumers' Imperium: The Global Production of American Domesticity, 1865–1920*. Chapel Hill: University of North Carolina, 2007.

Hosley, William. *The Japan Idea: Art and Life in Victorian America*. Hartford, CT: Wadsworth Atheneum, 1990.

Horowitz, Helen Lefkowitz. *Alma Mater: Design and Experience in the Women's Colleges from Their Nineteenth-Century Beginnings to the 1930s*. Amherst: University of Massachusetts, 1993.

Hurn, Ethel Alice. *Wisconsin Women in the War Between the States*. Madison: Wisconsin History Commission, 1911.

Iriye, Akira, ed. *Mutual Images: Essays in American–Japanese Relations*. Cambridge, MA: Harvard University, 1975.

Jacob, Kathryn Allamong. *Capital Elites: High Society in Washington, D.C., After the Civil War*. Washington, DC: Smithsonian Institution, 1994.

Jacobson, Matthew Fry. *Barbarian Virtues: The United States Encounters Foreign People at Home and Abroad, 1876–1917*. New York: Hill and Wang, 2000.

Janke, Lucinda Prout. *A Guide to Civil War Washington, D.C.* Charleston, SC: History Press, 2013.

Jefferson, Roland M. and Alan E. Fusonie. *The Japanese Flowering Cherry Trees of Washington, D.C.: A Living Symbol of Friendship*. National Arboretum Contribution No. 4. Washington, DC: U.S. Department of Agriculture, 1977.

Jenkins, Mark Collins. *National Geographic: 125 Years*. Washington, DC: National Geographic, 2013.

Johnson, Robert Underwood. *Remembered Yesterdays*. Boston: Little, Brown, 1923.

Jonnes, Jill. *Urban Forests: A Natural History of Trees and People in the American Landscape*. New York: Viking, 2016.

Junior League of Washington. Thomas Froncek, ed. *The City of Washington: An Illustrated History*. New York: Alfred A. Knopf, 1979.

Keene, Donald. *Emperor of Japan: Meiji and His World, 1852–1912*. New York: Columbia University, 2002.

Kerr, Douglas and Julia Kuehn, eds. *A Century of Travels in China: Critical Essays on Travel Writing From the 1840s to the 1940s*. Hong Kong: Hong Kong University, 2007.

Kohler, Sue and Pamela Scott, eds. *Designing the Nation's Capital: The 1901 Plan for Washington, D.C.* Washington, DC: U.S. Commission of Fine Arts, 2006.

Leech, Margaret. *Reveille in Washington, 1860–1865*. New York: Grosset & Dunlap, 1941.

Lerner, Gerda. *The Creation of Feminist Consciousness: From the Middle Ages to Eighteen-Seventy*. New York: Oxford University, 1993.

Lutes, Jean Marie. *Front-Page Girls: Women Journalists in American Culture and Fiction, 1880–1930*. Ithaca, NY: Cornell University, 2006.

Martin, Geoffrey J. *American Geography and Geographers*. New York: Oxford University, 2015.

Mattox, Henry E. *Twilight of Amateur Diplomacy: The American Foreign Service and Its Senior Officers in the 1890s*. Kent, OH: Kent State University, 1989.

McCarry, Charles, ed. *From the Field: A Collection of Writings from* National Geographic. Washington, DC: National Geographic, 1997.

McClellan, Ann. *The Cherry Blossom Festival: Sakura Celebration*. Piermont, NH: Bunker Hill, 2005.

Meyer, Karl E. and Shareen Blair Brysac. *The China Collectors: America's Century–Long Hunt for Asian Art Treasures*. New York: Palgrave/MacMillan, 2015.

Miles, Sara. *Discourses of Difference: An Analysis of Women's Travel Writing and Colonialism*. London and New York: Routledge, 1991.

Miller, Iris. *Washington in Maps, 1606–2000*. New York: Rizzoli, 2002.

Mollenhoff, David. *Madison: A History of the Formative Years*, 2nd ed. Madison: University of Wisconsin, 2004.

Morgan, Susan. *Place Matters: Gendered Geography in Victorian Women's Travel Books About Southeast Asia*. New Brunswick, NJ: Rutgers University, 1996.

Mott, Frank Luther. *A History of American Magazines, 1741–1930*. Cambridge, MA: Harvard University Press, c. 1958–1968.

Newman, Cathy. *Women Photographers at National Geographic*. Washington, DC: National Geographic, 2000.

Nimura, Janice P. *Daughters of the Samurai: A Journey from East to West and Back*. New York: W. W. Norton, 2015.

Noonan, Mark J. *Reading* The Century Illustrated Monthly Magazine*: American Literature and Culture, 1870–1893*. Kent, OH: Kent State University, 2010.

O'Toole, Patricia. *The Five of Hearts: An Intimate Portrait of Henry Adams and His Friends, 1880–1918*. New York: Ballantine, 1990.

Passonneau, Joseph R. *Washington through Two Centuries: A History in Maps and Images*. New York: Monacelli Press, 2004.

Poole, Robert. *Explorers House: National Geographic and the World It Made*. New York: Penguin, 2004.

Pratt, Mary Louise. *Imperial Eyes: Travel Writing and Transculturation*. 2nd ed. London and New York: Routledge, 2008.

Preston, Diana. *The Boxer Rebellion: The Dramatic Story of China's War on Foreigners That Shook the World in the Summer of 1900*. New York: Walker, 1999.

Reps, John W. *Washington on View: The Nation's Capital Since 1790*. Chapel Hill: University of North Carolina, 1991.

Reischauer, Edwin O. *Japan: The Story of a Nation*. New York: Alfred A. Knopf, 1974. [Revised ed.]

Ritchie, Donald A. *Press Gallery: Congress and the Washington Correspondents*. Cambridge, MA: Harvard University, 1991.

Rosenstone, Robert A. *Mirror in the Shrine: Encounters with Meiji Japan*. Cambridge, MA: Harvard University, 1988.

Rothenberg, Tamar. *Presenting America's World: Strategies of Innocence in* National Geographic Magazine, *1888–1945*. Burlington, VT: Ashgate, 2007.

Russell, Mary. *The Blessings of a Good Thick Skirt: Women Travellers and Their World*. London: Collins, 1986.

Rutkow, Eric. *American Canopy: Trees, Forests and the Making of a Nation*. New York: Scribner, 2012.

Rydell, Robert W. *All the World's a Fair: Visions of Empire at American International Expositions, 1876–1916*. Chicago: University of Chicago, 1984.

Sabin, Burritt. *A Historical Guide to Yokohama*. Yokohama: Yurindo, 2002.

Samuels, Gayle Brandow. *Enduring Roots: Encounters with Trees, History, and the American Landscape*. Rutgers, NJ: Rutgers University, 1999.

Schlereth, Thomas J. *Victorian America: Transformations in Everyday Life, 1876–1915*. New York: Harper Perennial, 1991.

Schlesinger, Arthur. M., Jr., ed. *The Almanac of American History*. New York: G. P. Putnam's, 1983.

Schulten, Susan. *Geographical Imagination in America, 1880–1950*. Chicago: University of Chicago, 2001.

Schulten, Susan. *Mapping the Nation: History and Cartography in Nineteenth-Century America*. Chicago: University of Chicago, 2012.

Scott, Anne Firor. *Making the Invisible Woman Visible*. Urbana: University of Illinois, 1984.

Scott, Pamela. *Capital Engineers: The U.S. Army Corps of Engineers in the Development of Washington, D.C., 1790–2004*. Alexandria, VA: Office of History, U.S. Army Corps of Engineers, 2005.

Seale, William. *The Imperial Season: America's Capital in the Time of the First Ambassadors, 1893–1918*. Washington, DC: Smithsonian Institution, 2013.

Shaffer, Marguerite. *See America First: Tourism and National Identity, 1880–1940*. Washington, DC: Smithsonian Institution, 2001.

Shannon, Anne. *Finding Japan: Early Canadian Encounters with Asia*. Vancouver, BC: Heritage House, 2012.

Sidmore, Daniel Howard. "Eliza Ruhamah Scidmore: More Than a Footnote in History." Master's thesis, Benedictine University, 2000.

Silber, Nina. *Daughters of the Union: Northern Women Fight the Civil War*. Cambridge, MA: Harvard University, 2005.

Skidmore, Warren. *Thomas Skidmore (Scudamore), 1605–1684, of Westerleigh, Gloucestershire, and Fairfield, Connecticut: His Ancestors and His Descendants to the Ninth Generation*. Akron, OH: Skidmore, 1980.

Smith-Rosenberg, Carroll. *Disorderly Conduct: Visions of Gender in Victorian America*. New York: Alfred A. Knopf, 1985.

Sterry, Lorraine. *Victorian Women Travelers in Meiji Japan: Discovering a "New" Land*. Folkestone, UK: Global Oriental, 2009.

Stevenson, Louise. *The Victorian Homefront: American Thought and Culture, 1860–1880*. New York: Twayne, 1991.

Stone, Daniel. *The Food Explorer: The True Adventures of the Globe–Trotting Botanist Who Transformed What America Eats*. New York: Dutton, 2018.

Taft, Mrs. William Howard. *Recollections of Full Years*. New York: Dodd, Mead, 1914.

Tebbel, John. *Compact History of the American Newspaper*. New York: Hawthorn, 1963.

Thurin, Susan Schoenbauer. *Victorian Travelers and the Opening of China, 1842–1907*. Athens: Ohio University, 1999.

Thwaites, Reuben. *The Story of Madison*. Thwaites, 1900.

Twain, Mark and Charles Dudley Warner. *The Gilded Age: A Tale for Today*, 1873. [Reprint: New York and London: Harper & Brothers, 1904.]

Wagner, Margaret E., et al. *Library of Congress Civil War Desk Reference*. New York: Simon and Schuster, 2002.

West, Nancy. *Kodak and the Lens of Nostalgia*. Charlottesville: University of Virginia, 2000.

Winkle, Kenneth J. *Lincoln's Citadel: The Civil War in Washington, D.C.* New York: W. W. Norton, 2013.

Worster, Donald. *A Passion for Nature: The Life of John Muir*. New York: Oxford University, 2008.

Yoshihara, Mari. *Embracing the East: White Women and American Orientalism*. Oxford and New York: Oxford University, 2003.

Yuzo, Kato, ed. *Yokohama, Past and Present*. Yokohama: Yokohama City University, 1990.

Ziparo, Jessica. *This Grand Experiment: When Women Entered the Federal Workforce in Civil War-Era Washington, D.C.* Chapel Hill: University of North Carolina, 2017.

Index

Note to Index: ERS = Eliza Ruhamah Scidmore; *f* refers to an Illustration